Green Energy and Technology

Climate change, environmental impact and the limited natural resources urge scientific research and novel technical solutions. The monograph series Green Energy and Technology serves as a publishing platform for scientific and technological approaches to "green"—i.e. environmentally friendly and sustainable—technologies. While a focus lies on energy and power supply, it also covers "green" solutions in industrial engineering and engineering design. Green Energy and Technology addresses researchers, advanced students, technical consultants as well as decision makers in industries and politics. Hence, the level of presentation spans from instructional to highly technical.

Indexed in Scopus.

Indexed in Ei Compendex.

Mayken Espinoza-Andaluz • Jordy Santana-Villamar • Brayan Ordóñez-Saca • Carlos Vallejo-Cervantes • Luis Rodríguez-Álava
Editors

Congress on Research, Development and Innovation in Renewable Energies

Selected Papers from CIDiER 2024

Springer

Editors
Mayken Espinoza-Andaluz
Escuela Superior Politécnica del Litoral
Guayaquil, Ecuador

Jordy Santana-Villamar
Escuela Superior Politécnica del Litoral
Guayaquil, Ecuador

Brayan Ordóñez-Saca
Escuela Superior Politécnica del Litoral
Guayaquil, Ecuador

Carlos Vallejo-Cervantes
Escuela Superior Politécnica del Litoral
Guayaquil, Ecuador

Luis Rodríguez-Álava
Escuela Superior Politécnica del Litoral
Guayaquil, Ecuador

ISSN 1865-3529 ISSN 1865-3537 (electronic)
Green Energy and Technology
ISBN 978-3-031-88994-3 ISBN 978-3-031-88995-0 (eBook)
https://doi.org/10.1007/978-3-031-88995-0

This Springer imprint is published by the registered company Springer Nature Switzerland AG
The registered company address is: Gewerbestrasse 11, 6330 Cham, Switzerland

Preface

The effects of climate change have grown over time, such as the increase in global temperature, shifting weather seasons, and rise of meteorological phenomena worldwide. Every change is accelerated by the significant greenhouse gas emissions released to the atmosphere using fossil fuels in the energy sector. Due to this, international alliances, such as the European Green Deal, have arisen to face the challenges of changing to a sustainable, clean, and economically viable energy matrix. In addition, one of the main concerns of the United Nations 2030 Agenda for Sustainable Development is to promote sustainable initiatives related to developing renewable energies, such as SDG 7: Affordable and Clean Energy and SDG 13: Climate Action. Therefore, disseminating scientific and technical works that help meet the objectives of sustainable development and, in turn, help mitigate the effects of climate change is essential. Based on the success of CIDiER 2021, CIDiER 2022, and CIDiER 2023, we are pleased to bring the selected papers from the fourth meeting of the Congress on Research, Development, and Innovation in Renewable Energies, CIDiER 2024, held on September 23–24, 2024, in Guayaquil, Ecuador.

This book looks at biomass, wind energy, solar energy, hydrogen energy, geothermal energy, tidal energy, computational modeling, control systems, and artificial intelligence applied to renewable energy studies. This 2024 Congress aimed to spread theoretical and experimental study results and applications related to relevant topics in renewable energies and generate a space of multidisciplinary interaction to strengthen and establish new contact networks in the academic and research community. Participants from all disciplines related to renewable energy and several Latin American countries contributed to this important event. This edited book examines some essential topics related to renewable energies, such as developing and studying new materials, process design and simulation, computational modeling, energy efficiency, and chemical and mechanical analysis. Finally, the editors of this book would like to thank the Springer editorial team and all contributing authors for their commitment and collaborative effort in making this book.

Guayaquil, Ecuador — Mayken Espinoza-Andaluz
Guayaquil, Ecuador — Jordy Santana-Villamar
Guayaquil, Ecuador — Brayan Ordóñez-Saca
Guayaquil, Ecuador — Carlos Vallejo-Cervantes
Guayaquil, Ecuador — Luis Rodríguez-Alava

Contents

About the Editors

Mayken Espinoza-Andaluz is a full professor at Escuela Superior Politécnica del Litoral (ESPOL) and co-supervisor of PhD students in heat transfer and energy systems at Lund University. He earned his PhD in heat transfer from the Department of Energy Sciences at Lund University. His areas of research include renewable energy, fuel cells, proton-exchange membrane fuel cells (PEMFCs), solid oxide fuel cells (SOFCs), physical and chemical phenomena, energy efficiency, lattice Boltzmann methods (LBMs), and computational modeling. He is a reviewer for several international journals, including the *International Journal of Energy Research*, *International Journal of Hydrogen Energy*, and *Computers & Mathematics with Applications*. He is the second vice president of the Ecuadorian Hydrogen Association and the current official speaker on Hydrogen and Fuel Cells from ESPOL.

Jordy Santana-Villamar is a mechanical engineer who completed his Master's in Materials Science and Engineering. He conducted a research stay at Instituto de Carboquímica in Zaragoza, Spain, majoring in vanadium-redox flow batteries and hydrogen fuel cell stack operation. Currently, he holds a research position and leads experimental research on polymer electrolyte fuel cells at ESPOL, Ecuador. His expertise also encompasses composite materials, particularly epoxy/graphite composites for bipolar plates in fuel cells, and development of biodegradable chitosan-based materials for food packaging. As the editor of the conference proceedings for the Congress on Research, Development, and Innovation in Renewable Energies (CIDiER), he plays a role in advancing research on renewable energies. He has published over 15 scientific works on PEM fuel cell experimentation and epoxy/graphite composites in top academic journals. Additionally, he recently received the Best Oral Presentation Award at CEEGE 2024 International Conference in Los Angeles, USA, and achieved second place in the Falling Walls Lab Colombia

2024 international competition. He has also secured funding for the research project "Study and Optimization of Wave Energy Systems for Application on the Coasts of Santa Elena, Ecuador" and serves as codirector of the project "Theoretical and Applied Research in Hydrogen Fuel Cells to Promote Their Application in the Decarbonization of Strategic Sectors."

Brayan Ordóñez-Saca received his Bachelor's Degree in Mechanical Engineering in 2022 from ESPOL. He is currently working as a research assistant in the Laboratory of Renewable Energy Sources in the same institution. He also has experience in the production of energy through generator sets. Furthermore, he developed an investigation about the modeling of a hybrid dryer at laboratory scale. He has participated in a research project about Study of the Fluid Dynamic Behavior of a Hybrid Dryer. Actually, his research interests spin around fuel cells, specifically proton-exchange membrane fuel cell, and their performance study applying electrochemical tests. On the other hand, he is now working on the development of expanded graphite-resin composite material with secondary fillers for bipolar plates. Recently, he has been working as an editor of the fourth Congress on Research, Development, and Innovation in Renewable Energies (CIDiER), Guayaquil, Ecuador.

Carlos Vallejo-Cervantes received his Bachelor's Degree in Mechanical Engineering in 2020 from ESPOL, Guayaquil, Ecuador. Currently, he works as a research technician in Renewable Energy Laboratory, a division of Renewable and Alternative Energy Center. He was the student representative to the university superior council. He has an extensive experience in the field of mechanical design as well as modeling. In the current work, he has been working on fuel cells and specifically on bipolar plates composed of graphite and resin. Previously, he was a contractor for one of the most important beer companies in the country. His work in this company has a great industrial impact both nationally and in the region.

Luis Rodríguez-Alava is a distinguished student at the Faculty of Mechanical Engineering and Production Sciences at ESPOL. He served as a teaching assistant for electricity and magnetism physics for a semester, demonstrating exceptional teaching skills and understanding of concepts. Additionally, he has made significant contributions as a research assistant in the Physics Laboratory, participating in the preparation of a conference paper presented at GIREP-EPEC 2023, which focused on the study of the magnetic force generated by an electromagnet. He received recognition for second place in the Science and Mathematics Fair organized by the polytechnic community for his research on the electromagnetic force of the electromagnet. Currently, he is developing a manuscript on vibratory systems in MATLAB and has joined the Laboratory of Renewable Energy Sources as a research assistant, working on a project funded by CEDIA. His dedication and academic excellence reflect his commitment to research and scientific advancement in engineering.

Part I
Solar Energy

Design of a Solar Greenhouse with an Integrated Aquaponic System

Gustavo Bohórquez, Catherine Crispo, and María E. Álvarez

1 Introduction

Agriculture is a fundamental strategy for fighting hunger. However, this activity represents the most significant proportion of human land use, with approximately 5000 Mha, equivalent to 38% of the earth's land surface [1]. In addition, modern food systems rely heavily on fossil fuels, which consume about 30% of the world's available energy and generate more than 20% of global greenhouse gas emissions [2]. To address the exponential growth of the world's population and the resulting food insecurity, agricultural production will need to double in the next 30 years. This would directly impact pollution and greenhouse gas emissions [3]. Solar greenhouses are enclosures where food is grown by providing a suitable environment [4]. Greenhouses receive the necessary sunlight for photosynthesis; however, each type of plant requires specific environments for its best production. They require moderate temperature, light, carbon dioxide, oxygen, mineral nutrients, and the movement of air and water [4]. Thus, temperature is a dominant environmental factor in plant growth, which is why, to be optimal, it must be high, but at the same time, it must ensure that it does not cause changes in the metabolism of the species. Therefore, light intensity and its spectral distribution are essential for plant growth and development. Consequently, the design of a greenhouse depends mainly on the type of plants to be grown inside, the light (intensity and angle of radiation), heat requirements, local climate, and construction materials.

Solar greenhouses employ specially designed structures for cultivating plants and vegetables while harnessing solar energy. The authors detailed methods for designing solar greenhouses for sustainable year-round food production, emphasizing

---

G. Bohórquez · C. Crispo · M. E. Álvarez (✉)
Universidad Metropolitana, Caracas, Miranda, Venezuela
e-mail: mealvarez@unimet.edu.ve

M. Espinoza-Andaluz et al. (eds.), *Congress on Research, Development and Innovation in Renewable Energies*, Green Energy and Technology,
https://doi.org/10.1007/978-3-031-88995-0_1

essential factors such as sun orientation and thermal resistance of greenhouse roofs and wall materials [5]. On the other hand, aquaponic systems focus on profitability and waste treatment by combining hydroponic crops with aquatic species breeding. This approach proves to be ecologically effective [6]. Aquaponic systems thus meet existing needs by maximizing environmental utilization with minimal waste generation, balancing human requirements and biosphere capacity [7].

The authors [8] designed and constructed a small-scale aquaponic system with a cultivation area of 1 m^2, employing a media bed configuration, combining white-stemmed chard (*Beta vulgaris* var. *cicla*) and Nile tilapia (*Oreochromis niloticus*). After 63 days of growth, they assessed system productivity by measuring leaf length, plant height, leaf count, total biomass of the chard, and fish weight and size. Results showed that the proposed system is a food production alternative capable of reducing environmental impact and enhancing productivity. The system demonstrated plant performance similar to that of traditional soil-based cultivation methods. This study proposes the design of a solar greenhouse integrated with an aquaponic system, considering the climatic conditions of Caracas (Venezuela) and the results from the developed experimental aquaponic prototype.

2 Solar Greenhouse Design

First, the authors selected the location for the design of the solar greenhouse, and fundamental environmental parameters were recorded, such as solar path, elevation angles, and azimuth angles for the solstices and equinoxes at noon. The aquaponic system integrated into the greenhouse was based on the experimental prototype developed by [8]. Therefore, white-stemmed chard and Nile tilapia were selected for the aquaponic system, considering their symbiotic relationship as plant and animal species. The media bed technique was chosen due to its simplicity and high oxygenation of species. Additionally, the substrate acts as a biological and mechanical filter. Furthermore, the aquaponic system requires the following main components: fish tank, growing beds, water supply reservoir, aeration system, and water flow system. Additional design elements for the rest of the greenhouse included construction materials, greenhouse orientation, rainwater harvesting system, roof tilt angle, and solar energy supply system.

2.1 Location and Environmental Considerations

The southern rooftop of the Corimón building at Universidad Metropolitana was chosen as the location for the solar greenhouse, with an area of approximately 216 m^2 (21.4 m × 10.1 m). Using the SunCalc.org platform [9], the sun path diagram for the selected area throughout the year was obtained. Additionally, elevation and azimuth angles for the solstices and equinoxes at noon were recorded, describing the

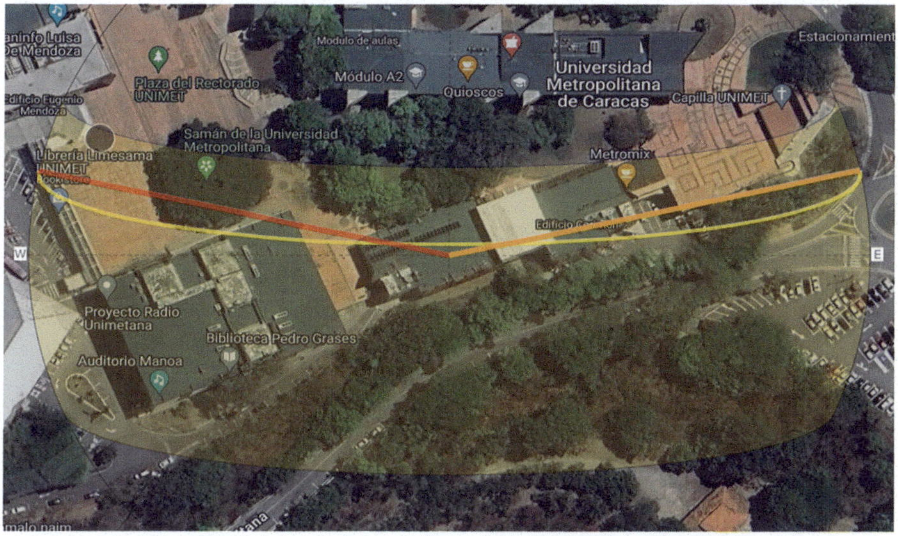

Fig. 1 Sun path diagram on the rooftop of the Corimón building, Universidad Metropolitana

Table 1 Position of the sun at UNIMET for the solstices and equinoxes (noon) [9]

Date	Elevation angle	Azimuth angle
March 20, 2023	76.36	140.35
June 21, 2023	75.34	27.15
September 23, 2023	78.30	155.18
December 21, 2023	55.52	169.78

sun's position for these dates. Figure 1 and Table 1 depict the sun's path and position on the rooftop of the Corimón building. Conducting the study at this location ensures minimal shadow interference from other structures that could affect greenhouse performance. It was found that the average sun elevation angle during the coldest months (December and January) is approximately 58°.

2.2 *Aquaponic System to Be Implemented in the Greenhouse*

The aquaponic system to be integrated into the solar greenhouse was based on the previous work by [8]. Due to the available space on the rooftop, scaling up was implemented, establishing a prototype four times larger. Each box will have a cultivation area of 1.0 m^2, aimed at standardizing the average plant distance to 20 cm, as this spacing resulted in the highest productivity, according to [8]. Considering the four boxes, there are 25 green leaf plants per box totaling 100 plants across 4.0 m^2. Table 2 outlines the general characteristics of the components of this larger-scale system.

Table 2 Description of the components of the aquaponics system

Component	Characteristic	Value
Fish tank	Volume (L)	1000
Growing beds	Quantity	4
	Total cultivation area (m^2)	4.0
Plants	Quantity	100
Fish feeding	Daily rate (g/day)	160

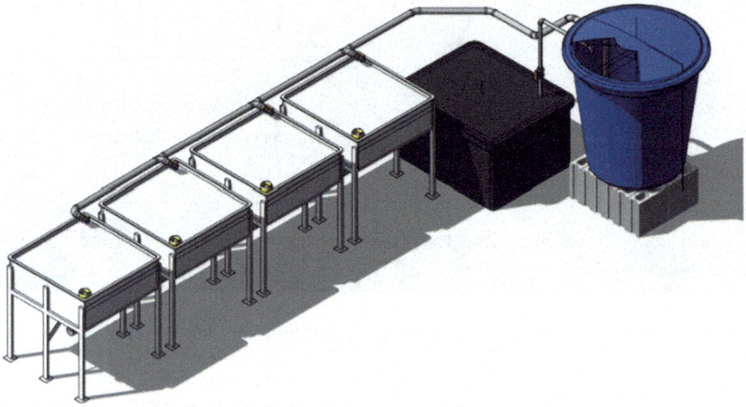

Fig. 2 Aquaponic system to be implemented in the greenhouse

The fish tank capacity was established at 1000 L, deemed suitable for the cultivation area and the number of plants specified [10]. A double-layer conical tank was selected for this purpose. Additionally, each growing bed had a bell siphon for water circulation to the sump. The sump tank, made from 40 mm thick PVC material, was chosen to hold a maximum capacity of 780 L of water. Figure 2 illustrates the isometric diagram of the proposed system.

For the water recirculation system, PVC schedule 40 pipes were selected based on their diameters and lengths. The 3/4 in. pipes draining the growing beds are not connected directly to the 3 in. pipe returning to the sump; instead, they discharge into this pipe through holes to ensure proper operation of the bell siphons. Additionally, models of submersible pumps MULTI 5800, FP0S1250X, and UTP017-1 from brands SICCE, Pentair Flotec, and Acquaer were evaluated. Operating points were calculated for each proposed pump. Accordingly, the SICCE MULTI 5800 pump was chosen for use in the aquaponic system, as it exceeds the minimum required flow rate (2000 L/h), has the lowest nominal power consumption at 0.090 kW, and requires the least hydraulic power at its operating point.

For the aeration required in the fish tank, the Aqua Miracle ACO004 air pump was selected. This pump delivers a flow rate of 4543 L/h, approximately 4.5 times the tank's total volume, ensuring adequate aeration for the aquatic species present.

2.3 Construction Details: Dimensions and Orientation

Figure 3 depicts the dimensions of the proposed solar greenhouse, which features a length-to-width ratio of 2:1 designed to maximize the roof area facing south.

Roof Pitch Angle of the Greenhouse

Based on the average elevation angles in December and January, it was determined that the ideal roof angle to achieve a solar incidence angle of 0° during these months is 32°. However, this inclination results in a solar incidence angle greater than 45° during the summer solstice, which reduces light transmission. Therefore, a roof pitch of 22.6° was selected to ensure the solar incidence angle remains below 45° at the lowest and highest sun elevations. This angle guarantees optimal light transmission throughout the year and maximizes solar exposure during the colder months.

Additionally, the roof should face south to receive maximum sunlight. However, the building's rooftop deviates 21° toward the east. In this context, it was decided that the greenhouse should have the same orientation, as this deviation still allows for around 90% of sunlight compared to a fully south-facing orientation. Moreover, this orientation prevents overheating in the greenhouse during the late afternoon and enhances heating capacity in the early morning hours [5].

Construction Materials

The materials for glazing the greenhouse were selected based on the following variables: light transmission, thermal resistance, lifespan, and light behavior. The wall material was determined based on the recommended thermal resistance for the climatic conditions of this area, as established by [5]. Table 3 presents the detailed properties of each material studied for the selection of glazing, both for the roof and the south-facing wall of the greenhouse. Additionally, polyethylene has a minimal lifespan, which would result in frequent replacement and continuous expenditure.

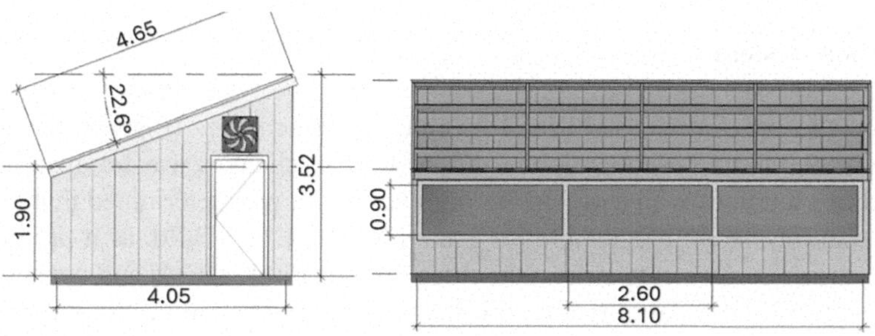

Fig. 3 Dimensions of the solar greenhouse (in meters)

Table 3 Properties of greenhouse glazing materials

Material		Light transmission (%)	Thermal resistance (*R*-factor) (ft^2 °F h/BTU)	Service light (years)	Light behavior
Glass	Single crystal (3 mm)	93	0.950	25+	Transparent
	Double crystal	81	2.084	25+	Transparent
	Double crystal, Low E	79	3.846	25+	Transparent
Polycarbonate	Single layer	91	0.830	10–15	Translucent
	Double layer (8 mm)	81	1.724	10–15	Translucent
	Triple layer (16 mm)	75	2.500	15–20	Translucent
	Five layers (20 mm)	60	3.570	15–40	Translucent
Polyethylene	Single film	90	0.830	2–4	Translucent

On the other hand, double-pane Low-E glass and triple-layer polycarbonate (16 mm) strike a better balance between light transmission and thermal resistance. However, triple-layer polycarbonate was chosen for the roof because its translucency allows for diffuse light transmission, which is advantageous for crop growth [5].

Regarding vertical glazing on the south wall of the greenhouse, the goal is to maximize light transmission during the colder months when the solar elevation angle is lower, resulting in a reduced angle of incidence. In this regard, it was determined that low-emissivity double-pane glass (Low-E) should be used because it offers high thermal resistance and excellent light transmission. As for the greenhouse walls, considering the recommended thermal resistance value of between 14 and 20 ft^2 °F h/BTU for areas with annual minimum temperatures between 10 and 20 °C [5], ISOPANEL structural panels were selected. These panels consist of a core of self-extinguishing expanded polystyrene (EPS) covered on both sides with galvanized steel sheets. Specifically, a 10 cm thick panel with a density of 20 kg/m^3 was chosen, providing a thermal resistance of 15.2 ft^2 °F h/BTU.

Ventilation System

Based on the previously established greenhouse dimensions (Fig. 3), the internal volume is 90.0 m^3. Consequently, a value of 1.50 air changes was set, resulting in a required extraction flow rate of 4767 ft^3/min (CFM). It was determined that the most suitable fan for use in the greenhouse is the FREDIVE AFP3-20, as it provides the highest extraction flow rate among the three options (5000 CFM) and consumes the least energy (1/2 HP). Additionally, this fan is single-phase, simplifying its wiring to the photovoltaic solar energy system compared to a three-phase fan. It is important to note that this fan should operate with a thermostat to prevent greenhouse overheating during warmer months. Furthermore, considering the fan diameter of 20 in., 4 square intake shutters with a side length of 14.5 in. were installed, providing an air intake area of 0.543 m^2.

3 Photovoltaic Solar Energy System

3.1 Required Energy

In Table 4, the power requirements for each of the electrical equipment used in the greenhouse are shown. The daily consumption was determined based on the equipment used in the aquaponics system and the greenhouse, including their lighting. This formed the basis for calculating the daily energy needs, using the global efficiency value *R* described in Eq. (1) [11]. Typical coefficient values are shown in Table 5. Thus, the daily energy requirement was obtained using Eq. (2)

$$R = (1 - k_b - k_i - k_r - k_v) \cdot (1 - k_a \cdot N/P_d) \tag{1}$$

Applied Eq. (2) and assuming modern regulators, inverters, and new stationary batteries with a 3-day autonomy, the overall efficiency obtained is 0.731.

$$E_{\text{daily}} = \frac{\text{Daily consumption}}{R} \cdot (1 + \text{Safety coefficient}) \tag{2}$$

where *E* is the daily energy requirement (kWh), *R* is the daily efficiency factor (dimensionless), and *S* is the safety coefficient (dimensionless). Consequently, the daily energy requirement for the solar greenhouse is 9.57 kWh.

Table 4 Daily energy consumption of solar greenhouse equipment

Component	Quantities	Power (kW)	Working hours per day	Consumption per day
Water pump	1	0.090	24	2.16
Air pump	1	0.058	10	0.580
Fan	1	0.373	5	1.87
Fan thermostat	1	0.003	24	0.0720
Led lamps (lighting)	6	0.024	8	1.15
Simultaneous power (kW)		**0.668**	**Total (kWh)**	**5.83**

Table 5 Coefficient for calculating overall efficiency

k_b	k_i	k_r	k_v	k_a	N	P_d
0.05	0.05	0.1	0.05	0.005	3	0.6

3.2 Peak Sun Hours (PSH)

According to the irradiation data collected from the NASA POWER platform [12] for the location of Universidad Metropolitana, the highest average annual solar irradiation can be obtained with a panel tilt angle equal to the university's latitude (10.5°). Accordingly, June has the lowest peak sun hours (5.09 PSH). Therefore, it was selected to estimate the number of solar panels.

3.3 Photovoltaic Solar Panels, Controllers, and Batteries

For the photovoltaic solar energy system, the SG340P polycrystalline solar panel model from Peimar, an Italian brand available locally, was selected. This model offers a nominal power of 340 Wp. The voltage and current at maximum power are 38.3 V and 8.88 A, respectively. With a value of 5.09 PSH for the study, the daily energy generated per panel (according to Eq. 3) is:

$$E_{\mathrm{GM}} = 38.3V \cdot 8.88A \cdot 5.09h = 1.73\ \mathrm{kWh/panel} \tag{3}$$

The number of solar panels will be (Eq. 4):

$$\text{Number of solar panels} = \frac{9.57\ \mathrm{kWh}}{1.73\ \mathrm{kWh/panel}} = 5.53\ \text{panels} \approx 6\ \text{panels} \tag{4}$$

The daily energy generated by each panel was determined to be 1.73 kWh. Accordingly, the requisite number of solar panels is 5.53, nearly equivalent to 6. The configuration of these panels was chosen so that three strings of two modules in series are associated in parallel, thus occupying an area of 11.6 m^2. From the established configuration of the solar panels and given the short-circuit current per panel (9.51 A) and open-circuit voltage (46.7 V), it can be determined that the maximum current (28.5 A) and maximum voltage (93.5 V) are obtained. In this context, the BlueSolar MPPT 150/60 charge controller was selected, which would work at a system voltage of 24 V and support a maximum voltage of 150 V and a load current of 60 A. Considering that the daily consumption of the solar greenhouse is 5.82 kWh, it was obtained that the bank capacity required for a maximum depth of discharge of 60%, a system voltage of 150 V, and a system current of 60 A. A system voltage of 24 V and 3 days of autonomy is 1214 Ah. Thus, using the RITAR RA12-200 battery model, which has a voltage of 12 V and a capacity of 200 Ah, a bank of 14 batteries is required, which consists of a parallel connection of seven series of two batteries. In Table 4, it was determined that the equipment belonging to the greenhouse needs a simultaneous power of 0.668 kW. When adding a 20% safety factor, a value of 0.802 kW was obtained. Consequently, it was established that the inverter suitable for these conditions is the Phoenix 24/1200 model, which has a nominal output power of 1.00 kW and tolerates a peak power of 2.20 kW, thus being ideal to support the starting currents of the air pump and the greenhouse fan.

3.4 *Rainwater Harvesting System*

The average annual precipitation at the Universidad Metropolitana was obtained using NASA POWER [12]. From it, knowing the effective catchment area and a runoff coefficient of 0.80 [13], the volume of water that could be collected annually was determined. Considering a daily water loss of 3% of the water system [14] and the time desired to reserve water, the required storage capacity was obtained, which served as a factor for selecting the proposed tanks. Subsequently, the isometry of the pipes for rainwater harvesting was made through Autodesk Revit. According to [12], the average annual rainfall at the Universidad Metropolitana is 735 mm, equivalent to L/m^2. Thus, considering that the catchment area corresponds to the greenhouse roof of 38.6 m^2, the volume of water possible to collect per year (L) was obtained to be 22,701 L. Considering that the amount of water present in the aquaponic system is about 1600 L, it is possible to estimate the daily loss of this system, taking as a reference a maximum percentage of 3% of the total water volume [14]. Thus, establishing that it is desired to have a water reserve for February and March (60 days) since these are the months with less rainfall [12], the minimum storage capacity that would be required was determined to be 2880 L. Consequently, it was decided that three tanks of 1000 L each would be used to have a final storage capacity of 3000 L. Thus, Fig. 4 shows the rainwater harvesting system. It should be pointed out that the 4-in. pipe belonging to section B corresponds to the first water pipe, which filters the initial dirt carried by the rainwater. Finally, after finalizing the

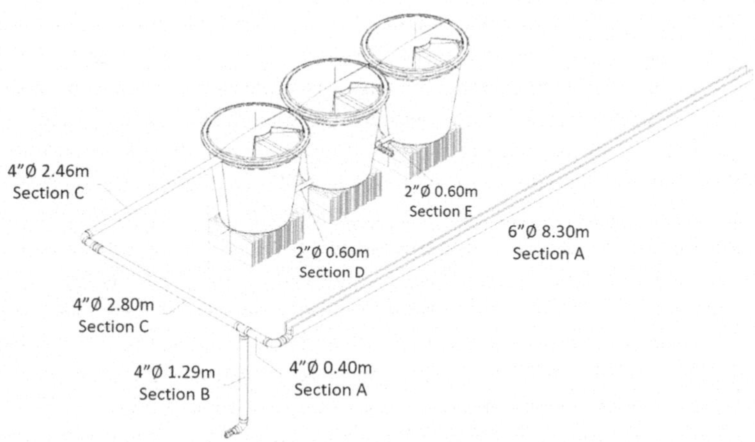

Fig. 4 Isometry of the rainwater harvesting system detailed

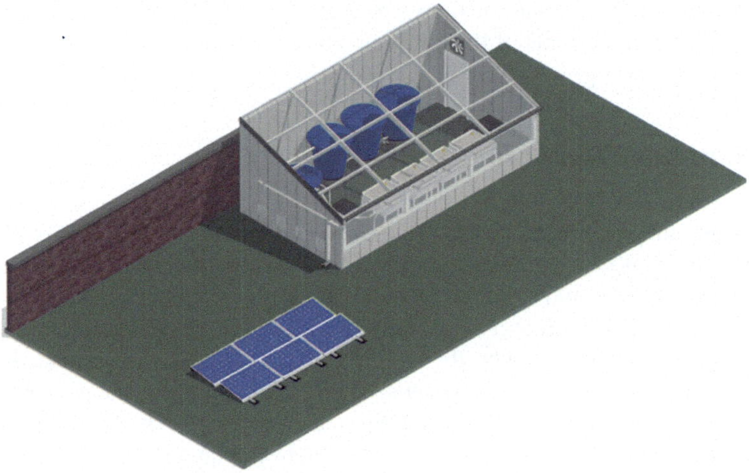

Fig. 5 Isometry of the solar greenhouse layout

design of the greenhouse, the photovoltaic system, and rainwater harvesting, Fig. 5 shows the distribution on the building's roof.

4 Conclusion

The present work proposes the design of a solar greenhouse as a sustainable alternative for food production. The system considers integrating an aquaponic method for food production based on the experimental prototype previously tested by the authors. It is based on the city of Caracas's environmental conditions and is located at the Universidad Metropolitana headquarters on the Corimón Building's rooftop. The greenhouse's dimensions are 8.10 m × 4.05 m and 3.52 m maximum height. The aquaponic system can grow 100 plants over a total area of 1 m^2. The design ensures optimal light transmission for the plant species throughout the year through a roof pitch of 22.6° and adequate thermal insulation and ventilation to prevent overheating in the warmer months. For the greenhouse's autonomous and sustainable operation, the design incorporates a photovoltaic solar energy system with six panels (2.04 kWp), meeting the greenhouse's daily energy demand of 9.57 kWh. Additionally, the design includes a rainwater harvesting system with a storage capacity of 3000 L. This enables the greenhouse to meet its energy and water requirements independently, without relying on external sources. The electrical autonomy of the greenhouse is ensured for at least 3 days through a 14-battery bank. The results of this study provide the basis for constructing a solar greenhouse. The proposal demonstrates an attractive possibility for food production by leveraging the natural symbiotic relationship between fish and plants and optimizing the construction materials to establish optimal growing conditions.

References

1. Food and Agriculture Organization of the United Nations (FAO): Agricultural land use by the numbers. https://www.fao.org/sustainability/news/detail/en/c/1274219/#:~:text=Globally%20agricultural%20land%20area%20is,and%20pastures)%20for%20grazing%20livestock
2. Food and Agriculture Organization of the United Nations (FAO): Ensure access to affordable, secure, sustainable and modern energy for all. https://unstats.un.org/sdgs/report/2017/goal-07/
3. Institute of Agrarian and Environmental Issues & UNICEF: Global food crisis, challenges and solutions
4. Garg, H.: Advances in Solar Energy Technology. D. Reidel Publishing Company, Dordrecht (1987)
5. Schiller, L., Plinke, M.: The Year-Round Solar Greenhouse: How to Design and Build a Net-Zero Energy Greenhouse. New Society Publishers, Gabriola Island, BC (2017)
6. Bonilla, L., Casabianca, L., Montaño, J., Pantoja, S., Rada, D., Salcedo, C.: Design and construction of a prototype of an aquaponic system for the use and treatment of fish farming waste from the Hacienda La Cosmopolitana, Restrepo—Meta. Revista de Tecnología (2015)
7. Goddek, S., Joyce, A., Kotzen, B., Dos-Santos, M.: Aquaponics and global food challenges. In: Aquaponics Food Production Systems. Springer, Cham (2019)
8. Bohórquez, G., Crispo, C.: Solar greenhouse: study of the integration of an aquaponic system for food production. Caracas (2023)
9. Hoffmann, T.: SunCalc—sunrise, sunset, shadow length, solar eclipse, sun position, sun phase, sun height, sun calculator, sun movement, map, sunlight phases, elevation, photovoltaic system, photovoltaic. http://suncalc.net/
10. López, J.: Aquaponic cultivation: specialized guide. https://cifalmalaga.org/web/wp-content/uploads/2020/04/2019.11.07-LIBRO-ACUAPONIA.pdf
11. Alvarado, J.: Design and Calculation of a Stand-Alone Photovoltaic System. Universidad Politécnica de Madrid, España (2018)
12. NASA POWER: Prediction of Worldwide Energy Resources (POWER). https://power.larc.nasa.gov/ (2020)
13. Adler, I., Carmona, G., Bojalil, J.: Manual de captación de aguas de lluvia para centros urbanos. International Renewable Resources Institute Mexico. https://docplayer.es/7110890-Manual-de-captacion-de-aguas-de-lluvia-para-centros-urbanos-ilan-adler-gabriela-carmona-jose-antonio-bojalil.html
14. Somerville, C., Cohen, M., Pantanella, E., Stankus, A., Lovatelli, A.: Small-scale aquaponics food production—integrated fish and plant culture. FAO Fisheries and Aquaculture Technical Paper No. 589. FAO, Rome (2022)

Machine Learning for the Prediction of Perovskite Solar Cell Performance: A Brief Review

Clara Rojas-Rincón, Yeraldin Vélez-Galvis, Mónica Botero-Londoño, Erick Reyes-Vera, and Alexander Sepúlveda

1 Introduction

In recent years, there has been considerable progress in the study, design, and production of perovskite solar cells (PSCs). PSCs are currently regarded as one of the best alternatives to silicon solar cells due to their high absorption coefficient, low cost, fabrication on flexible and curved substrates, ease of production, tunable transparency, and low levelized cost of electricity (LCOE) [1, 2]. However, perovskite solar cells'efficiency and stability strongly depend on the physicochemical features of the perovskite material. These qualities include the perovskite film's crystalline structure, chemistry, and morphology. This significantly affects the short-circuit current density (J_{sc}), open-circuit voltage (V_{oc}), fill factor (FF), and efficiency (η), parameters that allow the performance of the cells to be evaluated.

Solar radiation absorption, charge carrier movement and separation, and energy transfer to the charge are all necessary for a solar cell to function [3]. The geometrical configuration of the cell, the materials used in the various layers, the process, and the synthesis parameters for each layer all affect the effectiveness of PSCs [4]. The basic architecture of a perovskite solar cell is depicted in Fig. 1. The device's active layer is perovskite (ABX3), which absorbs photons from the sun and creates photogenerated carriers. It is the cell's most important layer, and it is what gives

C. Rojas-Rincón · M. Botero-Londoño · A. Sepúlveda (✉)
School of Electrical, Electronics and Telecommunications Engineering, Universidad Industrial de Santander, Bucaramanga, Colombia
e-mail: alexander.sepulveda@saber.uis.edu.co

Y. Vélez-Galvis · E. Reyes-Vera
Department of Electronics and Telecommunications, Instituto Tecnológico Metropolitano, Medellín, Colombia

M. Espinoza-Andaluz et al. (eds.), *Congress on Research, Development and Innovation in Renewable Energies*, Green Energy and Technology,
https://doi.org/10.1007/978-3-031-88995-0_2

Fig. 1 Configuration of a perovskite solar cell

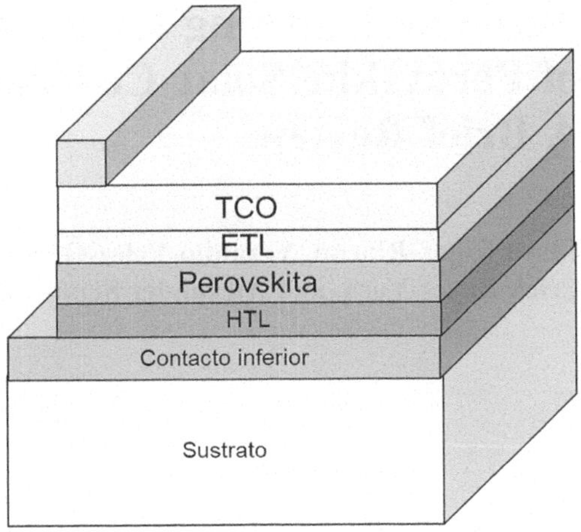

devices and technology their given names. Each cell's layer can be made from various materials and synthesis methods.

Machine learning (ML) applications in materials science are fast expanding due to their ability to provide interpretable data that can assist in designing and manufacturing perovskite solar cells [5, 6]. This research area can give a more accurate assessment of factors that affect PSC performance characteristics, as in [4, 7], including V_{oc}, J_{sc}, FF, and η. In addition, ML can find complicated patterns and correlations between material qualities and cell performance by analyzing vast amounts of experimental and simulated data using sophisticated algorithms [8, 9]. This not only streamlines the development process by lowering the time and costs associated with traditional experimentation, but it also allows for creativity in designing new structures and material compositions that maximize the efficiency and stability of perovskite solar cells.

Some researchers have proposed machine learning algorithms for analyzing data and determining the impact of geometrical, optical, thermal, and electrical variables in perovskite solar cells [6, 10]. These algorithms predict whether the planned cell will result in a better-performing device by generating high V_{oc}, J_{sc}, FF, and η values. Thus, this method could transform the modeling of perovskite solar cells, lowering experimental research expenses and analysis times. Furthermore, it would enable cells to be built with exceptional efficiency without producing many samples [6, 11].

In this manuscript, the authors present a review of the state-of-the-art in machine learning methods applied to predicting perovskite solar cell performance. The review focuses on the machine learning approaches used to model and predict the behavior of perovskite solar cells, encompassing recent studies and emerging trends. It is acknowledged that there are related articles in the state-of-the-art; however, this document provides a taxonomy of machine learning algorithms applied to predicting perovskite solar cells. We also discuss the challenges and limitations associated with integrating machine learning into this field.

2 Computational Modeling of Perovskite Solar Cells

Computational modeling of PSCs is a growing research subject due to its enormous potential for studying the behavior and performance of these solar cells. Computational techniques enable the identification of features that restrict their efficiency and longevity. Furthermore, these tools enable the exploration of various structures, materials, and film thicknesses, among others. They allow the study of their electrical, thermal, and optical behavior without the need to manufacture them, significantly lowering research costs and implementation times. This allows for more efficient design before fabrication [12].

There are numerous approaches to the computational modeling of PSCs. The primary techniques comprise the finite element method (FEM) [13], drift-diffusion modeling (DDP) [14, 15], the transfer matrix method [16], ray-tracing privative [16], and theoretical modeling [17]. The 3D finite element method (FEM) is well-suited for representing intricate shapes and several physical phenomena such as optics, heat, and electricity. However, it requires significant computational resources. This option has proven to be highly effective, as models utilizing this methodology consistently align more closely with real-world conditions [13]. The DDP method solves the drift-diffusion and Poisson equations to determine the electrical potential and ensure accurate precise charge transport. Despite its complexity and parameter dependence, the perovskite absorber incorporates mobile ions. The optical transfer matrix method, employed in software such as SETFOS, enhances light absorption by computing the reflection and transmission in thin films. However, it may not encompass all intricate optical phenomena. Ray tracing is a technique that traces the path of light through a device. It combines with the transfer matrix method to accurately account for large-scale textures. While it offers high precision, it is computationally intensive. The field of theoretical modeling aims to analyze the influence of flaws and various circumstances on the performance of devices, offering a basic yet simplified comprehension of intricate processes. Nevertheless, it requires approximations, and the outcomes are typically less precise than experimental results.

Multiple open-source software applications facilitate basic simulations of perovskite solar cells, such as WXAMPS, SCAPS-1D, and GVPDM. These computational tools focus mainly on the electrical processes inside the cell structure. A descriptive analysis of some of these simulation tools is presented in Table 1. For example, SCAPS-1D is one of the most widely used tools by researchers due to its ability to efficiently model solar cells [18, 19]. This tool also allows for incorporating material absorption coefficient curves and simulating defects, which enriches the accuracy of the results toward more realistic conditions. Concerning commercial software such as COMSOL Multiphysics, TCAD, PC1D, and SETFOS, COMSOL has also been employed more since it allows the integration of several physical models (thermal, optical, semiconductor, etc.), which makes it more effective in terms of design because it is more similar to what was obtained in the experimental phase. It even allows the use of doped materials.

Table 1 Comparative table between the different software for PSC simulations

Name	License	Use	Physical study
COMSOL Multiphysics [20]	Privative	Modeling of 1D, 2D, and 3D geometries	Optical, electrical, and thermal
SCAPS-1D [21]	Free	Modeling of 1D geometries	Electric
GVPDM [22]	Free	Modeling of 1D, 2D, and 3D geometries	Optical and electrical
WXAMPS [23]	Free	Microelectronic and photonic modeling	Electric
TCAD-SILVACO [24]	Privative	Semiconductor modeling	Electrical, optical, and mechanical
PC1D [25]	Free with restrictions	Modeling of 1D geometries	Electric

3 Data for the Machine Learning Activity

Data is the unprocessed information that machine learning algorithms use to build models and make predictions about the efficiency of solar cells, filled factors, or other intriguing parameters. These input data could come from experimental results [26] or simulations [27].

Simulated data provide a regulated and comprehensive dataset for training algorithms, as exemplified by [28]. They employed SCAPS to generate input and output parameters. Similarly, in [29], Shrivastav and collaborators utilized the ML algorithms to predict the performance of PSC by using simulated data obtained from SCAPS.

On the other hand, experimental data is commonly acquired from scientific research articles in which writers have conducted the manufacturing of the perovskite cells and measured the output characteristics. This task can be done manually (as in [26, 30, 31]) or using natural language processing methods.

Thus, machine learning and deep learning techniques with high-fidelity data make it possible to quickly evaluate and enhance PSC designs. By utilizing these algorithms, researchers can effectively investigate a wide range of design possibilities, determine the most favorable material characteristics, and forecast the influence of different manufacturing conditions on the performance of solar cells [32, 33].

Some projects collect data from PSCs. In particular, Li et al. [34] collected data from 333 perovskite solar cells obtained from 200 scientific papers. Information about the perovskite layer composition, band gap, and the energy difference concerning the hole transport layer (ΔH) and electron transport layer (ΔL) was carefully retrieved. This dataset is expanded in a subsequent study [31] by including information about those observations' thickness and grain size. However, in both circumstances, the available data is limited. By contrast, Jacobson and collaborators [30] introduce a perovskite dataset with over 400 variables collected from over 42,000 perovskite solar cell records between 2003 and 2020.

4 Perovskite Solar Cells Performance Prediction Using Machine Learning

The general model for estimating the efficiency of perovskite solar cells (PSCs) using machine learning follows several critical steps. First, experimental or simulated data on solar cells' physical, chemical, and geometrical properties are collected. These data are then divided into training and testing sets. Next, an appropriate machine learning algorithm is selected, and the model is trained using the training dataset. After this, the model is tested using the testing dataset. Subsequently, the results are validated and evaluated using appropriate metrics to ensure the model's accuracy. Finally, the model's hyperparameters are optimized to maximize its performance and improve its predictive capacity [35].

In recent years, machine learning (ML) approaches have proven to be effective in modeling the performance of PSCs and providing valuable insights into the physical, chemical, and electrical phenomena involved. These methodologies make the perovskite cell design process faster and more efficient thanks to their ability to identify complex patterns and make predictions [36]. Moreover, they allow for the rapid evaluation of synthesis parameters (geometrical, chemical, electrical, etc.) and their impact on PSC performance (efficiency, stability, fill factor, among others).

Several key activities can be highlighted within the field of ML applied to PSCs: predicting properties such as bandgap [37–39]; performance prediction [7, 26]; inferring relevant descriptors [4, 7, 31]; stability analysis [40–42]; and explaining relationships between variables [37].

Machine learning presents several methodologies. The first is the pure prediction method, following the current trend of using "black box" models, in which even the creator cannot explain the generated predictions [43]. Another approach is based on explainable models that clarify the connections and rationales behind the predictions, commonly referred to as explainable artificial intelligence (X-AI) or explainable machine learning (X-ML) [37]. Additionally, physically informed machine learning (PIML) integrates physical principles into machine learning algorithms, aiming to enhance the precision and comprehensibility of the predictive models [44].

4.1 Data-Driven Models

Machine learning has demonstrated its efficacy in forecasting the performance of PSCs. Several ML approaches, including K-nearest neighbors (KNN), support vector regression (SVR), random forests (RF), artificial neural networks (ANN), gradient-boosted decision trees (GBDT), and XGBoost, have been used for this task. While RF, GBDT, and XGBoost methods have been found to have superior predictive performance, these models are considered "black boxes" because they do not explicitly incorporate knowledge of the physical laws governing the phenomenon from a PV material science perspective [7, 45, 46].

In [28], the authors employed a collection of supervised machine learning methods to construct a model that represents the connection between the input and output parameters given by SCAPS. The objective of this technique was to determine the input parameters that have a major impact on the efficiency of PSCs. In [29], Shrivastav and collaborators utilized the XGBoost algorithm to predict the performance of PSC. The algorithm was trained by using simulated data obtained from SCAPS. Their model yielded a coefficient of determination (R^2) of 99.99% and a mean squared error (MSE) of 0.0006, demonstrating exceptional precision and reliability in the predictions. XGBoost is known for its capacity to handle intricate, non-linear connections within the data. This makes it very suitable for predicting the performance of perovskite solar cells, which entails multiple interactions between materials and device topologies.

4.2 Explainable Machine Learning

Explainable machine learning (X-ML) seeks to determine the synthesis parameters that are most important and relevant. A common approach in this category is symbolic regression (SR), which attempts not only to predict but also to provide understandable explanations of the relationships between variables. For example [37], introduces an approach based on symbolic regression to model the bandgap of lead-free inorganic perovskites. This approach was based on the selection and combination of physicochemical descriptors, using mathematical operators to generate a high-dimensional feature space of 300,000 descriptors. The predictive model developed increased the performance measure given by the correlation coefficient from 81.6% to 92.4% and reduced the root mean square error (RMSE) from 0.89 to 0.59 eV.

4.3 Informed Machine Learning

The integration of theoretical physical knowledge with machine learning is useful for improving the accuracy and reliability of the results obtained. Incorporating physical knowledge (by integrating differential equations) provides a solid basis for developing more robust predictive models, improving the extrapolation capability of the models concerning existing empirical approaches [47].

Physically informed neural networks (PINNs) have emerged as a tool that combines physical principles with ML. PINNs have the ability to integrate complex physical laws into ML models. It could provide a deep understanding of how synthesis parameters affect performance, which could lead to significant advances [44]. This approach has been applied to other optoelectronic materials, such as tantalum and aluminum. In the case of tantalum, PINNs have enabled accurate large-scale simulations, while, for aluminum, interatomic potentials based on physical bond orders have been developed, reproducing first-principles energies and predicting various physical properties [47].

5 Conclusion

This review has demonstrated that machine learning techniques are valuable tools for driving the development of perovskite solar cells, offering opportunities to design more efficient and cost-effective solar technologies. For instance [28], achieved a coefficient of determination (R^2) of 99.99% using SCAPS-generated data to model PSC performance, demonstrating high precision. Similarly [29], employed the XGBoost algorithm, significantly reducing the mean squared error (MSE) in predictions. However, while "black box" models have proven effective in making predictions, their opacity limits the understanding of the physical and chemical principles that affect the efficiency of perovskite solar cells. To overcome this limitation, approaches such as explainable artificial intelligence (X-AI) and physically informed machine learning (PIML) are being developed, which promise to improve the accuracy, generalization, and interpretability of machine learning models.

Despite the growing availability of machine learning methods and fundamental knowledge about perovskite solar cells, these two approaches have been developed in isolation. This has hindered the integration of physical principles into data-driven models, representing a significant gap in current research. The lack of studies explicitly incorporating the principles of light-to-electric energy conversion into machine learning models highlights an opportunity for the development of more accurate and explainable predictions in this field.

Stability remains one of the major challenges for perovskite solar cells, and although researchers are exploring various strategies to improve it, few studies have used machine learning tools to address this issue. Additionally, the application of generative models offers promising potential for the creation of new solar cell structures, facilitating their simulation and validation and expanding the exploration of innovative perovskite material configurations.

Acknowledgments We thank MINCIENCIAS-Colombia for the financial support provided under contract 2022-0724, call 890, with resources administered by ICETEX-Colombia.

References

1. Yoo, J.J., Seo, G., Chua, M.R., Park, T.G., Lu, Y., Rotermund, F., Kim, Y.-K., Moon, C.S., Jeon, N.J., Correa-Baena, J.-P., Bulović, V., Shin, S.S., Bawendi, M.G., Seo, J.: Efficient perovskite solar cells via improved carrier management. Nature. **590**, 587–593 (2021). https://doi.org/10.1038/s41586-021-03285-w

2. Zhu, P., Chen, C., Dai, J., Zhang, Y., Mao, R., Chen, S., Huang, J., Zhu, J.: Toward the commercialization of perovskite solar modules. Adv. Mater. (2024). https://doi.org/10.1002/adma.202307357
3. Zhang, B., Xu, W., Peng, L., Li, Y., Zhang, W., Wang, Z.: Nature-inspired interfacial engineering for energy harvesting. Nat. Rev. Electr. Eng. **4**, 218–233 (2024). https://doi.org/10.1038/s44287-024-00029-6
4. Vélez, J., Botero Londoño, M.A., Sepulveda, A.: Measurement of information content of perovskite solar cell's synthesis descriptors related to performance parameters. Emerg. Mater. **7**, 1961–1968 (2024). https://doi.org/10.1007/s42247-024-00667-4
5. Makke, N., Chawla, S.: Interpretable scientific discovery with symbolic regression: a review. Artif. Intell. Rev. **57**, 2 (2024). https://doi.org/10.1007/s10462-023-10622-0
6. Yan, W., Liu, Y., Zang, Y., Cheng, J., Wang, Y., Chu, L., Tan, X., Liu, L., Zhou, P., Li, W., Zhong, Z.: Machine learning enabled development of unexplored perovskite solar cells with high efficiency. Nano Energy. **99**, 107394 (2022). https://doi.org/10.1016/j.nanoen.2022.107394
7. Lu, Y., Wei, D., Liu, W., Meng, J., Huo, X., Zhang, Y., Liang, Z., Qiao, B., Zhao, S., Song, D., Xu, Z.: Predicting the device performance of the perovskite solar cells from the experimental parameters through machine learning of existing experimental results. J. Energy Chem. **77**, 200–208 (2023). https://doi.org/10.1016/j.jechem.2022.10.024
8. David, T.W., Anizelli, H., Jacobsson, T.J., Gray, C., Teahan, W., Kettle, J.: Enhancing the stability of organic photovoltaics through machine learning. Nano Energy. **78**, 105342 (2020). https://doi.org/10.1016/j.nanoen.2020.105342
9. Hui, Z., Wang, M., Yin, X., Wang, Y., Yue, Y.: Machine learning for perovskite solar cell design. Comput. Mater. Sci. **226**, 112215 (2023). https://doi.org/10.1016/j.commatsci.2023.112215
10. Mahmood, A., Wang, J.-L.: Machine learning for high performance organic solar cells: current scenario and future prospects. Energy Environ. Sci. **14**, 90–105 (2021). https://doi.org/10.1039/D0EE02838J
11. Le Corre, V.M., Stolterfoht, M., Perdigón Toro, L., Feuerstein, M., Wolff, C., Gil-Escrig, L., Bolink, H.J., Neher, D., Koster, L.J.A.: Charge transport layers limiting the efficiency of perovskite solar cells: how to optimize conductivity, doping, and thickness. ACS Appl. Energy Mater. **2**, 6280–6287 (2019). https://doi.org/10.1021/acsaem.9b00856
12. Liu, F., Zhu, J., Wei, J., Li, Y., Lv, M., Yang, S., Zhang, B., Yao, J., Dai, S.: Numerical simulation: toward the design of high-efficiency planar perovskite solar cells. Appl. Phys. Lett. **104** (2014). https://doi.org/10.1063/1.4885367
13. Zandi, S., Razaghi, M.: Finite element simulation of perovskite solar cell: a study on efficiency improvement based on structural and material modification. Sol. Energy. **179**, 298–306 (2019). https://doi.org/10.1016/j.solener.2018.12.032
14. Hossain, M.K., Arnab, A.A., Samajdar, D.P., Rubel, M.H.K., Hossain, M.M., Islam, M.R., Das, R.C., Bencherif, H., Rahman, M.F., Madan, J., Pandey, R., Bhattarai, S., Amami, M., Dwivedi, D.K.: Design insights into La2NiMnO6-based perovskite solar cells employing different charge transport layers: DFT and SCAPS-1D frameworks. Energy Fuel. **37**, 13377–13396 (2023). https://doi.org/10.1021/acs.energyfuels.3c02361
15. Hossain, M.K., Ishraque Toki, G.F., Samajdar, D.P., Rubel, M.H.K., Mushtaq, M., Islam, M.R., Rahman, M.F., Bhattarai, S., Bencherif, H., Mohammed, M.K.A., Pandey, R., Madan, J.: Photovoltaic performance investigation of Cs3Bi2I9-based perovskite solar cells with various charge transport channels using DFT and SCAPS-1D frameworks. Energy Fuel. **37**, 7380–7400 (2023). https://doi.org/10.1021/acs.energyfuels.3c00540
16. Gohri, S., Madan, J., Pandey, R.: Highly efficient 3D–2D perovskite tandem solar cells: a combined ray tracing and transfer matrix-based simulation study. J. Electron. Mater. **53**, 1508–1516 (2024). https://doi.org/10.1007/s11664-023-10890-0
17. De Los, M., Santos, I., Cortina-Marrero, H.J., Ruíz-Sánchez, M.A., Hechavarría-Difur, L., Sánchez-Rodríguez, F.J., Courel, M., Hu, H.: Optimization of CH3NH3PbI3 perovskite solar

cells: a theoretical and experimental study. Sol. Energy. **199**, 198–205 (2020). https://doi.org/10.1016/j.solener.2020.02.026
18. Banik, S., Das, A., Das, B.K., Islam, N.: Numerical simulation and performance optimization of a lead-free inorganic perovskite solar cell using SCAPS-1D. Heliyon. **10**, e23985 (2024). https://doi.org/10.1016/j.heliyon.2024.e23985
19. Elsayed, M.R.A., Elseman, A.M., Abdelmageed, A.A., Hashem, H.M., Hassen, A.: Synthesis and numerical simulation of formamidinium-based perovskite solar cells: a predictable device performance at NIS-Egypt. Sci. Rep. **13** (2023). https://doi.org/10.1038/s41598-023-37018-y
20. Makableh, Y.F., Abu Awad, I., Hassan, W., Aljaiuossi, G.: Enhancement of the thermal properties of heterojunction perovskite solar cells by nanostructured contacts design. Sol. Energy. **202**, 204–209 (2020). https://doi.org/10.1016/j.solener.2020.04.002
21. Ompong, D., Clements, M.: Optimization of formamidinium-based perovskite solar cell using SCAPS-1D. Results Opt. **14**, 100611 (2024). https://doi.org/10.1016/j.rio.2024.100611
22. MacKenzie, R.C.I., Shuttle, C.G., Chabinyc, M.L., Nelson, J.: Extracting microscopic device parameters from transient photocurrent measurements of P3HT:PCBM solar cells. Adv. Energy Mater. **2**, 662–669 (2012). https://doi.org/10.1002/aenm.201100709
23. Hossain, M.K., Arnab, A.A., Das, R.C., Hossain, K.M., Rubel, M.H.K., Rahman, M.F., Bencherif, H., Emetere, M.E., Mohammed, M.K.A., Pandey, R.: Combined DFT, SCAPS-1D, and wxAMPS frameworks for design optimization of efficient Cs2BiAgI6-based perovskite solar cells with different charge transport layers. RSC Adv. **12**, 34850–34873 (2022). https://doi.org/10.1039/D2RA06734J
24. Chabane, H., Dehimi, L., Bencherif, H., Rao, S., Benhaya, A., Guenifi, N., Sasikumar, P., Younsi, Z., Shahatha, S.H., Mohammad, M.R., Alotaibi, N.H., Haldhar, R., Hossain, M.K.: Optimized Al0.25Ga0.75 as solar cell performance using a new approach based on hybridizing Silvaco TCAD simulator with real coded genetic algorithm. J. Opt. (2024). https://doi.org/10.1007/s12596-024-01969-w
25. Yadav, C., Kumar, S.: Numerical simulation of novel designed perovskite/silicon heterojunction solar cell. Opt. Mater. (Amst.). **123**, 111847 (2022). https://doi.org/10.1016/j.optmat.2021.111847
26. Odabaşı Özer, Ç., Yıldırım, R.: Performance analysis of perovskite solar cells in 2013–2018 using machine-learning tools. Nano Energy. **56**, 770–791 (2019). https://doi.org/10.1016/j.nanoen.2018.11.069
27. Wu, T., Wang, J.: Global discovery of stable and non-toxic hybrid organic-inorganic perovskites for photovoltaic systems by combining machine learning method with first principle calculations. Nano Energy. **66**, 104070 (2019). https://doi.org/10.1016/j.nanoen.2019.104070
28. Rauf, A., Sarker, S., Shafiqul Islam, M., Al Jame, H., Ahsan, S., Tohidul Islam, M., Shahriyar Nishat, S., Md. Shorowordi, K., Carbonara, J., Ahmed, S.: Effect of bromide incorporation on the electronic and photovoltaic properties of Sn-based perovskite devices: a multiscale investigation utilizing first principles approach and numerical simulation, aided by machine learning models. Sol. Energy. **253**, 375–388 (2023). https://doi.org/10.1016/j.solener.2023.02.034
29. Shrivastav, N., Madan, J., Pandey, R.: Predicting photovoltaic efficiency in Cs-based perovskite solar cells: a comprehensive study integrating SCAPS simulation and machine learning models. Solid State Commun. **380**, 115437 (2024). https://doi.org/10.1016/j.ssc.2024.115437
30. Jacobsson, T.J., Hultqvist, A., García-Fernández, A., et al.: An open-access database and analysis tool for perovskite solar cells based on the FAIR data principles. Nat. Energy. **7**, 107–115 (2022). https://doi.org/10.1038/s41560-021-00941-3
31. Vélez, J., Sepúlveda, F., Botero, M., Otalora, C., Camacho, C.: Absorber layer thickness as a new feature in statistical learning tools of perovskite solar cells. J. Appl. Res. Technol. (2023). https://doi.org/10.22201/icat.24486736e.2023.21.5.2057
32. Hasanzadeh Azar, M., Aynehband, S., Abdollahi, H., Alimohammadi, H., Rajabi, N., Angizi, S., Kamraninejad, V., Teimouri, R., Mohammadpour, R., Simchi, A.: SCAPS empowered machine learning modelling of perovskite solar cells: predictive design of active layer and hole transport materials. Photonics. **10**, 271 (2023). https://doi.org/10.3390/photonics10030271

33. Shah, S.A.A., Sayyad, M.H., Khan, K., Guo, K., Shen, F., Sun, J., Tareen, A.K., Gong, Y., Guo, Z.: Progress towards high-efficiency and stable tin-based perovskite solar cells. Energies (Basel). **13**, 5092 (2020). https://doi.org/10.3390/en13195092
34. Li, J., Pradhan, B., Gaur, S., Thomas, J.: Predictions and strategies learned from machine learning to develop high-performing perovskite solar cells. Adv. Energy Mater. **9** (2019). https://doi.org/10.1002/aenm.201901891
35. Goodfellow, I., Bengio, Y., Courville, A.: Deep Learning. The MIT Press (2016)
36. Le Corre, V.M., Sherkar, T.S., Koopmans, M., Koster, L.J.A.: Identification of the dominant recombination process for perovskite solar cells based on machine learning. Cell Rep. Phys. Sci. **2**, 100346 (2021). https://doi.org/10.1016/j.xcrp.2021.100346
37. Baloch, A.A.B., Albadwawi, O., AlShehhi, B., Alberts, V.: Bandgap model using symbolic regression for environmentally compatible lead-free inorganic double perovskites. In: 2022 IEEE 49th Photovoltaics Specialists Conference (PVSC), pp. 452–455. IEEE (2022). https://doi.org/10.1109/PVSC48317.2022.9938842
38. Gladkikh, V., Kim, D.Y., Hajibabaei, A., Jana, A., Myung, C.W., Kim, K.S.: Machine learning for predicting the band gaps of ABX3 perovskites from elemental properties. J. Phys. Chem. C. **124**, 8905–8918 (2020). https://doi.org/10.1021/acs.jpcc.9b11768
39. Stanley, J., Gagliardi, A.: Machine learning bandgaps of inorganic mixed halide perovskites. In: 2018 IEEE 18th International Conference on Nanotechnology (IEEE-NANO), pp. 1–4. IEEE (2018). https://doi.org/10.1109/NANO.2018.8626420
40. Mammeri, M., Dehimi, L., Bencherif, H., Pezzimenti, F.: Paths towards high perovskite solar cells stability using machine learning techniques. Sol. Energy. **249**, 651–660 (2023). https://doi.org/10.1016/j.solener.2022.12.002
41. Odabaşı, Ç., Yıldırım, R.: Machine learning analysis on stability of perovskite solar cells. Sol. Energy Mater. Sol. Cells. **205**, 110284 (2020). https://doi.org/10.1016/j.solmat.2019.110284
42. Graniero, P., Khenkin, M., Köbler, H., Hartono, N.T.P., Schlatmann, R., Abate, A., Unger, E., Jacobsson, T.J., Ulbrich, C.: The challenge of studying perovskite solar cells' stability with machine learning. Front. Energy Res. **11** (2023). https://doi.org/10.3389/fenrg.2023.1118654
43. Cheng, L., Zang, H., Wei, Z., Zhang, F., Sun, G.: Evaluation of opaque deep-learning solar power forecast models towards power-grid applications. Renew. Energy. **198**, 960–972 (2022). https://doi.org/10.1016/j.renene.2022.08.054
44. Zichi, L., Liu, T., Drueke, E., Zhao, L., Xu, G.: Physically informed machine-learning algorithms for the identification of two-dimensional atomic crystals. Sci. Rep. **13** (2023). https://doi.org/10.1038/s41598-023-33298-6
45. Liu, Y., Yan, W., Han, S., Zhu, H., Tu, Y., Guan, L., Tan, X.: How machine learning predicts and explains the performance of perovskite solar cells. Sol. RRL. **6** (2022). https://doi.org/10.1002/solr.202101100
46. Salah, M.M., Ismail, Z., Abdellatif, S.: Selecting an appropriate machine-learning model for perovskite solar cell datasets. Mater. Renew. Sustain. Energy. **12**, 187–198 (2023). https://doi.org/10.1007/s40243-023-00239-2
47. Mayr, F., Harth, M., Kouroudis, I., Rinderle, M., Gagliardi, A.: Machine learning and optoelectronic materials discovery: a growing synergy. J. Phys. Chem. Lett. **13**, 1940–1951 (2022). https://doi.org/10.1021/acs.jpclett.1c04223

The Performance Ratio of a Grid-Connected Photovoltaic System: An Experimental and Calculative Analysis

Juan Carlos Solano and Juan José Arciniega

1 Introduction

The performance ratio (PR) of a photovoltaic (PV) system—regardless of its configuration—is a crucial parameter for assessing the efficiency of PV systems. By understanding the PR value, it is possible to simulate and predict the PV production of a system before installation. The PR provides a measure of the energy production of a PV system relative to its theoretical maximum potential, accounting for various system losses. Knowing the PR is essential for the proper sizing of PV systems to ensure realistic production estimates [1–4].

Ideally, a loss-free PV system would have a PR of 1 (or 100%). However, overall losses are influenced by various factors that decrease efficiency, such as temperature, shading, soiling, DC/AC conversion losses, cable losses, inverter losses, and system inefficiencies. Understanding where these losses occur and improving the PR of PV systems is essential to optimizing performance and ensuring reliable energy production for users.

PR values typically range between 70% and 90% for well-maintained systems, with higher values indicating more efficient performance. For example, a study of PV systems in the Midlands region of the UK recorded an average monthly PR of 85.74%, with seasonal variations affecting the PR due to changes in solar irradiance and ambient temperature [5]. Similarly, a simulated study of a 5 MW grid-connected PV system in Indonesia revealed that the PR varied between 80% and 86%, influenced by high ambient temperatures and system configuration [4]. Factors such as module temperature, irradiance levels, and shading significantly influence PR values in different geographical locations and system configurations [1, 3,

J. C. Solano (✉) · J. J. Arciniega
Facultad de la Energía, Universidad Nacional de Loja, Loja, Ecuador
e-mail: juan.solano@unl.edu.ec

M. Espinoza-Andaluz et al. (eds.), *Congress on Research, Development and Innovation in Renewable Energies*, Green Energy and Technology,
https://doi.org/10.1007/978-3-031-88995-0_3

Table 1 Summary of PR values from various PV system studies

Author(s)	Type of PV system	PR value (%)
[7]	Rooftop PV and floating PV	76.39–82.69
[6]	Various PV systems	70–90
[5]	Residential PV	88.12–85.74
[4]	Grid-connected (tropical climate)	80–86
[2]	Clustered PV	79.3
[8]	Ground-mounted installation	80–85
[9]	Residential rooftop installation	75
[10]	Hybrid system	78–82
[11]	Industrial installation	70–75
[3]	Building-integrated PV system	85–90

6]. Table 1 summarizes the PR values from various studies, highlighting the diversity in system performance as a function of geographical and environmental factors.

Analysis of existing literature reveals several key relationships and trends in the study of PV system performance. Many studies emphasize the impact of environmental conditions such as temperature, irradiance, and shading. For example, Dhimish's research highlights the influence of thermal variations and hot spots, with higher PR values observed in regions with stable weather conditions [5]. Conversely, research by Ueda et al. demonstrates the effects of grid voltage and snowfall on PV systems [2].

The calculation of the PR of PV systems relies on various methodologies designed to measure and analyze the system's efficiency under real-world conditions. One commonly used methodology is a simulation with specialized software such as PVsyst, which models the behavior of the PV system under different environmental conditions. Syahindra et al. [7] used *PVsyst* to evaluate the PR of PV systems installed in electric vehicle charging stations, validating these results with precise mathematical calculations. Similarly, Karunanithi et al. [6] explored the use of machine learning algorithms to estimate PR, evaluating different models such as Random Forest and AdaBoost, finding that these approaches can improve the accuracy of performance estimates.

Another important methodology includes the analysis of experimental data and the correction of environmental factors such as temperature and irradiance. Dhimish [5] conducted an exhaustive study of 8000 PV installations in England, adjusting the monthly PR using a model developed by the US National Renewable Energy Laboratory (NREL) to correct for variations in temperature and irradiance. Kumara et al. [4] performed simulations and long-term monitoring of a 5 MW grid-connected PV system in Indonesia, using synthetic meteorological data to calculate the monthly and annual PR of the system. These methodologies allow for a detailed and accurate evaluation of PR, providing valuable insights for optimizing the design and operation of PV systems.

The tilt angle of solar panels also impacts the amount of energy they produce. To maximize energy production, the tilt angle should be such that the panels are oriented as perpendicularly as possible to the sun throughout the day. In general, the tilt angle should be equal to the site's latitude to maximize the annual energy yield [12]. For example, a study in Thailand (latitude 14°) found that a PV system with a tilt angle of 15° achieved the highest yield (PR = 0.7) [13]. The performance ratio of a PV system is also affected by the orientation (azimuth) of the panels. It is recommended that PV panels should always be oriented to the earth's equator [14].

Despite these data, significant gaps exist in the reviewed studies, particularly regarding the long-term degradation of the PR and the impact of new PV technologies. Future research should focus on these areas to provide a more complete understanding of PV system performance characteristics and identify strategies to improve the PR. Similarly, numerous efforts are being made to integrate advanced monitoring techniques for estimating the PR, such as machine learning algorithms. Although studies are limited, they show promising potential for improving system efficiency and reliability [6].

This paper contributes to an experimental study aimed at calculating the PR of a 5 kWp grid-connected photovoltaic system installed at the National University of Loja in Ecuador. The system monitors various parameters every 5 min, such as irradiance, DC and AC power, ambient and cell temperature, wind speed, and PV production. Over the past 4 months (February, March, April, and May), data collection has enabled a detailed analysis of the system's PR, providing valuable information for optimizing the sizing of future PV installations [15]. This approach allows for a more accurate determination of the PR, contributing to the development of more precise models for predicting PV system performance under varying environmental conditions.

The remainder of this chapter is structured as follows: Section 2 presents the methodology used for data collection and analysis. Section 3 discusses the results, including the implications of the findings on the PR of grid-connected PV systems. Finally, Sect. 4 summarizes the findings and presents the main conclusions.

2 Methodology

This research focuses on the acquisition and processing of experimental data from a 4.85 kWp grid-connected PV system installed at the National University of Loja, Ecuador. The system comprises 12 photovoltaic panels, each with a power rating of 405 W, an efficiency of 21%, and a tilt angle of 12°. In our study conducted in the City of Loja-Ecuador, with a latitude around −4° and close to zero latitude, the tilt angle does not considerably affect the energy production, so the tilt angle was not considered for calculating the PR. Data were collected from February to May 2024. Figure 1 provides an aerial photograph of the PV system and a schematic of power exchange with the electrical grid and building consumption.

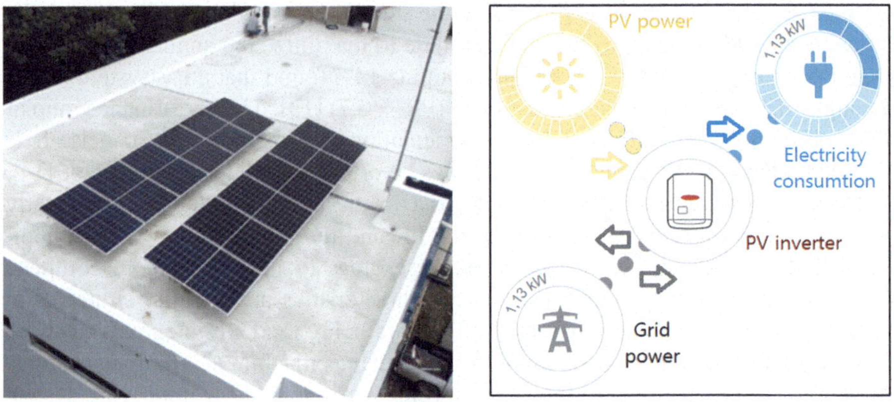

Fig. 1 (Left) Aerial view of the 4.85 kWp photovoltaic system at UNL, with 12 panels of 405 W each, at a tilt angle of 12°. (Right) Data diagram showing the power exchange between the PV system, the consumer, and the grid

2.1 *Data Collection*

The variables recorded every 5 min include irradiance, ambient temperature, PV panel temperature, wind speed, DC power, and AC power. These are all variables that the system is recording every 5 min. The sampling time can be selected through the configuration of the measurement equipment. Our study used the shortest possible sampling time, i.e., 5 min, in order to obtain more accurate results in the calculations. However, the proposed methodology can be adjusted to other sampling times as long as they are factors of 60 min.

The aim of data processing is to determine the characteristic PR of the PV system, identify DC/AC conversion losses, and compare these results with other studies in the scientific literature. The main variables were recorded from 5 February 2024 to 31 May 2024.

Once the PR is known, it is possible to estimate or predict the solar energy production of other systems simply by knowing their irradiance, nominal power, and panel efficiency. Figure 2 shows an overview of the data processing and the results obtained.

2.2 *Measured Variables*

Irradiance (I_{rrad}) [W/m^2]

Measured in watts per square meter, irradiance represents the power per unit area captured by the radiation sensor positioned at the same inclination as the panels.

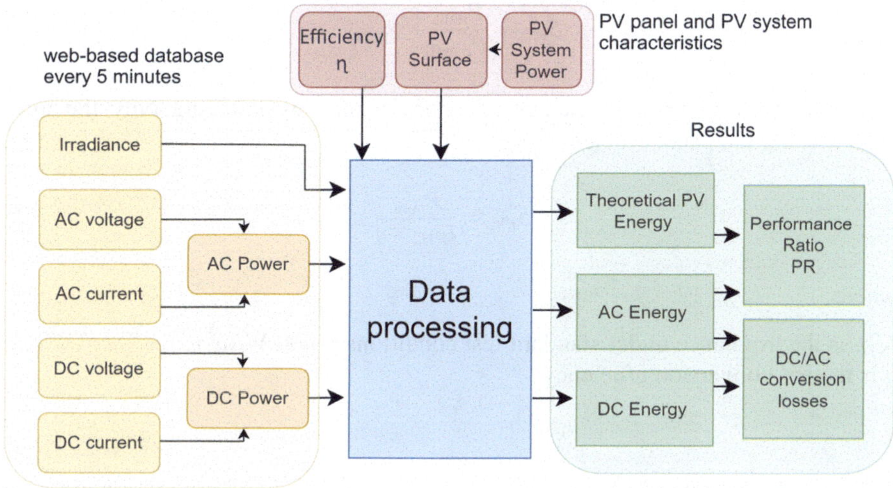

Fig. 2 Variables considered for data processing and analysis. In yellow are the variables recorded every 5 min. In red are the PV system specifications. In green are the results obtained

Direct Current Power (P_{dc}) [W]

DC power is calculated by multiplying DC voltage by DC current. This power is obtained directly from the PV panels before entering the inverter.

Alternating Current Power (P_{ac}) [W]

AC power is obtained by multiplying AC voltage by AC current. This power is measured at the inverter's output, connecting directly with building consumption and the electrical grid.

By knowing the DC and AC power, it is possible to calculate conversion losses. Similarly, knowing the AC power and the incident irradiance on the system, the characteristic performance of our PV system can be determined.

2.3 *PV System Specifications*

PV System Surface Area (S_{PV}) [m^2]

S_{PV} is calculated based on the number of PV panels and their individual power ratings (P_{panel}). The panel's length and width are obtained from the specifications sheet to calculate the individual area S_{panel}. Subsequently, the total installed PV power (P_{nom}) determines the surface area.

$$S_{PV} = \frac{P_{nom}}{P_{panel}} \times S_{panel} \tag{1}$$

Alternatively, the surface area can be approximated by knowing only the total power, using the panel efficiency (η).

$$S_{PV} \approx \frac{P_{nom}}{I_{STC} \times \eta} \tag{2}$$

where

I_{STC} is the irradiance under standard test conditions (1000 W/m^2)
η is the PV conversion efficiency

2.4 Data Processing

Sampling Interval (t) [min]

For mathematical data processing, the time interval in which the data are recorded is necessary. In this study, the interval is 5 min, but the methodology can accommodate any value as long as t is a factor of 60.

Number of Days (d)

The analysis spans d days. For monthly analyses, the number of days of the specific month must be counted. For yearly analysis, d could be 365.

Total Number of Samples (N)

Knowing the data sampling interval and the days, the total number of samples can be calculated for energy calculations.

$$N = \frac{60}{t} \times 24 \times d \tag{3}$$

2.5 Calculations

Irradiation ($I_{irradiation}$) [kWh/m^2]

Irradiation is the energy per unit area captured by the panels over a specific time period, measured in kWh/m^2. It sums irradiance over time (Eq. 4). Unlike irradiance, irradiation is always associated with a certain time in which the irradiance is summed. This time can be hourly, daily, weekly, monthly, yearly, etc. For this study, the analysis time is monthly.

$$I_{\text{irradiation}} = \left(\frac{t}{60}\right)\left(\frac{1}{1000}\right)\sum_{i=1}^{N} I_{\text{irradiance}_i} \tag{4}$$

Monthly Average Daily Irradiation (I_{daily})

It represents the monthly irradiation divided by the number of days in the month. This is often the way irradiation is presented in software or databases. It is useful to calculate the energy produced by a PV system, as it provides information on how much irradiation is accumulated in a day at a specific site. It is not precisely a daily irradiation but averages all the daily irradiations of a given month. In case Eq. (4) has been calculated on a monthly basis, the I_{daily} will be:

$$I_{\text{daily}} = \frac{I_{\text{irradiation}}}{d} \tag{5}$$

Theoretical PV Energy Output ($E_{\text{PV, th}}$) [kWh]

The theoretical PV energy output under ideal conditions is calculated by multiplying irradiation, surface area, and panel efficiency.

$$E_{\text{PV,th}} = I_{\text{irradiation}} \times S_{\text{PV}} \times \eta \tag{6}$$

DC PV Energy Output ($E_{\text{PV, dc}}$) [kWh]

The energy obtained at the panels' output is calculated using DC power.

$$E_{\text{PV,dc}} = \left(\frac{t}{60}\right)\left(\frac{1}{1000}\right)\sum_{i=1}^{N} P_{\text{dc}_i} \tag{7}$$

AC PV Energy Output ($E_{PV, ac}$) [kWh]

The energy output in AC, measured at the inverter's output, is calculated similarly to DC power.

$$E_{PV,ac} = \left(\frac{t}{60}\right)\left(\frac{1}{1000}\right)\sum_{i=1}^{N} P_{ac_i} \tag{8}$$

Conversion Losses (L_{conv}) [%]

Conversion losses in the inverter are expressed as a percentage of the energy lost during the conversion process.

$$L_{conv} = \left(1 - \frac{E_{PV,ac}}{E_{PV,dc}}\right) \times 100\% \tag{9}$$

Performance Ratio (PR)

The PR represents the percentage of theoretical energy that is effectively used by the PV system, indicating the overall system efficiency (Eq. 10). A value close to 1 indicates a correct functioning of all elements of the system, while low values may indicate problems at some point in the whole energy conversion process.

$$\text{PR} = \frac{E_{PV,ac}}{E_{PV,th}} \times 100\% \tag{10}$$

Predicted Energy ($E_{PV, predicted}$) [kWh]

The results allow for the formulation of a prediction equation for solar energy that a PV system can deliver over a specified period (Eq. 11).

$$E_{PV,predicted} = I_{irradiation} \times S_{PV} \times \eta \tag{11}$$

Equation (11) can be adapted based on I_{daily} (see Eq. 5).

$$E_{PV,predicted} = I_{daily} \times S_{PV} \times \eta \times d \tag{12}$$

3 Results and Discussion

3.1 Results

All formulas indicated in the Methodology were calculated using the following data:

1. Sampling interval (t): 5 minutes
2. Nominal power (P_{nom}): 4.85 kWp
3. PV system surface area (S_{PV}): 24 m^2
4. PV conversion efficiency (η): 21%

For this analysis, d corresponds to 25 days for February (4 days without data recording), 30 days for April, and 31 days for March and May. Table 2 presents the results of the total monthly irradiation and monthly average daily irradiation calculated for the months from February to May (see Eqs. 4 and 5).

According to Eqs. (6)–(8), Table 3 shows the theoretical energy output, DC energy output, and AC energy output.

Finally, Table 4 provides the conversion losses and PR values obtained from the calculations of Eqs. (9) and (10).

On the one hand, Fig. 3 illustrates the relationship of power over time, clearly showing how the estimated theoretical power progressively decreases. Initially, this reduction is reflected in the direct current (DC) power, and finally, after accounting for conversion losses, the actual system power, which is the alternating current (AC) power, is obtained. On the other hand, Fig. 4 depicts the accumulation of these losses as an average over the analysis period.

Table 2 Total monthly irradiation and monthly average daily irradiation

Month	Total monthly irradiation ($I_{irradiation}$) [kWh/m^2]	Monthly average daily irradiation (I_{daily}) [kWh/m^2]
February	76.09	3.04
March	92.25	2.98
April	73.33	2.44
May	96.80	3.12

Table 3 Theoretical energy output, DC energy output, and AC energy output

Month	Theoretical energy PV output ($E_{PV,\ th}$) [kWh]	DC energy PV output ($E_{PV,\ dc}$) [kWh]	AC energy PV output ($E_{PV,\ ac}$) [kWh]
February	383.50	347.16	328.76
March	464.96	438.39	414.63
April	369.60	330.60	312.22
May	487.89	419.23	396.33

Table 4 Conversion losses and performance ratio

Month	Conversion losses (L_{conv}) [%]	Performance ratio (PR) [%]
February	5.30	85.73
March	5.42	89.18
April	5.56	84.48
May	5.46	81.23
Average	5.44	85.15

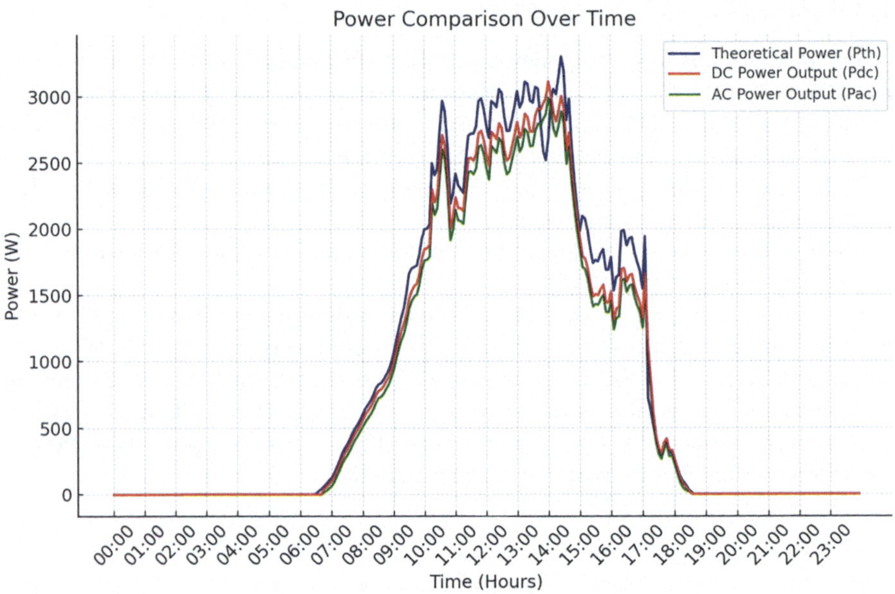

Fig. 3 Power outputs for a particular day

3.2 *Discussion*

The average PR obtained in this study is 85.15%, which falls within the normal range when compared to other studies and the values presented in Table 1 of the literature review. This value indicates that 14.85% of the theoretical energy is lost within the PV system, considering 100% as the energy that should be theoretically harnessed given the levels of irradiation. Within this 14.85%, the conversion losses in the inverter average 5.44%. The remaining 9.41% is attributed to losses related to cables and connectors, temperature-induced losses, and soiling, which can cause hot spots on the panels, leading to reduced photovoltaic production.

Comparing the monthly PR values with those from other regions and systems, our results show a high efficiency, particularly in March, which recorded the highest PR of 89.18%. This high performance is consistent with other well-maintained systems,

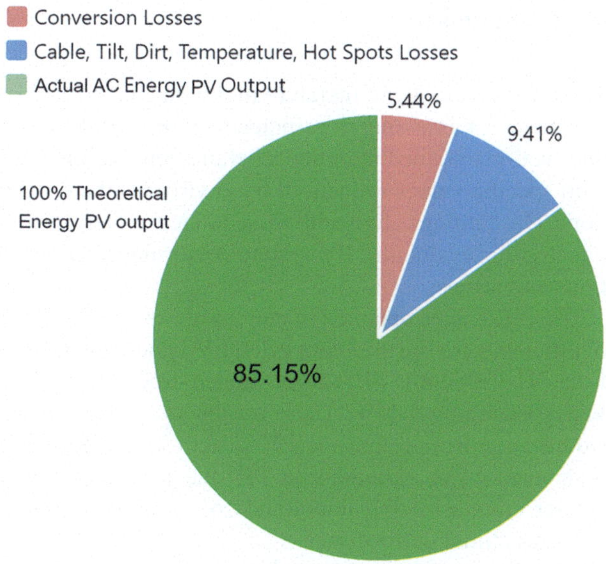

Fig. 4 Proportion of losses and effective energy use of the PV system

as discussed in the literature, where PR values typically range from 70% to 90% for systems under optimal conditions.

For instance, Dhimish [1] study in the Midlands region of the UK observed an average monthly PR of 85.74%, which is comparable to our findings. Similarly, Kumara et al. [2] reported PR values between 80% and 86% for a grid-connected PV system in Indonesia, influenced by high ambient temperatures and system configuration. These results underscore the significant influence of environmental factors such as temperature, irradiance, and shading on PR values.

In this study, the fluctuations in PR can be attributed to variations in irradiance and temperature across different months. The highest irradiance in May (96.80 kWh/m^2) corresponds to a lower PR (81.23%), indicating the potential impacts of higher temperatures on panel efficiency. Conversely, the lower irradiance in April (73.33 kWh/m^2) with a PR of 84.48% suggests more stable weather conditions.

The conversion losses observed are consistent with typical inverter efficiencies, highlighting the importance of selecting high-efficiency inverters to minimize energy losses. The average conversion loss of 5.44% aligns with findings from other studies, indicating that while inverter losses are significant, they are within the expected range for modern PV systems.

Overall, this study's results contribute valuable insights into the performance of grid-connected PV systems in Ecuador, demonstrating that with appropriate maintenance and optimal configuration, these systems can achieve high efficiency levels comparable to those reported in other regions. Future research should focus on long-term monitoring to assess the degradation of PR over time and explore the impact of new PV technologies on system performance.

4 Conclusion

The review of existing literature highlights the critical role of the PR as an essential parameter for assessing the efficiency of photovoltaic (PV) systems. Studies indicate that well-maintained systems generally achieve PR values between 70% and 90%, with specific values influenced by environmental conditions and system configurations. Our findings align with these ranges, underscoring the robustness of the PR as a metric for evaluating PV system performance across different geographical locations and setups.

The PR values obtained in our study average 85.15%, demonstrating a high level of efficiency for the grid-connected PV system installed at the National University of Loja. The monthly PR values ranged from 81.23% to 89.18%, with conversion losses averaging 5.44%. These results are consistent with those reported in similar studies, confirming that the PV system is performing effectively within expected parameters. The variations in PR due to changes in irradiance and temperature further emphasize the importance of continuous monitoring and maintenance to sustain optimal performance.

Continuous monitoring and analysis of the PR are crucial for optimizing the performance of PV systems. Understanding the factors that affect PR, such as temperature, shading, and soiling, allows for timely interventions to mitigate losses and enhance system efficiency. Regular monitoring helps identify potential issues early, ensuring that the system operates at its maximum potential and providing reliable energy production for users.

Finally, knowing the PR is invaluable in the design phase of PV systems. It enables accurate predictions of the real energy output that can be expected, facilitating better planning and sizing of installations. By incorporating the PR into simulations and models, designers can provide realistic production estimates, ensuring that the PV systems meet energy demands effectively. This predictive capability is essential for optimizing investment returns and achieving energy sustainability goals.

Acknowledgments The authors acknowledge the support of the "Universidad Nacional de Loja" by means of the research project: 34-DI-FEIRNNR-2021 "*Desarrollo de un sistema de soporte de decisiones para el autoconsumo fotovoltaico en el Ecuador.*"

References

1. Yang, Q., Xu, L., Xia, L., Chen, Z., Zhang, S., Zhou, R.: Performance ratio test and uncertainty evaluation of photovoltaic power generation system based on linear regression. In: 5th IEEE Conference on Energy Internet and Energy System Integration: Energy Internet for Carbon Neutrality, EI2 2021, pp. 1911–1916 (2021). https://doi.org/10.1109/EI252483.2021.9713184
2. Ueda, Y., Kurokawa, K., Itou, T., Kitamura, K., Miyamoto, Y., Yokota, M., Sugihara, H.: Performance ratio and yield analysis of grid connected clustered PV systems in Japan. In:

Conference Record of the 2006 IEEE 4th World Conference on Photovoltaic Energy Conversion, WCPEC-4, pp. 2296–2299 (2006). https://doi.org/10.1109/WCPEC.2006.279631
3. Cha, H.-L., Ko, J.-W., Lim, J.-R., Kim, D.-K., Ahn, H.-K.: Performance ratio of crystalline Si and triple junction a-Si thin film photovoltaic modules for the application to BIPVs. Trans. Electr. Electron. Mater. **18**, 30–34 (2017). https://doi.org/10.4313/TEEM.2017.18.1.30
4. Kumara, I.N.S., Ashari, M., Sampeallo, A.S., Pawitra, A.A.G.A.: Simulated energy production and performance ratio of 5 MW grid-connected photovoltaic under tropical savannah climate in Kupang Timor Island of Indonesia. Int. J. Eng. Technol. Innov. **7**, 117–129 (2017)
5. Dhimish, M.: Thermal impact on the performance ratio of photovoltaic systems: a case study of 8000 photovoltaic installations. Case Stud. Therm. Eng. **21** (2020). https://doi.org/10.1016/j.csite.2020.100693
6. Karunanithi, M., Thilak, S.J., Rizvi, A.A., Khan, T.A., Ahmad, I., Kouatly, R.: Performance ratio estimation of solar power plants using machine learning algorithms. In: 2023 IEEE 64th International Scientific Conference on Information Technology and Management Science of Riga Technical University, ITMS 2023—Proceedings (2023). https://doi.org/10.1109/ITMS59786.2023.10317646
7. Syahindra, K.D., Ma'Arif, S., Widayat, A.A., Fauzi, A.F., Setiawan, E.A.: Solar PV system performance ratio evaluation for electric vehicles charging stations in transit oriented development (TOD) areas. In: E3S Web of Conferences (2021). https://doi.org/10.1051/e3sconf/202123102002
8. Ghiani, E., Pilo, F., Cossu, S.: On the performance ratio of photovoltaic installations. In: 2013 IEEE Grenoble Conference PowerTech. PowerTech (2013). https://doi.org/10.1109/PTC.2013.6652489
9. Nakamura, Y., Chantana, J., Takeguchi, K., Kawano, Y., Nishimura, T., Minemoto, T.: Long-term examinations up to 20 years of performance ratio and degradation rate of different-type photovoltaic modules at central part of Japan. Renew. Energy Focus. **44**, 98–105 (2023). https://doi.org/10.1016/j.ref.2022.12.003
10. Syahputra, E., Fajarin, R.A., Setiawan, E.A.: Characteristic analysis of photovoltaic on-grid system in tropical region for weather-corrected performance ratio calculation method implementation. In: E3S Web of Conferences (2018). https://doi.org/10.1051/e3sconf/20186701025
11. Sakarapunthip, N., Chenvidhya, D., Chuangchote, S., Kirtikara, K., Chenvidhya, T., Onreabroy, W.: Effects of dust accumulation and module cleaning on performance ratio of solar rooftop system and solar power plants. Jpn. J. Appl. Phys. **56**, 08ME02 (2017). https://doi.org/10.7567/JJAP.56.08ME02
12. Khoo, Y.S., Nobre, A., Malhotra, R., Yang, D., Ruther, R., Reindl, T., Aberle, A.G.: Optimal orientation and tilt angle for maximizing in-plane solar irradiation for PV applications in Singapore. IEEE J. Photovolt. **4**, 647–653 (2014). https://doi.org/10.1109/JPHOTOV.2013.2292743
13. Singh, H., Sirisamphanwong, C., Santhi Rekha, S.M.: Effect of tilt and azimuth angle on the performance of PV rooftop system. Appl. Mech. Mater. **839**, 159–164 (2016). https://doi.org/10.4028/www.scientific.net/amm.839.159
14. Kokate, A., Wagh, M.: Experimental analysis of performance ratio of solar rooftop photovoltaic system (SRTPV) for various roof orientation and tilt. J. Phys. Conf. Ser. **1172**, 012067 (2019). https://doi.org/10.1088/1742-6596/1172/1/012067
15. Solano, J.C., Herrera, V., Ordóñez, Á., Caraballo, M., Lozano, A.: Economic analysis of residential photovoltaic self-consumption in Ecuador: simulation tool. In: Green Energy and Technology, pp. 57–69. Springer Science (2024). https://doi.org/10.1007/978-3-031-52171-3_4

Techno-Economic Analysis of an Off-Grid Solar PV System for Rural Electrification in La Virginia, Ecuador

Milton Cuenca, Maria Torres Sanchez, Jean Muñoz Campoverde, Ivonne Bermello Castro, and Carlos A. Cuenca

1 Introduction

Access to electricity is important for socioeconomic development and for improving the quality of life in rural communities. However, at present there are still many areas of Ecuador that do not have electricity supply through electricity distribution companies [1]. One of the best solutions to this major problem is the use of isolated photovoltaic systems (off-grid), which proves to be a viable and sustainable alternative to provide electricity in rural areas [2, 3] to meet the sustainable development goal (SDG) 7 "affordable and clean energy" and SDG 11 "sustainable cities and communities" [4].

The present project shows a technically and economically feasible solution that will help to eliminate the lack of electric energy in the houses of the sector "La Virginia" in the city of Quevedo, Ecuador, which consists of an off-grid photovoltaic system to supply electricity to the aforementioned area [5]. The city of Quevedo, located in the province of Los Rios where the study was conducted, has average radiation levels of 3040 to 31,500 Wh/m^2/day [6], which makes it a favorable area for the use of a solar photovoltaic system. Also, the angle of the sun's rays helps to capture solar energy more efficiently [7].

The development of the research shows the analysis of solar radiation data for the area, load demand calculations, system sizing using PVsyst software, and a

M. Cuenca (✉) · M. Torres Sanchez · J. Muñoz Campoverde · I. Bermello Castro
Faculty of Engineering Science, Universidad Técnica Estatal de Quevedo, Quevedo, Ecuador
e-mail: mcuenca@uteq.edu.ec

C. A. Cuenca
Facultad de Ingeniería Mecánica y Ciencias de la Producción, Escuela Superior Politécnica del Literal, ESPOL, Guayaquil, Ecuador

Facultad del Mar y Medio Ambiente, Universidad del Pacífico, Guayaquil, Ecuador

M. Espinoza-Andaluz et al. (eds.), *Congress on Research, Development and Innovation in Renewable Energies*, Green Energy and Technology,
https://doi.org/10.1007/978-3-031-88995-0_4

technical-economic feasibility evaluation for the implementation of the photovoltaic generation system [8]. Two configurations are evaluated: first, a centralized system to supply all the houses, and second, individual systems for each house. The design proposed in this work seeks to provide a functional solution for the families of the "La Virginia" sector through electricity self-supply. The electrical energy produced by this system will be environmentally friendly and will allow the implementation of technological development and innovation while improving the quality of life of the local population.

2 Methodology

This project was carried out using a methodology that combines quantitative and qualitative approaches to evaluate the technical and economic feasibility of an off-grid photovoltaic system in the "La Virginia" sector. The different stages of the development of this study are shown in the following sections.

2.1 Data Collection

The input data to know the different requirements needed for the study were obtained from different references, collecting the following information [9–12]. Information on the geographical location and climatic conditions of La Virginia was collected to understand the environmental factors influencing solar energy potential. This included data on temperature, humidity, and other relevant climatic variables that affect solar panel efficiency and energy generation. Additionally, the electricity demand of all dwellings in the sector was assessed.

Furthermore, solar radiation data was obtained using databases such as SOLCAST, which is important for estimating the potential solar energy that can be harnessed at the location. This data is a key factor in designing an efficient solar PV system. Information on the technical specifications and costs of photovoltaic components available in the local market was also gathered. This included data on solar panels, inverters, batteries, and other system components, which are essential for performing a cost-benefit analysis and ensuring the economic viability of the solar PV system.

It is important to mention SOLCAST data may not offer high-resolution, location-specific information, leading to inaccuracies in solar resource assessment due to reliance on satellite models with inherent uncertainties [13]. Temporal and spatial variability in solar radiation can also result in discrepancies in performance predictions, as historical data may not account for future climate changes [14].

To address these issues, it is crucial to validate SOLCAST data with local ground-based measurements, conduct sensitivity and scenario analyses to understand the impact of variable factors, use multiple data sources for cross-verification, and iteratively calibrate models with updated data.

2.2 Photovoltaic System Design

The PVsyst program was used for sizing and simulation of the system, considering some criteria to ensure the system meets the energy demands efficiently and cost-effectively [15–20].

Component selection begins with the analysis of solar irradiance data, which is critical in determining the potential solar energy available at the site. In this case, data from SOLCAST was utilized due to its comprehensive and up-to-date solar radiation data, which is essential for accurate system sizing. Load analysis is another factor to consider, as it involves an understanding of the daily energy consumption of households. This includes common household appliances, their power ratings, and usage hours, which help determine the electricity consumption for sizing the PV system components.

Component efficiency significantly impacts the overall system performance. High-efficiency components, such as monocrystalline panels with an efficiency of 21% or more [14], were selected to maximize energy production and minimize losses. The system's autonomy, or the number of days it can operate without sunlight, was set to 3 days, ensuring reliability during periods of low solar radiation. Economic considerations are also important to consider, focusing on minimizing costs while ensuring reliability. The cost analysis included the price of solar panels, batteries, inverters, and other components, as well as installation and maintenance costs.

The configuration of the panels involved arranging them in series and parallel to achieve the desired voltage and current levels. The battery bank was sized to provide 3 days of autonomy, using lithium batteries for their high energy density and long cycle life. The configuration was based on the calculated daily energy consumption and the desired autonomy. An inverter with a capacity slightly higher than the peak load was selected to handle potential surges in demand, ensuring the system can supply power to all connected loads without overloading the inverter.

Several assumptions were made during the design process. It was assumed that all homes have similar energy consumption patterns and appliances, simplifying the design process but potentially overlooking variations in individual household usage. The design also assumed average solar radiation levels throughout the year, which might not account for seasonal variations. This assumption is mitigated by the system's autonomy and battery storage capacity. The lifespan of components, particularly batteries and inverters, was assumed to be consistent with manufacturer specifications, impacting long-term economic analysis and maintenance planning.

Once these values were defined, two configurations were designed:

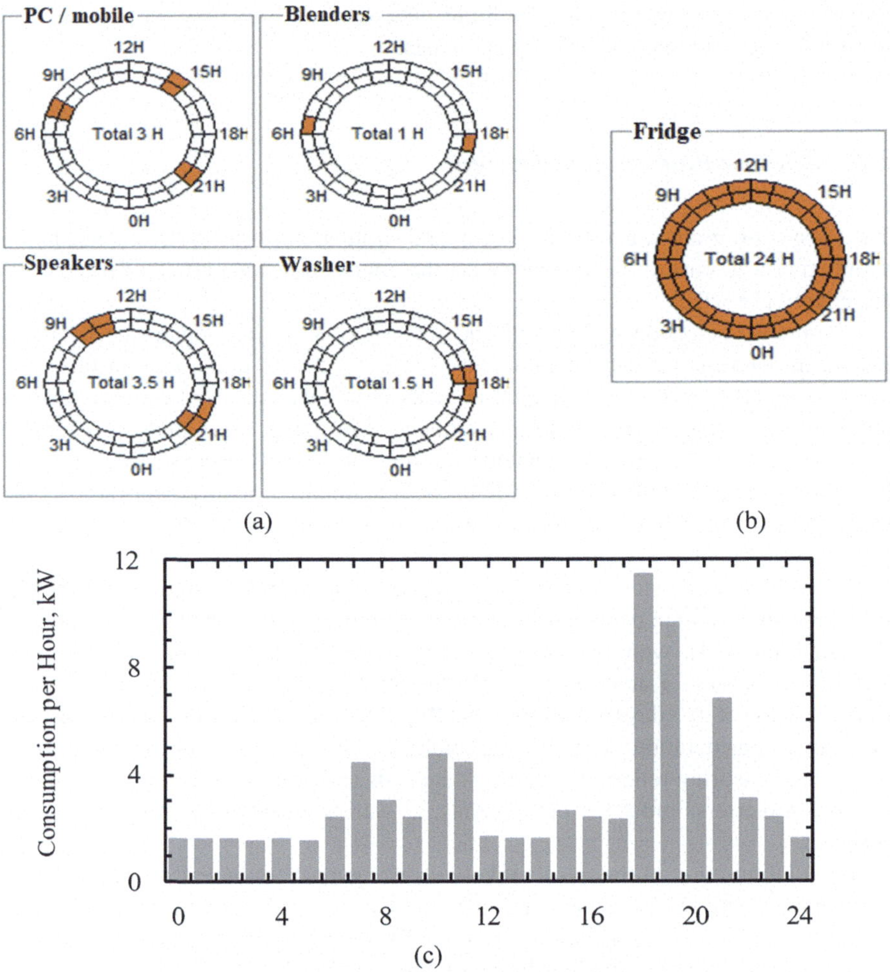

Fig. 1 Definition of daily consumption: (**a**) low consumption appliances, (**b**) high consumption appliances, (**c**) distribution of daily global consumption per hour

(a) Centralized system for all houses
(b) Individual systems for each house

Figure 1 shows the hourly distribution of the system. Here, it can be seen that each appliance has its own daily use schedule, and the refrigerator is the only appliance that remains connected 24 h a day. This type of distribution managed by the PVsyst program is practical and interactive since it allows one to mark the time from half an hour as a minimum of use represented as a square in the hourly circles. In this way, based on the hours marked, a daily load curve of the overall daily consumption is created.

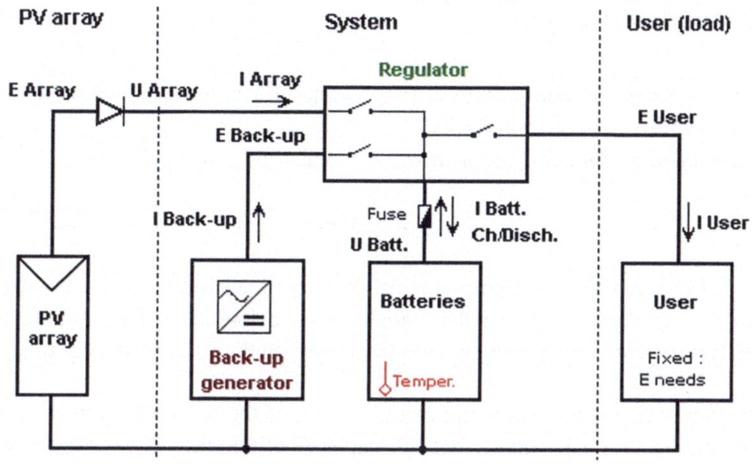

Fig. 2 Typical design of a stand-alone solar panel generation system

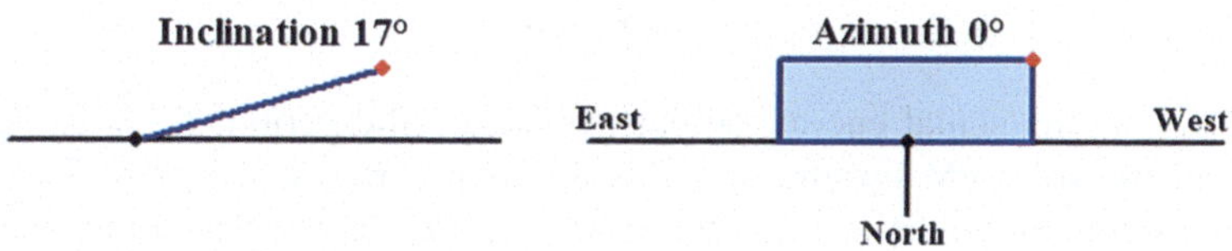

Fig. 3 Field parameters, orientation, and placement of solar panels

The scheme representing an off-grid photovoltaic system provided by the PVsyst software is shown in Fig. 2.

Quevedo is located at the geographical coordinates 1°03′18″ south latitude and 79°25′24″ west longitude [21, 22]. This location is critical in determining the optimal solar panel angle.

In general, to maximize solar energy collection throughout the year, it is recommended that the tilt angle of the solar panel be approximately equal to the latitude of the site [23]. However, since Quevedo is close to the Equator (latitude near 0°), this general principle must be adjusted.

For locations close to the Equator, such as Quevedo, a tilt angle slightly greater than the latitude is recommended to optimize solar gain and facilitate self-cleaning of the panels. Considering this, an optimal tilt angle for Quevedo would be in the range of 15° [24]. However, to ensure that the photovoltaic panel has a lower percentage of losses and a higher efficiency, a reference angle of 17° will be taken, as shown in Fig. 3.

2.3 Technical Analysis

The following technical parameters were evaluated for both configurations to ensure the system's efficiency and reliability.

The estimated annual energy production is calculated using Eq. (1) [25]:

$$E_{\text{prod}} = AG\eta \tag{1}$$

where A is the area of the solar panels in m^2, G is the solar irradiance in $kWh/m^2/day$, and η is the efficiency of the solar panels. This calculation determines the total energy that the system can produce annually based on the solar resources available at the site.

The capacity factor measures the actual output of the solar PV system compared to its potential output if it operated at full capacity all the time. It is calculated using Eq. (2) [26]:

$$\text{CF} = \frac{E_{\text{actual}}}{E_{\text{rated}}T} \tag{2}$$

where E_{actual} is the actual energy output over a specific period, E_{rated} is the rated capacity of the solar PV system, and T is the total period considered. This factor provides insight into how efficiently the system is operating relative to its maximum potential.

System performance is evaluated using the performance ratio (PR), which indicates the quality of the PV system irrespective of location and solar irradiance. The PR is calculated as shown in Eq. (3) [27]:

$$\text{PR} = \frac{E_{\text{actual}}}{E_{\text{ideal}}} \tag{3}$$

where E_{actual} is the actual energy output of the system, and E_{ideal} is the ideal energy output calculated as the product of the system's rated capacity, solar irradiance, and time. A higher PR indicates better system performance.

Detailed losses in the system can occur due to various factors such as shading, temperature effects, and inefficiencies in the inverter and other components. These losses are quantified to understand their impact on the overall system performance and to identify areas for improvement.

The solar fraction is the portion of the total energy demand met by the solar PV system whose value is calculated as shown in Eq. (4) [28]:

$$\text{ES} = \frac{E_{\text{solar}}}{E_{\text{total}}} \tag{4}$$

where E_{solar} is the energy supplied by the solar PV system, and E_{total} is the total energy demand. A higher solar fraction indicates a greater reliance on solar energy.

Load loss probability (LLP) is the probability that the system will not be able to meet the energy demand. It is determined by simulating the system's performance over time and calculating the frequency and duration of energy shortfalls.

2.4 *Economic Analysis*

This analysis is important for evaluating the economic feasibility and technical assumptions made in the project [29, 30]. It is focused on several financial parameters, including initial investment costs, operation and maintenance costs, revenues from energy sales, and financial indicators such as net present value (NPV), internal rate of return (IRR), and payback period. These parameters were selected because of the impact of the economic viability of the project.

Variations in initial investment costs can substantially affect the project's feasibility. A higher initial cost may extend the payback period and reduce the NPV, making the project less attractive to investors. Operation and maintenance costs, which are recurrent, can influence the long-term sustainability of the project. An increase in maintenance costs can decrease the IRR and extend the payback period, while efficient maintenance strategies can improve financial outcomes.

Revenues from energy sales are a critical factor in the project's profitability. If energy prices increase, the project's profitability improves, reflected in a higher NPV and IRR. The IRR is particularly sensitive to changes in cash flow patterns over the project's lifespan.

3 Results

This study provides a comparative analysis of centralized and individual solar PV systems for rural electrification in La Virginia, Ecuador. The centralized system involves a single, larger solar PV installation that supplies electricity to multiple households. This approach offers several advantages. Economies of scale can be achieved, leading to reduced costs per unit of electricity generated. Maintenance and operational tasks are centralized, which can simplify management and reduce costs. Additionally, the centralized system can support a larger energy storage capacity, enhancing reliability during periods of low solar irradiance.

However, centralized systems also have disadvantages. The initial investment cost is typically higher due to the scale of the infrastructure required. There is also a dependency on the central system's functionality; if the system fails, all connected households may experience power outages. Furthermore, the distribution of electricity from a central point to individual homes can lead to transmission losses, especially in geographically dispersed rural areas.

In contrast, individual systems are installed separately for each household. This configuration provides autonomy to each one, allowing them to manage their own energy consumption and storage. The initial investment for individual systems is generally lower per household compared to a centralized system. Additionally, the risk of a complete power outage is minimized since each system operates independently.

However, individual systems have their own drawbacks. They may not benefit from economies of scale, potentially leading to higher costs per unit of electricity generated. Maintenance can be more challenging as each system requires individual attention, which can increase overall maintenance costs.

The energy storage capacity is typically smaller, which may affect reliability during extended periods of low solar irradiance.

3.1 Case 1: Centralized System for All Homes

In this study case, a single centralized system will be created so that it can supply electricity to all the homes in the sector. Table 1 presents a summary of the nominal capacities at which the sized system must work to supply the load of all the homes that make up the "La Virginia" sector in the city of Quevedo. It must be considered that for the configuration of the modules there will be 8 × 8, that is, 8 in series and 8 in parallel.

Figure 4 shows the energy use per hour of the day with respect to the users and their stipulated load. Also, Table 2 shows that the demand calculated for all homes can be fully met during the system's operation period once it is implemented.

For this photovoltaic system to be implemented and put into operation, an economic analysis was carried out to determine how feasible it is to put the system into operation.

Figure 5 shows the cumulative cash flow during the years of operation of the system, in this case, 25 years (2023–2048); the red bars are the investments made, and the green bars are the profits generated after the operation of the PV system. Installation costs, panel purchases, batteries, controller, and maintenance labor are part of the investment.

The amount that would be disbursed for the PV system is USD 59,525.95, designed to supply all households together. The results of this economic analysis are based on the cumulative cash flow, considering that the system started operating

Table 1 Capabilities indicated by PVsyst software for Case 1

Items	Solar panels	Batteries
Modules	LR-66 HPH 500 M G2	KBM216_2P 13S 5000 Ah
Nominal power, kWp	32.0	–
MPP voltage, V	37.8	–
MPP current, A	13.2	–
Voltage, V	–	48
Total capacity, Ah	–	1200

Items	N°	Power, W	Use Hour/day	Energy Wh/day
Lamps (LED or fluorescent)	84	9W/lamp	7.0	5292
PC / mobile	48	80W/app	3.0	11520
Blenders	12	460W/app	1.0	5520
Fridge	12		24.0	38390
Washer	12		2.0	13500
Speakers	12	230W tot	3.5	9660
TV	12	60W tot	6.0	4320
Total				88202

(a)

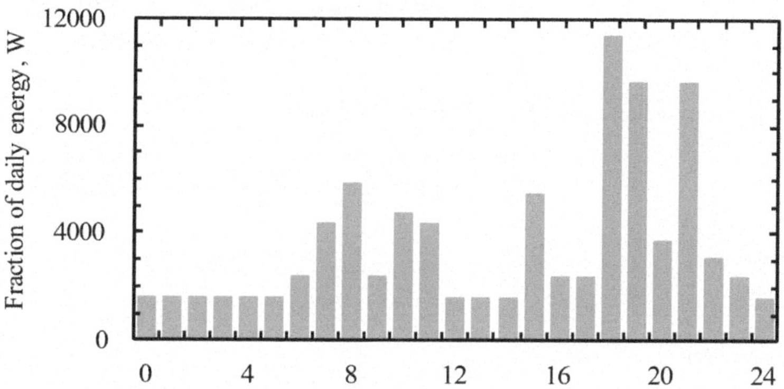

(b)

Fig. 4 User's electricity demand per day: (**a**) daily household consumers and (**b**) hourly distribution

Table 2 Energy balances in kWh for Case 1

Month	Global horizontal irradiation (kWh/m^2)	Global effective irradiance (kWh/m^2)	Available solar energy (kWh)	Unused energy (full battery) (kWh)	Energy supplied to the user (kWh)
January	121.6	105.9	3028	188	2734
February	115.6	104.5	3033	349	2470
March	146.1	138	4030	1212	2734
April	140.6	139.3	4055	1369	2646
May	116.2	118	3400	502	2734
June	99.7	102.5	2931	197	2646
July	96.9	97.7	2777	22	2734
August	113	112.2	3194	297	2734
September	122	116.8	3331	608	2646
October	114.8	105.3	2986	341	2734
November	116.8	103.3	2921	0	2646
December	118.4	102.3	2905	80	2734
Year	1421.7	1345.8	38,591.0	5165.0	32,192.0

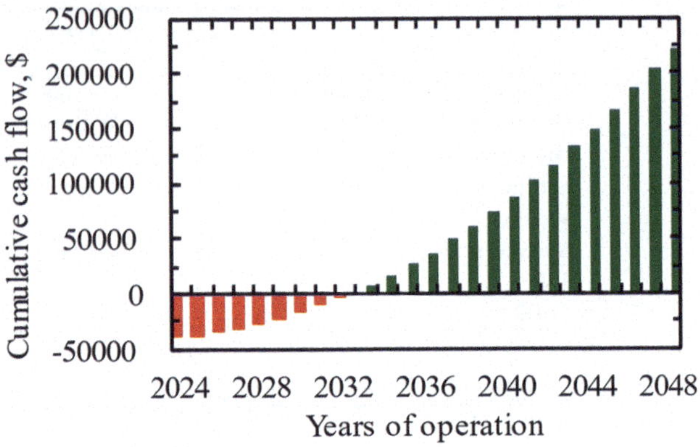

Fig. 5 Cumulative cash flow for Case 1

Table 3 Capacities indicated by PVsyst software for Case 2

Items	Solar panels	Batteries
Modules	LR5-66 HBD 500 M bifacial	KBM216_13S 60 Ah
Nominal power, kWp	3.00	–
MPP voltage, V	37.8	–
MPP current, A	13.2	–
Voltage, V	–	48
Total capacity, Ah	–	300

in the current year and show that the payback period for the investment is 13.4 years. From this point on, the income from profit will increase until 2048.

In addition, another parameter to consider is the annual operating cost defined at USD 3115.99. Considering this value and the socioeconomic profile of the target users, a tariff of USD 0.10 per kWh can be established. This rate, in comparison with the commercial rate for conventional energy in the residential sector of USD 0.092 per kWh, is competitive.

3.2 *Case 2: Individual Systems for Each House*

In this second case, it is proposed that a solar panel system be installed in each house to supply the average demand individually. Then, Table 3 shows the most relevant nominal design data for this case.

The energy needs for each month of the year, considering Case 2, are shown in Table 4. Here, it should be considered that the installation values cover two points: (1) the total installation cost and (2) the active amortization value.

For the total installation cost, a value of USD 11,530.94 is shown, which represents the total cost of installing the PV system for a single house, including the panels plus all the elements involved such as batteries, controller, inverter, wiring, etc., in addition to the values of all the elements of the set.

Table 4 Energy balances in kWh for Case 2

Month	Global horizontal irradiation (kWh/m^2)	Global effective irradiance (kWh/m^2)	Available solar energy (kWh)	Unused energy (full battery) (kWh)	Energy supplied to the user (kWh)
January	121.6	106.4	279.3	0.090	224.3
February	115.6	105.0	276.9	0.000	199.9
March	146.1	138.7	362.4	0.000	227.6
April	140.6	140	363.7	0.000	220.3
May	116.2	118.6	310.4	0.000	227.6
June	99.7	103	270.1	0.000	220.3
July	96.9	98.2	257.6	0.000	224.1
August	113.0	112.7	292.9	0.000	225.5
September	122.0	117.4	303.8	0.000	220.3
October	114.8	105.8	274.9	0.000	223.4
November	116.8	103.8	270.4	0.000	217.9
December	118.4	102.8	269	0.030	225.9
Year	1421.7	1352.4	3531.4	0.120	2657.1

On the other hand, depreciable assets show a value of USD 7554.28, which is defined by all the equipment or elements of the system that deteriorate over time due to use or different causes that may be natural or due to damage and repairs, causing their economic value to decrease.

The financing of this project can come from three main sources: loans, own funds, or state subsidies, in the case that the state consents to the aid. In this case, a combination of equity and state subsidy has been chosen. The proposed subsidy is equivalent to 30% of the total value of the entire SFV installation, which represents USD 3459.28, while the investment funds are equivalent to USD 8071.66. The sum of both amounts corresponds to the total installation costs.

The expenses related to the project involve three factors: the annual operating cost for the system, which amounts to 1003.71 USD/year; the loan annuity, which is zero dollars since the calculations were not estimated with loans; and the cost of energy used (LCOE), which is 0.60 USD/kWh, is not found in addition to the operating cost but is found in the expenses section. In other words, for this PV system, producing 1 kWh would cost 60 cents.

Finally, as for the return on investment, the net present value (NPV) was used to evaluate the financial viability of the project. Since the NPV calculated for this case corresponds to USD −5676.13, it indicates that the project, under current conditions, is not financially viable. This means that, instead of generating profits, the investment in the project would result in a net loss. This negative return expressed as a percentage corresponds to −70.3%, which allows us to deduce that this percentage of the initial investment, in present value terms, would be lost.

Figure 6 shows the cumulative cash flow in dollars over the 25-year life of the project, during which time there are only losses. The graph shows that, from the initial stage of the project, losses amount to approximately USD 8000 in the first year

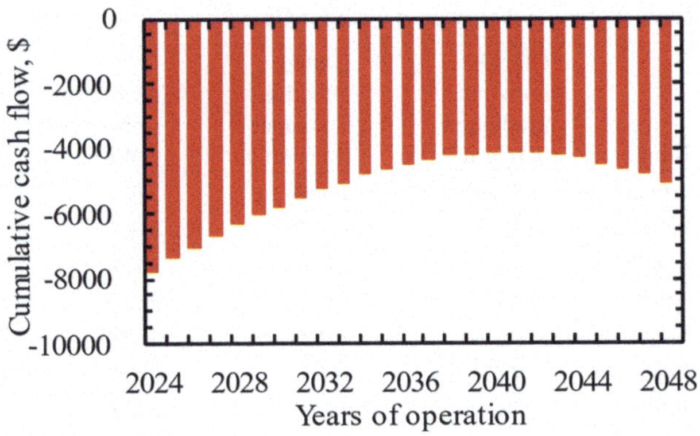

Fig. 6 Cumulative cash flow for Case 2

of operation. Finally, although in the following years the losses decreased, the project did not generate profits. It is important to mention that for a typical off-grid solar PV system, the installation cost can range from \$1000 to \$3000 per kilowatt (kW), translating to \$1 to \$3 million USD per MW. These costs are influenced by the local solar resource availability, which is relatively high in Ecuador due to its equatorial location, potentially reducing the cost per unit of electricity generated [8, 31].

On the other hand, for a natural gas combined cycle power plant, the capital costs for conversion to a biomethane-based bioenergy with carbon capture and storage system range from \$1.4 to \$2.1 per kilowatt (kW) of capacity. This translates to approximately \$1.4 to \$2.1 million USD per MW of capacity.

4 Conclusion

The present study demonstrates that both centralized and individual solar PV systems can effectively meet the energy demands of rural households, with each configuration offering unique advantages.

The centralized system benefits from economies of scale, reducing costs per unit of electricity generated, and simplifying maintenance and operational tasks. However, it presents challenges such as higher initial investment costs and potential transmission losses. On the other hand, individual systems provide autonomy to households and minimize the risk of complete power outages, although they may incur higher costs per unit of electricity generated and require more complex maintenance.

The design of the off-grid photovoltaic system for the "La Virginia" sector in Quevedo is viable and allows covering the energy demand of all the houses with an installed power per module of 32 kWp and a 48 V battery bank, considering an 8 × 8 configuration for Case 2 and presented as technically viable.

From the economic point of view, Case 1 (centralized system) shows a payback of the investment in approximately 13.4 years, with a positive NPV and an attractive IRR, which makes it financially viable. On the other hand, Case 2 (individual systems) proved to be less economically favorable, with a longer payback period and a negative NPV, suggesting that it is not a recommended option for this project.

Despite the positive outcomes, the study has several limitations. The analysis is based on average solar irradiance data, which may not account for seasonal variations and extreme weather conditions. Additionally, the financial analysis assumes stable energy prices and maintenance costs, which could fluctuate over time.

Future research should focus on incorporating more dynamic models that account for these variations and uncertainties. Further investigation into the integration of hybrid systems, combining solar PV with other renewable sources, could enhance reliability and resilience. Additionally, exploring community engagement and financing mechanisms could provide insights into optimizing the adoption and sustainability of renewable energy projects in rural areas.

References

1. Muñoz Campoverde, J.A., Torres Sanchez, M.I.: Factibilidad Tecnico-Economica De Un Sistema De Energia Solar Fotovoltaico ‘Off Grid’ Para Suministro Electrico Residencial En El Sector La Virginia De La Cuidad De Quevedo. Universidad Tecnica Estatal de Quevedo. https://repositorio.uteq.edu.ec/server/api/core/bitstreams/78521ef3-b0b1-4d44-a59d-ce2aa35d49da/content (2022)
2. Echegaray-Aveiga, R.C., Masabanda, M., Rodriguez, F., Toulkeridis, T., Mato, F.: Solar energy potential in ecuador. In: 2018 International Conference on eDemocracy & eGovernment (ICEDEG), April, pp. 46–51 (2018). https://doi.org/10.1109/ICEDEG.2018.8372318
3. Abdul-Ganiyu, S., Quansah, D.A., Ramde, E.W., Seidu, R., Adaramola, M.S.: Techno-economic analysis of solar photovoltaic (PV) and solar photovoltaic thermal (PVT) systems using exergy analysis. Sustain. Energy Technol. Assess. **47**, 101520 (2021). https://doi.org/10.1016/j.seta.2021.101520
4. Naciones Unidas: Objetivos y metas de desarrollo sostenible. Desarrollo Sostenible. https://www.un.org/sustainabledevelopment/es/objetivos-de-desarrollo-sostenible/ (2023). Accessed 15 Apr 2024
5. Chere-Quiñónez, B.F., Martínez-Peralta, A.J.: Technical-economic analysis of a grid-connected photovoltaic system Análisis técnico económico de un sistema fotovoltaico conectado a la red Resumen. Revista Científica Interdisciplinaria Investigación y Saberes. http://www.revistasdigitales.utelvt.edu.ec/revista/index.php/investigacion_y_saberes/article/view/246 (2024)
6. CONELEC: Atlas solar del Ecuador, pp. 1–51. CONELEC. http://www.conelec.gob.ec/archivos_articulo/Atlas.pdf (2008)
7. Ud-Din Khan, S., Wazeer, I., Almutairi, Z., Alanazi, M.: Techno-economic analysis of solar photovoltaic powered electrical energy storage (EES) system. Alexandria Eng. J. **61**(9), 6739–6753 (2022). https://doi.org/10.1016/j.aej.2021.12.025
8. Cuenca, A., Oña, C., Suquillo, I., Miniguano, H.: Design methodology of off-grid PV solar powered systems for rural areas in Ecuador. Rev. Técnica Energía. **20**(1), 43–51 (2023). https://doi.org/10.37116/revistaenergia.v20.n1.2023.537
9. Toledo-Orozco, M., Martinez, L., Quito, H., Quizhpi, F., Álvarez-Bel, C., Morales, D.: Methodology to determine the management of demand in recharging electric vehicles in

vertically integrated markets includes photovoltaic solar generation. Energies. **15**(24), 9590 (2022). https://doi.org/10.3390/en15249590
10. Shah, S.A.A., Das Valasai, G., Memon, A.A., Laghari, A.N., Jalbani, N.B., Strait, J.L.: Techno-economic analysis of solar PV electricity supply to rural areas of Balochistan, Pakistan. Energies. **11**(7), 1777 (2018). https://doi.org/10.3390/en11071777
11. Zhao, F., et al.: Estimating rural electric power consumption using npp-viirs night-time light, toponym and poi data in ethnic minority areas of China. Remote Sens. **12**(17), 1–20 (2020). https://doi.org/10.3390/rs12172836
12. AhmadImad, A.-S., Fatih, Y.: Modelling the use of PVSYST software for a stand-alone PV solar system 'off grid' with batteries by utilizing silicon hetero-junction technology (HJT) panels in Iraq/Basra. Al Rafidain J. Eng. Sci. **2**(1), 32–42 (2023). https://doi.org/10.61268/0hyjvn12
13. Solcast. Solar live & forecast accuracy. https://solcast.com/forecast-accuracy (2024)
14. Woods, T.N., Harder, J.W., Kopp, G., Snow, M.: Solar-cycle variability results from the solar radiation and climate experiment (SORCE) mission. Sol. Phys. **297**(4), 43 (2022). https://doi.org/10.1007/s11207-022-01980-z
15. Rarasati, A.D., Ardian, M.D., Setiawan, E.A., Mohammad, I.H.: Community behavior analysis of electricity consumption in rural areas (a case study of Batang Regency). IOP Conf. Ser. Mater. Sci. Eng. **550**(1), 012002 (2019). https://doi.org/10.1088/1757-899X/550/1/012002
16. Blennow, H.: Method for rural load estimations—a case study in Tanzania, Lund. http://citeseerx.ist.psu.edu/viewdoc/download?doi=10.1.1.620.1440&rep=rep1&type=pdf (2004)
17. Renné, D., George, R., Wilcox, S., Stoffel, T., Myers, D., Heimiller, D.: Solar resource assessment, Colorado. http://www1.eere.energy.gov/solar/pdfs/42301.pdf (2008)
18. Ríos, A., Guamán, J., Vargas, C., García, M.: Design, dimensioning, and installation of isolated photovoltaic solar charging station in Tungurahua, Ecuador. Int. J. Renew. Energy Res. **7**(1), 235–242 (2017). https://doi.org/10.20508/ijrer.v7i1.5340.g7005
19. Thanarak, P., Lhazom, K.: Economic feasibility analysis of off-grid PV systems for remote settlements in Bhutan. Int. J. Energy Econ. Policy. **11**(3), 206–214 (2021). https://doi.org/10.32479/ijeep.10665
20. Aboagye, B., Gyamfi, S., Puoza, J.C., Obeng, M.: Techno-economic feasibility analysis of solar photovoltaic system for single households in peri-urban areas Kumasi, Ghana. Int. J. Sustain. Energy Dev. **8**(1), 392–404 (2020). https://doi.org/10.20533/ijsed.2046.3707.2020.0047
21. Franco Castillo, B.L.: Efecto de diferentes dosis de fitohormona Cerone sobre el proceso de maduración en la variedad de Piña MD-2 o Golden Sweet en la zona de Quevedo. Universidad Tecnica Estatal de Quevedo. https://repositorio.uteq.edu.ec/items/443530cd-b0d4-4902-bd31-840ebeadb7f6 (2017)
22. Kartikasari, F.D., Tarigan, E., Irawati, F., Louk, M.H.L., Limanto, S., Asmawati, E.: Optimal solar panel tilt angle calculation and simulation in Indonesia: a Liu and Jordan sky isotropic model-based approach. Int. J. Sci. Res. Arch. **9**(2), 116–121 (2023). https://doi.org/10.30574/ijsra.2023.9.2.0517
23. Khan, M.S., Ramli, M.A.M., Sindi, H.F., Hidayat, T., Bouchekara, H.R.E.H.: Estimation of solar radiation on a PV panel surface with an optimal tilt angle using electric charged particles optimization. Electronics. **11**(13) (2022). https://doi.org/10.3390/electronics11132056
24. Kwon, T.M., Weidemann, R.: Advanced LED Warning Signs for Rural Intersections Powered by Renewable Energy. Minnesota Department of Transportation, Minnesota. http://www.cts.umn.edu/Publications/ResearchReports/pdfdownload.pl?id=1470%0A, http://www.dot.state.mn.us/research/TS/2011/2011-04.pdf%0A, https://trid.trb.org/view/1090557 (2010)
25. De Meij, A., et al.: Wind energy resource mapping of Palestine. Renew. Sustain. Energy Rev. **56**, 551–562 (2016). https://doi.org/10.1016/j.rser.2015.11.090
26. Mignacca, B., Locatelli, G.: Economics and finance of small modular reactors: a systematic review and research agenda. Renew. Sustain. Energy Rev. **118**, 109519 (2020). https://doi.org/10.1016/j.rser.2019.109519

27. Odeh, S.: Analysis of the performance indicators of the PV power system. J. Power Energy Eng. **6**(6), 59–75 (2018). https://doi.org/10.4236/jpee.2018.66005
28. Ge, T.S., Xu, J.C.: Review of solar-powered desiccant cooling systems. In: Advances in Solar Heating and Cooling, pp. 329–379. Elsevier (2016). https://doi.org/10.1016/B978-0-08-100301-5.00013-8
29. Oko, C., Diemuodeke, O., Omunakwe, N.F., Nnamdi, E.A.: Design and economic analysis of a photovoltaic system: a case study. Int. J. Renew. Energy Dev. **1**(3), 65–73 (2012). https://doi.org/10.14710/ijred.1.3.65-73
30. Shahzad, M.K., Zahid, A., ur Rashid, T., Rehan, M.A., Ali, M., Ahmad, M.: Techno-economic feasibility analysis of a solar-biomass off-grid system for the electrification of remote rural areas in Pakistan using HOMER software. Renew. Energy. **106**, 264–273 (2017). https://doi.org/10.1016/j.renene.2017.01.033
31. Arif, S., Taweekun, J., Ali, H.M., Ahmed, A., Bhutto, A.A.: Building resilient communities: techno-economic assessment of standalone off-grid PV powered net zero energy (NZE) villages. Heliyon. **9**(11), e21426 (2023). https://doi.org/10.1016/j.heliyon.2023.e21426

Part II
Biomass

Thermodynamic Assessment of a Biomass Gasification-Driven Tri-generation Chemical Loop Combustion Setup: A Colombian Case Study

Jhojan Zea, Brandon Martínez, Mario Luna-Del Risco, Lisandra Rocha-Meneses, Jorge Sierra, Hernando A. Yepes, Carlos E. Arrieta, Juan Pablo Gómez Montoya, Carlos A. Arredondo, and Esteban Vanegas

1 Introduction

The world faces increasing energy demand, coupled with the urgent need to reduce greenhouse gas emissions. Energy shortage, coupled with concerns about the environmental impacts of traditional energy sources, highlights the need for alternative energy solutions [1]. Biomass energy is a promising alternative, offering a sustainable source of energy with a lower environmental impact [2–4]. While traditional biomass conversion methods like direct combustion, gasification, and pyrolysis contribute to greenhouse gas emissions [5], chemical looping combustion (CLC) offers a novel approach. CLC separates oxygen transfer from the actual combustion of the fuel using a metal oxide as an oxygen carrier, which efficiently moves oxygen from the air to the fuel, producing higher-quality syngas with lower CO_2 levels

J. Zea · B. Martínez · M. L.-D. Risco · C. E. Arrieta · C. A. Arredondo · E. Vanegas (✉)
Grupo de Investigación en Energía—GRINEN, Universidad de Medellín, Medellín, Colombia
e-mail: evanegas901@soyudemedellin.edu.co

L. Rocha-Meneses
Biosystems Engineering, Institute of Forestry and Engineering, Estonian University of Life Sciences, Tartu, Estonia

J. Sierra
Department of Mechatronics Engineering—MATyER, Instituto Tecnológico Metropolitano, Medellin, Colombia

H. A. Yepes
Aplicaciones en Termofluidos, Energía y Nanomateriales Avanzados—ATENA, Departamento de Ingeniería Mecánica, Universidad Francisco de Paula Santander Ocaña, Ocaña, Colombia

J. P. Gómez Montoya
Departamento de Ingeniería Mecánica, Universidad Tecnológica del Perú, Lima, Peru

M. Espinoza-Andaluz et al. (eds.), *Congress on Research, Development and Innovation in Renewable Energies*, Green Energy and Technology,
https://doi.org/10.1007/978-3-031-88995-0_5

[6, 7]. Demonstrating greater efficiency in capturing nearly pure CO_2 compared to oxy-combustion and other CO_2 separation methods from flue gas, such as chemical and physical absorption techniques [8]. This makes CLC a promising technology for reducing emissions, especially when combined with renewable energy sources like biomass [9, 10]. Biomass can be used in CLC systems directly or by pre-gasifying it. Direct introduction has advantages, especially when compared to coal, due to biomass's higher content of volatile matter and its lower sulfur and ash content [11], but it faces challenges like material deactivation and separation issues [12, 13]. Gasification, however, transforms biomass into a cleaner, more reactive gas, leading to higher efficiencies and broader feedstock flexibility [3, 14–17]. The study by Penthor et al. [18] exemplifies this, where they successfully coupled a 100 kW gasifier with a 120 kW CLC pilot plant, using wood chips and standard wood pellets, achieving a near-complete conversion of CO and H_2. Fan et al. [19] highlighted the environmental benefits of this integration in a biomass and coal co-fueled system, reaching significant reductions in greenhouse gas emissions when replacing coal with biomass, while Spallina et al. [20] demonstrated high electric and CO_2 capture efficiencies in a coal gasification-CLC system.

CLC integrated with tri-generation systems offers the potential to produce cooling, heating, and power concurrently. Research by Rajabi et al. [21] evaluated CLC's application in multigeneration setups, emphasizing the superior electrical efficiency of fuel cell-based systems and significant CO_2 reduction. They noted that factors like oxygen carrier type and operational conditions critically influence system performance, with energy efficiency values ranging from 39% to 70% and exergy efficiency from 22.16% to 73.25%. Jiménez et al. [22] explored chemical loop hydrogen generation (CLHG), showcasing its potential for efficient trigeneration of energy, hydrogen, and process heating. Their thermodynamic analysis revealed an overall cycle efficiency between 63% and 67%, depending on process heating temperatures. Additionally, Segurado et al. [23] presented a case study on biomass gasification in a trigeneration plant, highlighting its environmental benefits through significant CO_2 emission reductions. They also identified challenges in implementation, noting lower efficiency compared to coal-based systems.

This research adapts an existing tri-generation system for hot climates like northern Colombia, incorporating pre-gasification of biomass before its integration into the CLC system. Our study includes a detailed thermodynamic analysis and sensitivity analysis to assess energy and exergy efficiencies, identifying areas of exergy loss and proposing optimization strategies. The findings demonstrate the feasibility of this adapted technology, contributing valuable insights for future development. This research underscores the importance of adapting existing technologies to local contexts [24, 25], playing a crucial role in the global transition toward sustainable energy solutions.

2 Case Study

Colombia, with a population of 45.5 million, is the 30th most populous country in the world. In recent years, the country has seen a significant increase in demand for refrigeration, especially in the hot and humid northern Caribbean region. For example, in Barranquilla, the average comfortable temperature is around 27.9 °C. This region experiences a higher demand for cooling energy compared to other parts of Colombia, particularly in tourism sectors [26]. Cartagena, for example, has a cooling demand of 47 MW, surpassing even larger cities like Bogota and Medellin [27]. This region is also experiencing rapid population growth, with a 1.1% increase from 1997 to 2017. Alongside this demographic expansion, economic growth, marked by a 3.7% average annual increase in real GDP [28], further amplifies the demand for refrigeration and air conditioning, driving overall energy consumption in Colombia. Between 2011 and 2016, air conditioning (AC) demand in Colombia increased by 66%, placing it among the top five largest demands in Latin America, with approximately 200,000–250,000 AC units sold each year [26].

While this study specifically focuses on northern Colombia, the system investigated is equally applicable to other regions with similar climates, economic trajectories, and abundant biomass resources. Many tropical countries experiencing population growth and economic expansion face a similar surge in demand for cooling and air conditioning systems. This trend is evident in regions like Southeast Asia, the Middle East, and parts of Africa and South America [29]. The Colombian government's policies, like Resolution 5926 of 2011 and Resolution 0549 of 2015, promoting eco-efficient buildings and sustainable construction [26], highlight the need for more effective energy policies and thermal comfort standards in rapidly developing areas. Furthermore, international agreements like the Kyoto Protocol (1997) and the Paris Agreement (2015), aimed at reducing greenhouse gas emissions and controlling the rise in global temperature, have significant implications for national policies. These agreements underscore the need for sustainable energy systems and thermal comfort standards in both existing and new construction. This research offers a potential path toward mitigating global carbon emissions, highlighting the broader environmental impact of this work.

3 System Description

Figure 1 illustrates the configuration of the modified system, which is based on the design proposed by Sun and Aziz [5]. This system is enhanced by the inclusion of a biomass gasification unit tailored for a locally available biomass feedstock, namely dried sugarcane bagasse from Colombia. A detailed discussion on the characteristics and significance of this biomass feedstock within the context of the system will be presented later in this chapter. The system effectively integrates biomass gasification with the CLC process to generate syngas (stream 6). It is composed of key units,

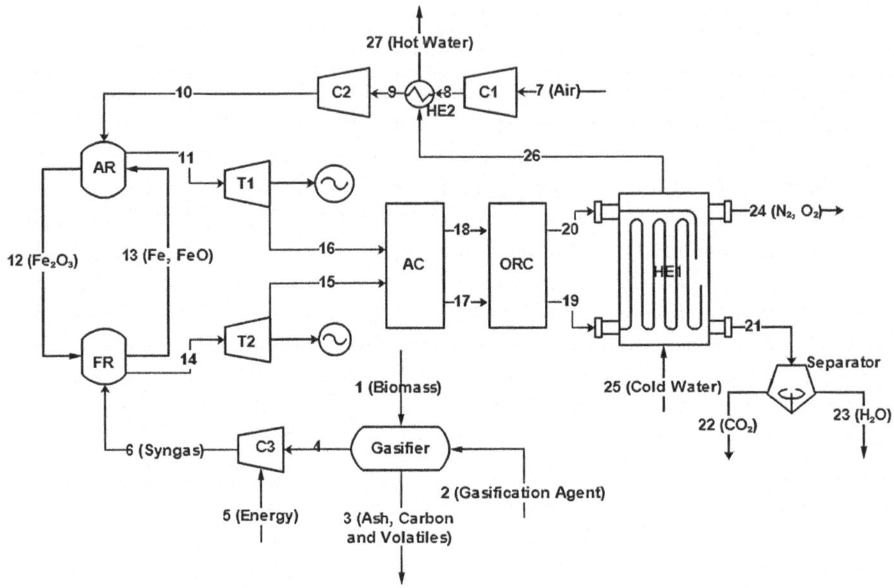

Fig. 1 Schematic diagram of the proposed system

including two gas turbines (GT), an absorption chiller (AC), an organic Rankine cycle (ORC), and a heat exchanger (HE1), all arranged to leverage the thermal gradient provided by the heat source.

The fuel reactor (FR) and air reactor (AR) initiate the process by producing high-temperature, high-pressure exhaust gases that drive the GTs (streams 11 and 14), facilitating the recovery of thermal energy to generate electricity. The subsequent use of an AC unit (stream 18) takes advantage of the remaining high temperatures of these gases [30]. Next, the ORC unit (stream 20) utilizes the AC's cooled exhaust gases for additional electricity generation, aligning with the system's heat output requirements. The gases proceed to HE1, where they are cooled further using water at 25 °C (stream 25) before being split (stream 21). The water, exiting HE1 at 65 °C (stream 26), is directed to the intercooler of the air compressor (HE2, stream 27), elevating its temperature to 75 °C. In addition to the heat produced, this process also cools the compressed air for the second air compressor stage, thereby boosting the system's overall efficiency [31].

3.1 *Biomass Gasification*

Dried sugarcane bagasse, a typical byproduct of sugar production [32], is utilized as the biomass feedstock in this system. Its disposal poses a significant challenge, but previous studies have explored its use for soil improvement and wastewater treatment [33]. Proximate analysis of sugarcane bagasse from Colombian agriculture reveals a moisture content of 5.25%, ash content of 3.90%, fixed carbon of 8.30%, and volatile material of 82.55% (wt%). Its calorific value is 16.92 MJ/kg. Ultimate analysis shows a composition of 46.60% carbon (C), 5.92% hydrogen (H), 0.14% oxygen (O), 43.05% nitrogen (N), and 0.09% sulfur (S) (wt% dry basis) [11]. Sugarcane bagasse (stream 1) is fed into the gasifier along with a gasifying agent (stream 2), initiating pyrolysis and gasification reactions. This process breaks down the biomass, primarily producing carbon monoxide (CO) and hydrogen (H_2) as syngas (stream 6), while ash (stream 3) is removed as a byproduct. The generated syngas is then pressurized (C3) to match or exceed the pressure required by the fuel reactor (FR), allowing it to flow into the CLC system (stream 4). Extensive research on sugarcane bagasse gasification, including the work by Lopes et al. [34], demonstrates promising results in synthesizing gas mixtures.

These studies highlight operating temperature as a crucial factor in maximizing CO/H_2 yield and minimizing CO_2 production. Circulating fluidized bed reactors using steam have been identified as the optimal technology for this process. Systems utilizing this technology have achieved hydrogen concentrations as high as 55.9% by volume and carbon monoxide concentrations up to 29% by volume.

3.2 *Chemical Looping Combustion System*

The chemical looping system under discussion employs a dual-reactor setup consisting of an FR and an AR. The syngas produced in the gasifier (stream 6) is channeled into the FR, which operates at a temperature of 950 °C and a pressure of 36 bars. Within the FR, the syngas reacts with the oxygen carrier (OC), iron oxide

Table 1 Reactions in the chemical looping system

Reactor	Reactions
Gasifier	Pyrolysis. Biomass $\rightarrow$ Char + Tar + Volatiles (CO, H_2, CH_4)
	Gasification. $C + H_2O \rightarrow CO + H_2/C + CO_2 \rightarrow 2CO$
FR	Reduction. $3Fe_2O_3 + CO \rightarrow 2Fe_3O_4 + CO_2$
	$3Fe_2O_3 + H_2 \rightarrow 2Fe_3O_4 + H_2O$
	$Fe_3O_4 + CO \rightarrow 3FeO + CO_2$
	$Fe_3O_4 + H_2 \rightarrow 3FeO + H_2O$
	$Fe_3O_4 + 4H_2 \rightarrow 3Fe + 4H_2O$
	$FeO + CO \rightarrow Fe + CO_2$
	$FeO + H_2 \rightarrow Fe + H_2O$
AR	Oxidization : $4Fe + 3O_2 \rightarrow 2Fe_2O_3$
	$4FeO + O_2 \rightarrow 2Fe_2O_3$
	$4Fe_3O_4 + O_2 \rightarrow 6Fe_2O_3$

(Fe_2O_3), resulting in the production of CO_2 and H_2O while also reducing the OC to ferrous oxide (FeO) and elemental iron (Fe) as outlined in Table 1.

The reduced OC (stream 13) is then transferred to the AR, where it undergoes reoxidation, predominantly being converted back to iron oxide (Fe_2O_3) at conditions of 1300 °C and 36 bars. Due to the exothermic nature of reactions in the AR as compared to the endothermic reactions in the FR, a thermal mediator, specifically alumina (Al_2O_3), is introduced to balance the heat exchange between the reactors, thus maintaining a self-regulating system [35, 36]. The oxygen carrier completes its cycle by returning to the FR for reutilization. The byproducts from this CLC process are an inert gas stream (stream 14), and a nitrogen-rich depleted air stream (stream 11), which are discharged from the system at a pressure of 36 bars but at varying temperatures of 950 and 1300 °C, respectively. In selecting an oxygen carrier, criteria such as environmental impact, oxygen transfer efficacy, high melting point, and cost-effectiveness were considered, leading to the choice of hematite (Fe_2O_3). Hematite is favored for its cost advantage over nickel-based alternatives and its established compatibility with biomass-derived syngas for CLC applications, as documented in Refs. [37, 38].

3.3 Tri-generation System

In this system, gases with no further chemical use, along with used air (streams 11 and 14), are directed through turbines to generate electricity. These streams, initially at high temperatures (up to 1300 °C) and pressures (36 bars), expand in the turbines, converting thermal energy into electrical power. While the gas streams (15 and 16) cool to 400–600 °C after the turbines, they still retain enough heat to

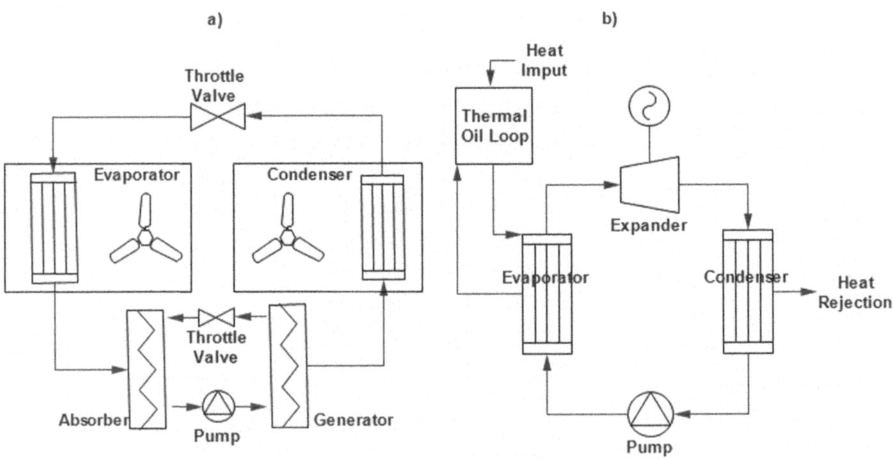

Fig. 2 Conceptual flow diagrams of (**a**) AC and (**b**) ORC

power a LiBr/H_2O absorption chiller (AC), providing cooling (see Fig. 2a). The AC system separates the warm diluted solution into refrigerant vapor and a concentrated solution. The vapor releases heat in the condenser, is depressurized, and then absorbs ambient heat in the evaporator, creating cooling and reverting to vapor. This vapor mixes with the concentrated solution from the generator, forming a diluted solution in the absorber, which is then pressurized for recirculation [39]. Exiting the AC, the cooled gas streams (17 and 18) enter the organic Rankine cycle (ORC) for further energy recovery (Fig. 2b).

These streams transfer heat to a low-boiling-point working fluid, causing it to vaporize. This vapor expands in an expander, converting thermal energy into mechanical energy. The exhaust gas is then cooled in a condenser, preparing the working fluid for the next cycle [40]. R245fa, a working fluid with a suitable critical temperature and low environmental impact, is chosen for its superior performance and compliance with regulatory standards. Compared to alternatives like R123, R245fa enhances energy and exergy efficiencies by up to 7% and has a lower global warming potential (GWP) [41].

This makes R245fa a more sustainable option for ORC systems [42]. The ORC process cools the flue gases (streams 19 and 20) to below 100 °C to prevent corrosion before they are used to heat cold water (stream 25) in HE1. The resulting hot water at 65 °C (stream 27) is further heated to 75 °C in the air compressor's intercooler (HE2).

The exhaust gas from the fuel reactor (FR) (stream 21) is processed in a separator, dividing it into CO_2 (stream 22) and H_2O (stream 23), allowing for potential carbon capture and storage and contributing to carbon-negative emissions. The final air stream (stream 24) is then released, completing the system's cycle.

4 Methodology

4.1 Simulation and Operating Parameters

The simulation of the system was performed in Aspen Plus V12.1 (Aspen Technology, Inc.). The property method selected for the simulation was the Peng-Robinson modified by Boston-Mathias (PR-BM), with the ELECTRNL method being used for the simulation of the AC. The simulation is done at a steady state with negligible losses in heat and pressure considered; also, all reactants are considered at chemical balance and phase equilibrium.

The stage of gasification of sugarcane bagasse is modeled using the Aspen Plus Ryield and RGibbs blocks at a temperature of 1100 °C and 1 atm [17]. The bagasse is subjected to a pyrolysis process in the Ryield block to devolatilize it without oxygen, giving rise to the conventional elements (C, H, O, N, S, ashes, carbon, and volatiles) [43]. In the RGibbs block, gasification reactions are carried out by chemical and phase equilibrium to minimize the total Gibbs free energy, subject to atomic equilibrium constraints. This divides the carbon between the gaseous phase,

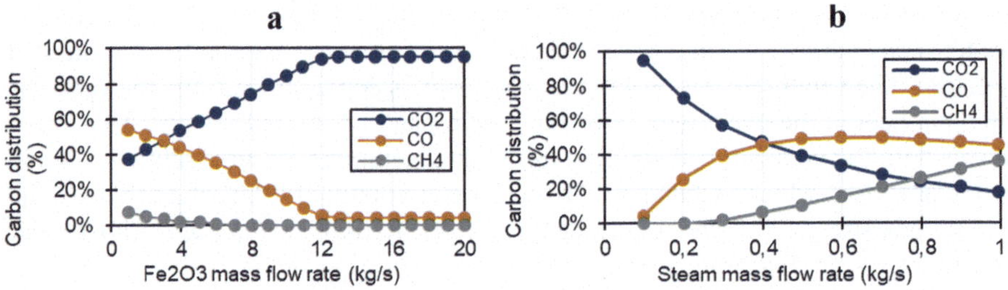

Fig. 3 (**a**) Effects of OC mass flow rate on carbon distribution. (**b**) Effects of steam mass flow rate on carbon distribution

Fig. 4 Effects of FR operating temperature on carbon distribution

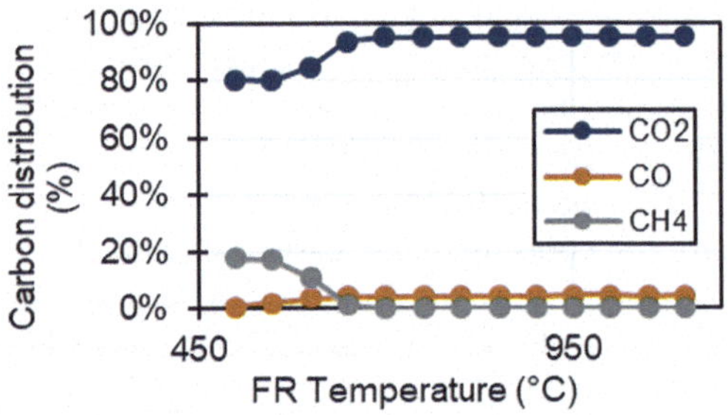

used for devolatilization and the solid phase (charcoal) that originates the coal gasification [44]. The by-products are ash, which is discarded, and syngas, which is then pressurized to the operating pressure of the FR by a compressor. The air reactor is also modeled as a single RGibbs block since the reactions that take place in the AR are fast and can be easily completed [5]. The fuel reactor's total heat duty is the sum of the reactors in the gasificator and the RGibbs fuel reactor.

A sensitivity analysis was performed for conditions in the CLC system, such as AR and FR temperatures, operating pressure, and OC mass feed rate, to select the most optimal parameters for the simulation, as the performance of the CLC system has a significant impact on the efficiency of the whole plant. The effects of the Fe_2O_3 and steam flow rates can be seen in Fig. 3a, b. From the sensibility analysis, it was concluded that the mass flow rate of Fe_2O_3 at the inlet of the fuel reactor should be higher than 12 kg/s to reduce the presence of CO and CH_4 at the outlet. Furthermore, the steam flow rate in the gasifier should be lower than 0.2 kg/s to combust the biomass completely.

As seen in Fig. 4, the temperature in the FR should be higher than 650 °C to achieve the complete combustion of the syngas generated in the gasifier.

Figure 5 shows that as the temperature in the AR increases, the energy required by the AR increases and so does its thermal load; on the other hand, the thermal load of the FR decreases as the AR temperature increases since the FR is provided with sensible energy by the OC that was heated in the AR [45]. Both reactors achieve a

Fig. 5 AR temperature's effects on the system's thermal load

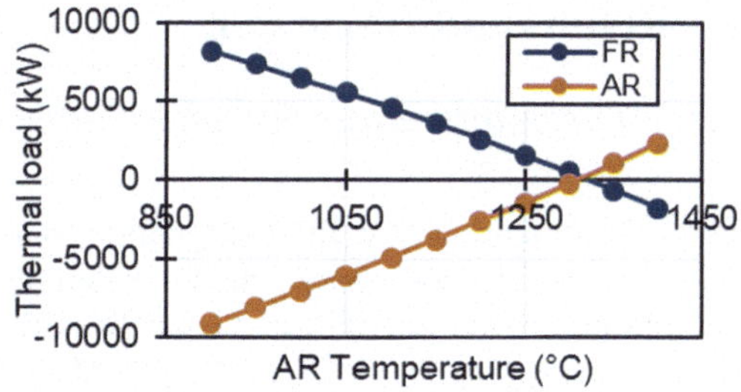

Table 2 Main design system parameters

Parameter	Value	Reference
Ambient-reference pressure (atm)	1	
Ambient-reference temperature (°C)	25	
Mass flow of biomass (kg/s)	1	
Air composition	79% N_2, 21% O_2 (Vol)	
OC composition	60% Fe_2O_3, 40% Al_2O_3	[46]
Gasification pressure (atm)	1	
Gasification temperature (°C)	1100	[46]
Operating pressure (atm)	36	
FR temperature (°C)	900	
AR temperature (°C)	1300	
Stream temperature (°C)	200	
Stream temperature (°C)	100	
Stream temperature (°C)	75	
Stream temperature (°C)	65	
AC COP	1.18	[47]
ORC efficiency (%)	15.76%	
Compressor and GT efficiency (%)	85%	[48]

net thermal load of 0 when the AR temperature is 1300 °C, in which the reactors are operating autothermally [5].

In the simulation, the methodology employs the MHeatx module to represent the heat exchanger in the ORC system, providing thermal energy to the AC. The system's COP is influenced by various factors, including the heat source temperature; for this study, the AC's COP was fixed at 1.18. The exit temperature of stream 18 from the heat exchanger is maintained at 200 °C. In addition, the ORC evaporator, modeled using the MHeatx module [41], sets the temperature of stream 20 at 100 °C after exchanging heat energy with the working fluid. The ORC cycle, utilizing R245fa, operates at pressures of 0.09 and 2.5 MPa, with an efficiency of 15.76%. This efficiency is unaffected by variations in the heat source temperature or the R245fa mass flow rate. The simulation also includes modeling of all compressors and turbines with isentropic modules at an efficiency of 85%. For detailed information on the remaining modules and their operational parameters, refer to Table 2.

5 Results and Discussion

5.1 Thermodynamic Results Under Design Conditions

For the comprehensive thermodynamic analysis, the simulation was conducted based on the operation parameters detailed in Table 2. Notably, under these design conditions, the syngas exiting the gasifier exhibits a molar composition comprising 51.22% hydrogen, 28.53% CO, 14.18% H_2O, 3.87% CO_2, and 2.17% N_2, alongside trace amounts of O_2 and methane.

The simulated system demonstrates an energy efficiency of 45.17%. This efficiency is distributed across electricity, cooling, and heating, with respective efficiencies of 19.74%, 13.68%, and 11.06%. The three turbines in the model collectively generate 3245.59 kW of power. Specifically, GT1 contributes 47.1%, GT2 accounts for 50.3%, and the ORC adds 2.6%, as illustrated in Fig. 6. Additionally, the AC produces 2258.81 kW of cooling power, while the heat exchangers contribute 1819.38 kW of heating power. The total output of the system is divided into 44.39% electricity, 30.76% cooling, and 24.86% heating, with 55% of this belonging to HE1 and 45% to HE2.

A detailed exergy analysis of these products is depicted in Fig. 7. The overall exergy efficiency of the system stands at 20.83%. Breaking this down, the exergy

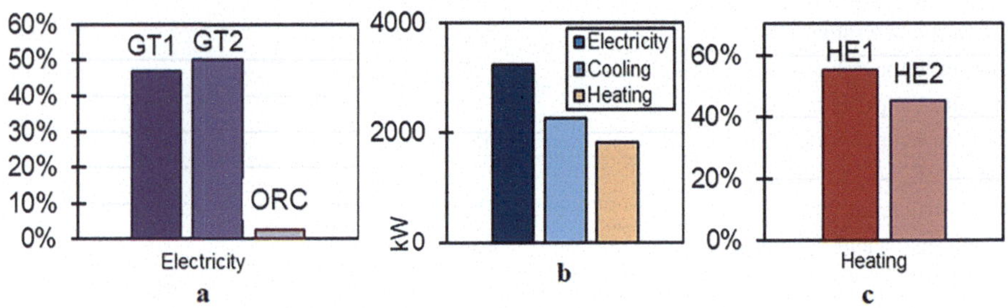

Fig. 6 Distribution of (**a**) electricity, (**b**) cooling, and (**c**) heating produced by the system

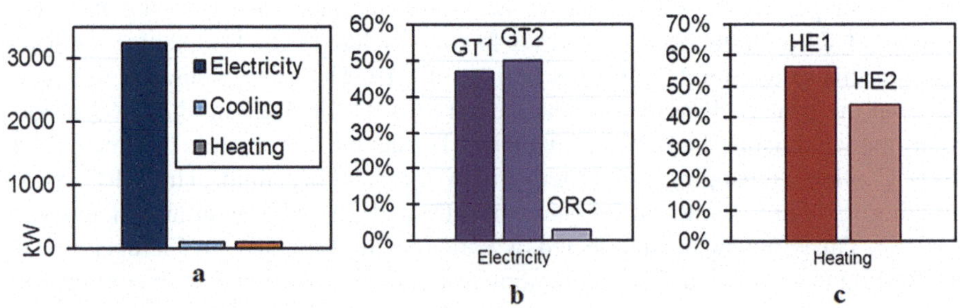

Fig. 7 Distribution of (**a**) cooling, (**b**) electricity, and (**c**) heating based on exergy

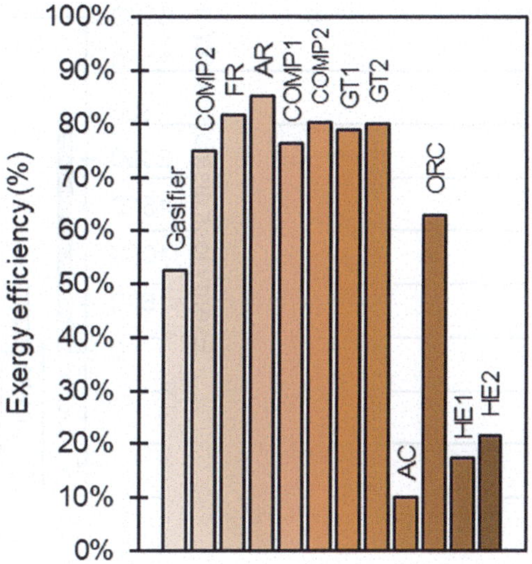

Fig. 8 Exergetic efficiency per block

efficiencies for electricity, cooling, and heating are 17.7%, 1.69%, and 1.43%, respectively. A significant portion of the exergy, amounting to 3235.8 kW, is derived from electricity. Specifically, GT1 contributes 47%, GT2 accounts for 50%, and the ORC adds 3%, while cooling and heating exergies contribute 99 kW and 88 kW, respectively, with 56% of the heating being generated by HE1 and 44% by HE2.

Figure 8 illustrates the exergy efficiency of each system component. The highest exergy efficiencies are recorded in the fuel and air reactors, achieving 81.65% and 85.28%, respectively. The three compressors and two gas turbines also exhibit high exergy efficiencies of over 75%. The ORC shows an exergy efficiency of 62.97%, surpassing its energy efficiency of 13.84%. The gasifier, with a moderate exergy efficiency of 52.53%, also plays a key role. Lower exergy efficiencies in the AC and the two heat exchangers, standing at 9.97%, 17.25%, and 21.36%, are attributed to the high-temperature differences during heat exchanges with the hot gases from the CLC system [5]. Figure 9 sheds light on the exergy destruction within each block. It reveals that the most significant exergetic losses occur in the gasifier and CLC section. The gasifier accounts for 2414.55 kW or 19.27% of the exergy loss, while the air reactor and fuel reactor are responsible for 3027.42 kW (24.2%) and 2791.13 kW (22.235%) of the losses, respectively. Combined, these elements contribute to 8230.13 kW or 65.7% of the total net exergy destruction. The remaining system components account for 4296.62 kW or 34.3% of the total exergy destruction, with the AC alone contributing to 1166.67 kW or 9.8% of this total.

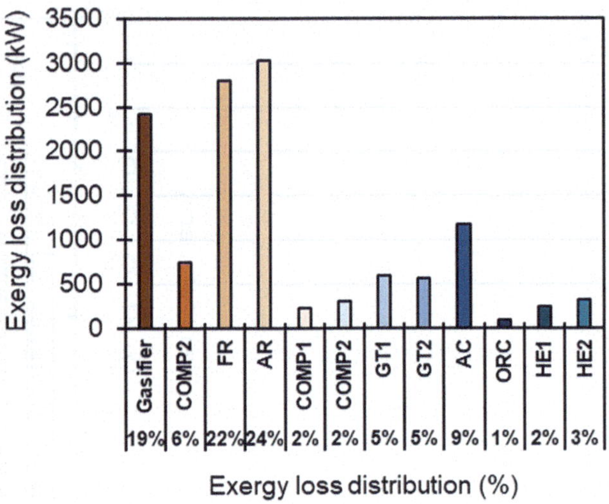

Fig. 9 Exergy destruction per block

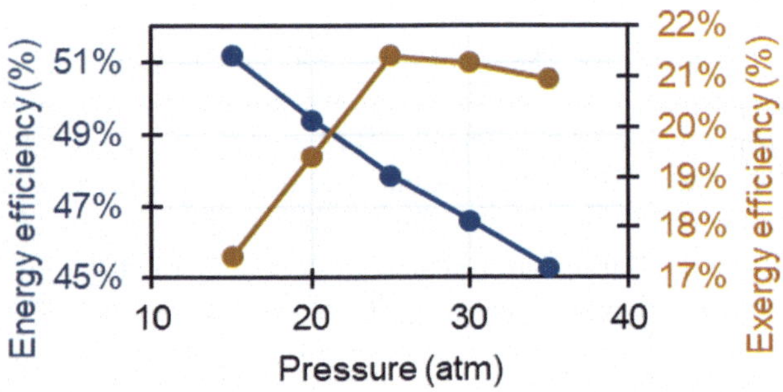

Fig. 10 Effects of operating pressure on exergy and energy efficiencies

5.2 *Sensitivity Analysis*

In a sensitivity analysis examining the effects of variables such as operating pressure, fuel reactor (FR) temperature, and stream 18 and 20 temperatures, significant impacts on the system's exergetic and energetic efficiencies were observed. For instance, an increase in operating pressure leads to a greater pressure drop in the turbines and a consequent reduction in the outlet temperature of the gas turbines. This change results in an enhancement of exergy efficiency from 17.43% at 15 bars to 21.4% at 25 bars, as illustrated in Fig. 10, primarily because the turbines work significantly contributes to the system's exergy. However, as the temperature of the expanded gases after the gas turbines decreases, the available energy for cooling, heating, and the ORC decreases, which causes the energy efficiency of the system to decrease from 51.20% at 15 bars to 45.28% at 35 bars.

The system's response to variations in FR temperature is also noteworthy. As shown in Fig. 11a, the energy efficiency reaches a high of 45.17% at 900 °C, declines to 43.73% at 950 °C and then increases again at 44.89% at 1100 °C; the

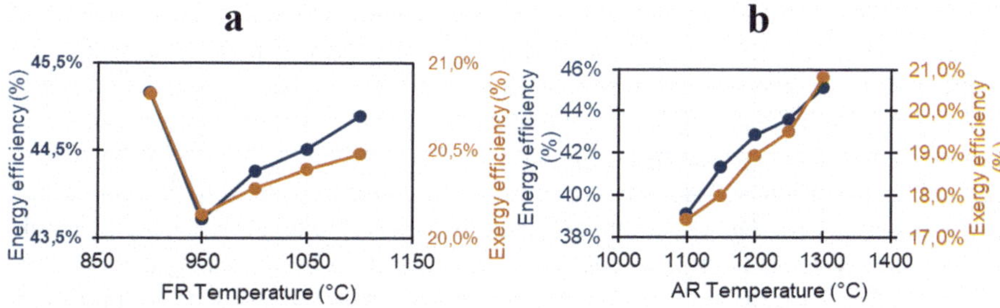

Fig. 11 (**a**) Effects of FR temperature in exergy and energy efficiencies. (**b**) Effects of AR temperature in exergy and energy efficiencies

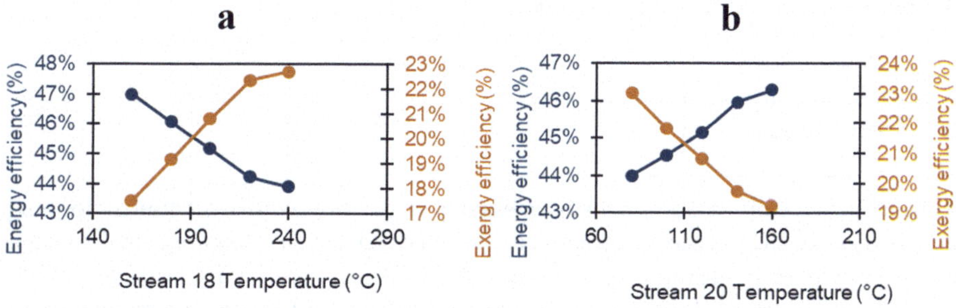

Fig. 12 (**a**) Effects of stream 18 temperature in exergy and energy efficiencies. (**b**) Effects of stream 20 temperature in exergy and energy efficiencies

exergy efficiency of the system decreases from a maximum of 20.83% at 900 °C to a minimum of 20.14% at 950 °C, to finally increase to 20.48% at 1100 °C. This fluctuation is attributed to changes in the support material and air flow rates within the FR, where while the temperature in the FR decreases, the amount of support material entering the FR increases, and the mass flow rate of air entering the FR decreases to maintain the heat transfer between both reactors; additionally, as the temperature in the FR increases, the work produces by the GT1 increases, while the increased temperature in the FR decreases the amount of energy of the depleted air coming from the AR, which decreases the work produced by the GT2. Moreover, as the air reactor (AR) temperature increases, the performance of the system improves, with energy and exergy efficiencies climbing from 39.12% and 17.43% at 1100 °C to 45.17% and 20.83% at 1300 °C, respectively, depicted in Fig. 11b. An increase in the AR temperature leads to a decrease in the flow rate of air to achieve the full oxidation of the reduced OC; this means less work in compressors 1 and 2, while the work produced in the gas turbine 2 increases due to the higher temperature of the depleted air exiting the air reactor; overall, high AR temperatures increase the network produced by the system and provide more exergy to harness in the GT2, resulting in an increase in energy and exergy efficiencies.

Additionally, the alterations in the temperatures of streams 18 and 20 show significant effects. An increase in stream 18's temperature decreases the system's cooling capacity and total production, reducing energy efficiency from 47.02% at 160 °C to 43.91% at 240 °C. On the other hand, the exergy efficiency of the system increases with the temperature of stream 18 from 17.52% at 160 °C to a maximum of 22.72% at 240 °C, as a higher temperature in stream 18 means more exergy to harness by the ORC, which has a higher overall exergy efficiency than the AC, as seen in Fig. 12a. Finally, Fig. 12b demonstrates the impact of stream 20s temperature on the system's efficiencies. As this temperature rises, the heating energy available for HE1 increases, reducing the ORC's work production but boosting the system's overall heating capacity and energy efficiency from 44.01% at 80 °C to 46.33% at 160 °C. However, this also leads to a decline in exergetic efficiency from 23.02% at 80 °C to 19.26% at 160 °C, as the increased heat transfer to the water and the greater temperature difference during this transfer escalate the generated entropy, thereby diminishing the overall exergetic efficiency.

6 Conclusion

This tri-generation system, which integrates a gasifier with a chemical looping combustion (CLC) process, effectively produces electricity, cooling, and heat. It employs two heat turbines and an organic Rankine cycle (ORC) for electricity generation, an air conditioning (AC) unit for cooling, and two heat exchangers for heat production. Simulations aimed at optimizing operational conditions demonstrated an overall energy efficiency of 45.17%, with specific efficiencies of 19.74% for electricity, 13.68% for cooling, and 11.06% for heating, resulting in 3245.59 kW of electricity, 2258.81 kW of cooling, and 1819.38 kW of heating power.

In addition to the energy efficiency metrics, the system achieved an exergetic efficiency of 20.83%. The analysis showed that the most of the exergy destruction occurred in the gasifier and the CLC section, accounting for 65.70% of total destruction, with the AC unit contributing 9.35%. This information is critical because it is an indication of areas where improvements could have an impact on overall system performance.

Sensitivity analysis showed that increasing operating pressure improves exergetic efficiency but reduces energetic efficiency. Optimal temperatures of 900 °C for the fuel reactor (FR) and 1300 °C for the air reactor (AR) yielded the highest combined exergetic and energetic efficiencies. Additionally, raising the temperature of combustion gases entering the ORC decreases exergetic efficiency but boosts energetic efficiency, while the opposite effect is observed for combustion gases entering the first heat exchanger (HE1).

These results underscore the potential of the proposed tri-generation system to provide multiple energy outputs simultaneously, making it a promising option for various industrial applications. Future research will aim to refine the system's design and assess its economic feasibility to further validate its practical application.

Acknowledgments The authors would like to thank Ecopetrol, Ministerio de Ciencia Tecnología e Innovación and Ministerio de Minas y Energía for their support in the development of the project "*Proceso integral para la obtención de hidrógeno de bajas emisiones a partir de biogás producido en plantas de tratamiento de aguas residuales usando transportadores sólidos de oxígeno de bajo costo*" Code 1115-929-93266 through the financing contract 274-2022 derived from the special cooperation agreement 753-2021 and to Universidades G8 Antioquia—Gobernación de Antioquia through the "Segunda convocatoria conjunta de proyectos de I+D+i en el marco de la agenda regional de I+D→i" program.

References

1. Mohamed, U., Zhao, Y., Yi, Q., Shi, L., Wei, G., Nimmo, W.: Evaluation of life cycle energy, economy, and CO_2 emissions for biomass chemical looping gasification to power generation. Renew. Energy. **176**, 366–387 (2021). https://doi.org/10.1016/j.renene.2021.05.067
2. Liu, Q., Hu, C., Peng, B., Liu, C., Li, Z., Wu, K., et al.: High H_2/CO ratio syngas production from chemical looping co-gasification of biomass and polyethylene with CaO/Fe_2O_3 oxygen carrier. Energy Convers. Manag. **199**, 111951 (2019). https://doi.org/10.1016/j.enconman.2019.111951
3. Zhao, X., Zhou, H., Sikarwar, V.S., Zhao, M., Park, A.-H.A., Fennell, P.S., et al.: Biomass-based chemical looping technologies: the good, the bad and the future. Energy Environ. Sci. **10**, 1885–1910 (2017). https://doi.org/10.1039/C6EE03718F
4. Haregu, S., Likna, Y., Tadesse, D., Masi, C.: Recent development of biomass energy as a sustainable energy source to mitigate environmental change. In: Ramanujam, P.K., Parameswaran, B., Bharathiraja, B., Magesh, A. (eds.) Bioenergy: Impacts on Environment and Economy, pp. 119–138. Springer Nature, Singapore (2023). https://doi.org/10.1007/978-981-99-3002-9_8
5. Sun, Z., Aziz, M.: Thermodynamic analysis of a tri-generation system driven by biomass direct chemical looping combustion process. Energy Convers. Manag. **244**, 114517 (2021). https://doi.org/10.1016/j.enconman.2021.114517
6. Abuelgasim, S., Wang, W., Abdalazeez, A.: A brief review for chemical looping combustion as a promising CO_2 capture technology: fundamentals and progress. Sci. Total Environ. **764**, 142892 (2021). https://doi.org/10.1016/j.scitotenv.2020.142892
7. Liu, G., Chang, C.C., Zhao, Y., Veksha, A., Giannis, A., Lim, T.T., et al.: Effect of alkali earth metal doping on the CuO/Al_2O_3 oxygen carrier agglomeration resistance during chemical looping combustion. J. Clean. Prod. **366**, 132970 (2022). https://doi.org/10.1016/j.jclepro.2022.132970
8. Zhou, L., Deshpande, K., Zhang, X., Agarwal, R.K.: Process simulation of chemical looping combustion using ASPEN plus for a mixture of biomass and coal with various oxygen carriers. Energy. **195**, 116955 (2020). https://doi.org/10.1016/j.energy.2020.116955
9. Mei, D., Soleimanisalim, A.H., Linderholm, C., Lyngfelt, A., Mattisson, T.: Reactivity and lifetime assessment of an oxygen releasable manganese ore with biomass fuels in a 10 kWth pilot rig for chemical looping combustion. Fuel Process. Technol. **215**, 106743 (2021). https://doi.org/10.1016/j.fuproc.2021.106743
10. Surywanshi, G.D., Pillai, B.B.K., Patnaikuni, V.S., Vooradi, R., Anne, S.B.: 4-E analyses of chemical looping combustion based subcritical, supercritical and ultra-supercritical coal-fired power plants. Energy Convers. Manag. **200**, 112050 (2019). https://doi.org/10.1016/j.enconman.2019.112050
11. Marrugo, G., Valdés, C.F., Chejne, F.: Characterization of Colombian agroindustrial biomass residues as energy resources. Energy Fuel. **30**, 8386–8398 (2016). https://doi.org/10.1021/acs.energyfuels.6b01596

12. Adánez-Rubio, I., Abad, A., Gayán, P., De Diego, L.F., García-Labiano, F., Adánez, J.: Biomass combustion with CO_2 capture by chemical looping with oxygen uncoupling (CLOU). Fuel Process. Technol. **124**, 104–114 (2014). https://doi.org/10.1016/j.fuproc.2014.02.019
13. Coppola, A., Scala, F.: Chemical looping for combustion of solid biomass: a review. Energy Fuel. **35**, 19248–19265 (2021). https://doi.org/10.1021/acs.energyfuels.1c02600
14. Labiano, F.G.: Estado actual y retos de los procesos chemical looping. Current status and challenges of chemical looping processes. https://www.gecarbon.org/boletines/articulos/Boletin_045-art3.pdf
15. Gogolev, I., Pikkarainen, T., Kauppinen, J., Linderholm, C., Steenari, B.-M., Lyngfelt, A.: Investigation of biomass alkali release in a dual circulating fluidized bed chemical looping combustion system. Fuel. **297**, 120743 (2021). https://doi.org/10.1016/j.fuel.2021.120743
16. Magdziarz, A., Dalai, A.K., Koziński, J.A.: Chemical composition, character and reactivity of renewable fuel ashes. Fuel. **176**, 135–145 (2016). https://doi.org/10.1016/j.fuel.2016.02.069
17. Bhui, B., Vairakannu, P.: Chemical looping and plasma technologies for gasification of coal and biomass. In: De, S., Agarwal, A.K., Moholkar, V.S., Thallada, B. (eds.) Coal and Biomass Gasification, pp. 499–520. Springer, Singapore (2018). https://doi.org/10.1007/978-981-10-7335-9_20
18. Penthor, S., Mayer, K., Kern, S., Kitzler, H., Wöss, D., Pröll, T., et al.: Chemical-looping combustion of raw syngas from biomass steam gasification—coupled operation of two dual fluidized bed pilot plants. Fuel. **127**, 178–185 (2014). https://doi.org/10.1016/j.fuel.2014.01.062
19. Fan, J., Hong, H., Zhu, L., Jiang, Q., Jin, H.: Thermodynamic and environmental evaluation of biomass and coal co-fuelled gasification chemical looping combustion with CO_2 capture for combined cooling, heating and power production. Appl. Energy. **195**, 861–876 (2017). https://doi.org/10.1016/j.apenergy.2017.03.093
20. Spallina, V., Romano, M.C., Chiesa, P., Lozza, G.: Integration of coal gasification and packed bed CLC process for high efficiency and near-zero emission power generation. Energy Procedia. **37**, 662–670 (2013). https://doi.org/10.1016/j.egypro.2013.05.154
21. Rajabi, M., Mehrpooya, M., Haibo, Z., Huang, Z.: Chemical looping technology in CHP (combined heat and power) and CCHP (combined cooling heating and power) systems: a critical review. Appl. Energy. **253**, 113544 (2019). https://doi.org/10.1016/j.apenergy.2019.113544
22. Jiménez Álvaro, Á., Urdiales Montesino, Á., Sánchez Orgaz, S., González Fernández, C.: Thermodynamic analysis of a dual power-hydrogen production system based on chemical-looping combustion. Energy. **137**, 1075–1085 (2017). https://doi.org/10.1016/j.energy.2017.03.132
23. Segurado, R., Pereira, S., Correia, D., Costa, M.: Techno-economic analysis of a trigeneration system based on biomass gasification. Renew. Sustain. Energy Rev. **103**, 501–514 (2019). https://doi.org/10.1016/j.rser.2019.01.008
24. Fan, L.-S., Chung, E.Y., Bayham, S.C., Kathe, M.V., Tong, A., Zeng, L.: Chemical looping combustion and gasification. In: Handbook of Clean Energy Systems, pp. 1–22. John Wiley & Sons, Ltd. (2015). https://doi.org/10.1002/9781118991978.hces150
25. Czakiert, T., Krzywanski, J., Zylka, A., Nowak, W.: Chemical looping combustion: a brief overview. Energies. **15**, 1563 (2022). https://doi.org/10.3390/en15041563
26. Medina, J.M., Rodriguez, C.M., Coronado, M.C., Garcia, L.M.: Scoping review of thermal comfort research in Colombia. Buildings. **11**, 232 (2021). https://doi.org/10.3390/buildings11060232
27. Ríos-Ocampo, J.P., Olaya, Y., Osorio, A., Henao, D., Smith, R., Arango-Aramburo, S.: Thermal districts in Colombia: developing a methodology to estimate the cooling potential demand. Renew. Sustain. Energy Rev. **165**, 112612 (2022). https://doi.org/10.1016/j.rser.2022.112612
28. Aguilera-Díaz, M.M., Reina-Aranza, Y.C., Orozco, A., Vega, J.Y., Barcos-Robles, R.: Evolución socioeconómica de la región Caribe colombiana entre 1997 y 2017. Banco de la República, Bogotá (2017). https://doi.org/10.32468/dtseru.258

29. Dong, Y., Coleman, M., Miller, S.A.: Greenhouse gas emissions from air conditioning and refrigeration service expansion in developing countries. Annu. Rev. Environ. Resour. **46**, 59–83 (2021). https://doi.org/10.1146/annurev-environ-012220-034103
30. Habibi, R., Pourfayaz, F., Mehrpooya, M., Kamali, H.: A natural gas-based eco-friendly polygeneration system including gas turbine, sorption-enhanced steam methane reforming, absorption chiller and flue gas CO_2 capture unit. Sustain. Energy Technol. Assess. **52**, 101984 (2022). https://doi.org/10.1016/j.seta.2022.101984
31. Boles, M.A., Yunus, A.Ç., Mehmet, K.: Termodinámica. McGraw Hill (2019)
32. Cadavez, C.C., De Souza-Santos, M.L.: Efficiency of a power generation alternative regarding the composition of feeding biomass-glycerol slurry; theoretical assessment. Energy. **214**, 118967 (2021). https://doi.org/10.1016/j.energy.2020.118967
33. Oruc, O., Dincer, I.: Evaluation of hydrogen production with iron-based chemical looping fed by different biomass. Int. J. Hydrog. Energy. **45**, 34557–34565 (2020). https://doi.org/10.1016/j.ijhydene.2020.04.119
34. Motta, I.L., Miranda, N.T., Maciel Filho, R., Wolf Maciel, M.R.: Sugarcane bagasse gasification: simulation and analysis of different operating parameters, fluidizing media, and gasifier types. Biomass Bioenergy. **122**, 433–445 (2019). https://doi.org/10.1016/j.biombioe.2019.01.051
35. Nandy, A., Loha, C., Gu, S., Sarkar, P., Karmakar, M.K., Chatterjee, P.K.: Present status and overview of chemical looping combustion technology. Renew. Sustain. Energy Rev. **59**, 597–619 (2016). https://doi.org/10.1016/j.rser.2016.01.003
36. Wang, X., Wang, X., Kong, Z., Shao, Y., Jin, B.: Auto-thermal operation and optimization of coal-fueled separated gasification chemical looping combustion in a pilot-scale unit. Chem. Eng. J. **383**, 123159 (2020). https://doi.org/10.1016/j.cej.2019.123159
37. Cheng, D., Yong, Q., Zhao, Y., Gong, B., Zhang, J.: Study on the interaction of the Fe-based oxygen carrier with ashes. Energy Fuel. **34**, 9796–9809 (2020). https://doi.org/10.1021/acs.energyfuels.0c01303
38. Lyngfelt, A., Brink, A., Langørgen, Ø., Mattisson, T., Rydén, M., Linderholm, C.: 11,000 h of chemical-looping combustion operation—where are we and where do we want to go? Int. J. Greenhouse Gas Control. **88**, 38–56 (2019). https://doi.org/10.1016/j.ijggc.2019.05.023
39. Ogidiama, O.V., Abu-Zahra, M.R.M., Shamim, T.: Techno-economic analysis of a poly-generation solar-assisted chemical looping combustion power plant. Appl. Energy. **228**, 724–735 (2018). https://doi.org/10.1016/j.apenergy.2018.06.091
40. Situmorang, Y.A., Zhao, Z., An, P., Rizkiana, J., Prakoso, T., Abudula, A., et al.: A small-scale power generation system based on biomass direct chemical looping process with organic rankine cycle. Chem. Eng. Process. Process Intensif. **163**, 108361 (2021). https://doi.org/10.1016/j.cep.2021.108361
41. Jiang, P., Mahmud Parvez, A., Meng, Y., Dong, X., Xu, M., Luo, X., et al.: Novel two-stage fluidized bed-plasma gasification integrated with SOFC and chemical looping combustion for the high efficiency power generation from MSW: a thermodynamic investigation. Energy Convers. Manag. **236**, 114066 (2021). https://doi.org/10.1016/j.enconman.2021.114066
42. Ozcan, H., Dincer, I.: Thermodynamic analysis of a combined chemical looping-based trigeneration system. Energy Convers. Manag. **85**, 477–487 (2014). https://doi.org/10.1016/j.enconman.2014.06.011
43. Tavares, R., Monteiro, E., Tabet, F., Rouboa, A.: Numerical investigation of optimum operating conditions for syngas and hydrogen production from biomass gasification using Aspen Plus. Renew. Energy. **146**, 1309–1314 (2020). https://doi.org/10.1016/j.renene.2019.07.051
44. Yu, J., Smith, JD.: Validation and application of a kinetic model for biomass gasification simulation and optimization in updraft gasifiers. Chem. Eng. Process. Process Intensif. 125: 214–226 (2018). https://doi.org/10.1016/j.cep.2018.02.003
45. Stenberg, V., Spallina, V., Mattisson, T., Rydén, M.: Techno-economic analysis of processes with integration of fluidized bed heat exchangers for H_2 production—part 2: chemical-looping combustion. Int. J. Hydrog. Energy. **46**, 25355–25375 (2021). https://doi.org/10.1016/j.ijhydene.2021.04.170

46. Miyahira, K., Aziz, M.: Hydrogen and ammonia production from low-grade agricultural waste adopting chemical looping process. J. Clean. Prod. **372**, 133827 (2022). https://doi.org/10.1016/j.jclepro.2022.133827
47. Wang, J., Fu, C.: Thermodynamic analysis of a solar-hybrid trigeneration system integrated with methane chemical-looping combustion. Energy Convers. Manag. **117**, 241–250 (2016). https://doi.org/10.1016/j.enconman.2016.03.039
48. Wang, S., Liu, C., Li, J., Sun, Z., Chen, X., Wang, X.: Exergoeconomic analysis of a novel trigeneration system containing supercritical CO_2 Brayton cycle, organic Rankine cycle and absorption refrigeration cycle for gas turbine waste heat recovery. Energy Convers. Manag. **221**, 113064 (2020). https://doi.org/10.1016/j.enconman.2020.113064

Experimental and Simulation Study of Convective Drying Process of Spent Coffee Grounds Including an Exergoeconomic Analysis

Diana L. Tinoco-Caicedo, Medelyne Mero-Benavides, Santiago Corral-Santos, and Ana M. Blanco-Marigorta

1 Introduction

The spent coffee grounds (SCGs) are the main waste of instant coffee factories and cafeterias [1]. From this agro-industrial waste, many value-added products such as biofuels, composites, and bioactive compounds have been proposed [2]. However, one of the main limiting factors of the conversion processes is the high operating cost of the drying process since it has a high energy consumption [3]. In this sense, the exergoeconomic analysis is an important decision-making tool to reduce production costs and increase process sustainability by decreasing the exergy destruction rate [4].

Many studies have been performed in different drying systems to quantify and identify the main components that cause low exergetic efficiency and high operational costs. Zeng et al. [5], by performing an exergoeconomic analysis, were able to recommend improvements in the drying chamber of an existing gas-type industrial

D. L. Tinoco-Caicedo (✉)
Facultad de Ciencias Naturales y Matemáticas (FCNM), Escuela Superior Politécnica del Litoral Ecuador, Guayaquil, Ecuador

Centro de Energías Renovables y Alternativas (CERA), Escuela Superior Politécnica del Litoral Ecuador, Guayaquil, Ecuador
e-mail: dtinoco@espol.edu.ec

M. Mero-Benavides · S. Corral-Santos
Facultad de Ciencias Naturales y Matemáticas (FCNM), Escuela Superior Politécnica del Litoral Ecuador, Guayaquil, Ecuador

A. M. Blanco-Marigorta
Department of Process Engineering, Universidad de Las Palmas de Gran Canaria, Las Palmas de Gran Canaria, Canary Islands, Spain

M. Espinoza-Andaluz et al. (eds.), *Congress on Research, Development and Innovation in Renewable Energies*, Green Energy and Technology,
https://doi.org/10.1007/978-3-031-88995-0_6

drying system for black tea since it had the lowest exergetic efficiency (19.87%) and the highest exergy destruction cost rate (18.5$/h). Singh et al. [6] analyzed the effect of the drying system design on the exergetic efficiency and concluded that when using single-stage spray drying system instead of a third stages system, the exergetic cost rate could be reduced from 322.9 to 43.7$/h, because the last one had the highest exergy consumption and the lowest exergetic efficiency. Atalay [7] identified that using electric energy instead of hybrid solar energy increases the exergy destruction cost by 48.6%.

Despite the relevance of the exergoeconomic analysis in the design and optimization of the drying systems, no exergoeconomic analyses of the drying process of SCGs based on real operational data have been found. Although some SCGs drying studies at temperatures from 100 to 250 °C [8] and 30 to 60 °C [9, 10] have been conducted, they do not evaluate the best-operating conditions to reduce the exergy destruction rate and the operating cost of the drying process.

In this context, this work presents an exergoeconomic analysis of a convective drying process of SCGs. The drying kinetic study and the characterization of the biomass were carried out by using SCGs from a factory located in Guayaquil, Ecuador. The drying process was simulated using Aspen Plus V12.1, and the obtained experimental data were used to validate the process simulation. The exergetic, economic, and exergoeconomic balances were performed using Engineering Equation Solver (ESS) software following the methodology from [4]. The exergy destruction rate, investment, and operational costs were determined for drying temperatures from 60 to 120 °C.

2 Materials and Methods

2.1 Experimental Studies

Characterization of SCGs

The SCGs were obtained from an instant coffee factory in Guayaquil, Ecuador. The samples were dried at 105 °C based on ISO 18123:2015. The proximate analysis was determined according to ISO 18122:2015. The ultimate analysis was performed by a CHNS Elemental Analyzer (Elementar). The particle size distribution was analyzed according to the ASTM D1511-10 procedure.

Drying Kinetic Study

The drying kinetic study was carried out at temperatures from 60 to 120 °C, since at higher temperatures, the degradation rate increases [11]. First, the dried SCGs were submerged in distilled water at room temperature during 2 h to reach their initial moisture [12]. Then, the SCGs were placed on a plate up to a thickness of 1.0 cm inside a convective dryer (UNB 500, Memmert, Germany). Periodically the samples were removed from the dryer, weighing them on digital balance to measure their average water content.

The critical moisture was determined as transition moisture, where the first constant drying rate deviation was observed. The equilibrium moisture was estimated as the constant behavior for an extended period [13]. The effective diffusivity was determined from the slope of the natural logarithm of the moisture ratio vs. the time according to Fick's second law of diffusion for the infinite dry plate [12]. The heat and mass coefficients were obtained from the dimensionless numbers based on Incropera et al. [14].

2.2 *Process Simulation*

The SCGs simulated drying system consists of an air preheater (HX-101) and a batch-type convection drying chamber (D-101). First, the flue gas flow (that exits the steam generator from the factory) at 300 °C (stream 6) with a mass composition of 18.5% CO_2, 6.8% H_2O, 7.09% N_2, and 3.8% O_2 enters the HX-101. Then, ambient air enters the HX-101 with a relative humidity of 73% at 25 °C (stream 1) and exits at a drying temperature (stream 2) varying between 60 and 120 °C. After that, the flue gases at the outlet of the HX-101 (stream 7) are discharged into the environment. Next, the hot air enters the D-101 to remove the water from the SCGs (stream 3), and the humid air exits from the dryer by stream 5. After drying, the dried SCGs (stream 4) are discharged.

The process simulation was performed in Aspen Plus V12.1. SCGs were simulated as unconventional components. The ultimate and proximate analysis results obtained in Sect. 2.1 were entered into the HCOALGEN and DCOALIGT models. First, the ideal thermodynamic model was used since it worked at atmospheric pressure, and the MIXCNCPSD for handling non-conventional solids configured the flow class. Next, the dryer was simulated using a convective dryer. Finally, the simulation was configured in batch mode, where the total drying time was set as the maximum iteration and operation time. Drying data obtained from the drying kinetic study such as heat and mass transfer coefficients, drying curve, equilibrium, and critical moisture were defined for the dryer.

2.3 *Model Validation*

The absolute moisture of the SCGs obtained in the simulation was compared with the experimental data for each drying air temperature. The same operating conditions were maintained to ensure the validity of the modeled process.

2.4 Exergoeconomic Analysis

The exergoeconomic analysis was developed according to the methodology offered by Bejan and Tsatsaronis [4] using Eqs. (1)–(6). Engineering Equations Solver (EES) software was used to determine the thermophysical properties for the calculation of enthalpy and entropy of each state. The enthalpy (h) and entropy (s) of SCGs were determined using the heat capacity taken from the literature [15].

$$e_{\mathrm{i}}^{\mathrm{PH}} = (h_{\mathrm{i}} - h_{\mathrm{o}}) - T_{\mathrm{o}}(s_{\mathrm{i}} - s_{\mathrm{o}}) \tag{1}$$

$$e_{\mathrm{i}}^{\mathrm{CH}} = \sum_{i} x_i e_{\mathrm{i}}^{\mathrm{ch}} + RT_{\mathrm{o}} \sum_{i} x_i \ln x_i \tag{2}$$

$$\dot{E}_{\mathrm{D,k}} = \dot{E}_{\mathrm{F,k}} - \dot{E}_{\mathrm{P,k}} \tag{3}$$

$$\dot{Z}_{\mathrm{k}} = \dot{Z}_{\mathrm{k}}^{\mathrm{O\&M}} + \dot{Z}_{\mathrm{k}}^{\mathrm{CI}} \tag{4}$$

$$\dot{C}_{\mathrm{F,k}} + \dot{Z}_{\mathrm{k}} = \dot{C}_{\mathrm{P,k}} \tag{5}$$

$$\dot{C}_{\mathrm{D,k}} = c_{\mathrm{F,k}} \dot{E}_{\mathrm{D,k}} \tag{6}$$

The economic analysis was carried out following the revenue requirement method. The economic parameters were taken from a previous study by [16]. The purchase equipment costs (PEC) of the HX-101 and the D-101 were determined using Eqs. (7) and (8) [17], respectively for 2010 (CEPCI de 532.9).

$$\mathrm{PEC}_{\mathrm{HX-101}} = 1900 + 2500 \cdot S \tag{7}$$

$$\mathrm{PEC}_{\mathrm{D-101}} = 10000 + 7900 \cdot S^{0.5} \tag{8}$$

where S is the area of each tray from the dryer (20 m^2) and the heat exchange area for the HX-101 (obtained in the simulator). The chemical engineering plant cost index (CEPCI) of 607.5 was used to adjust the PEC for 2019.

Table 1 shows the fuel exergy ($\dot{E}_{\mathrm{F,k}}$), product exergy ($\dot{E}_{\mathrm{P,k}}$), fuel cost ($\dot{C}_{\mathrm{F,k}}$), product cost ($\dot{C}_{\mathrm{P,k}}$), and auxiliary equations at the component level.

Table 1 Exergoeconomic balance

Component	$\dot{E}_{\mathrm{F,k}}$	$\dot{E}_{\mathrm{P,k}}$	$\dot{C}_{\mathrm{F,k}}$	$\dot{C}_{\mathrm{P,k}}$	Auxiliary equation
HX-101	$\dot{E}_6 - \dot{E}_7$	$\dot{E}_2 - \dot{E}_1$	$\dot{C}_6 - \dot{C}_7$	$\dot{C}_2 - \dot{C}_1$	$c_6 = 0$ $c_1 = 0$
D-101	$\dot{E}_2 - \dot{E}_4 + \dot{E}_3$	$\dot{E}_5$	$\dot{C}_2 - \dot{C}_4 + \dot{C}_3$	$\dot{C}_5$	$c_3 = 0$
Overall system	$\dot{E}_{\mathrm{F}} = (\dot{E}_6 - \dot{E}_7) + \dot{E}_3 + (\dot{E}_2 - \dot{E}_4)$				
	$\dot{E}_{\mathrm{P}} = \dot{E}_5$				

3 Results and Discussion

3.1 Experimental Studies

The ultimate and proximate analysis and the particle size distribution of SCGs are presented in Table 2.

The kinetic study of SCGs was carried out in a batch-type convective dryer at different drying temperatures between 60 and 120 °C.

Figure 1a shows the drying curves. It is observed that the drying time is reduced from 4.50 to 0.83 h when the temperature is increased from 60 to 120 °C. This reduction occurs because at high temperatures, the water evaporation rate increases (from 0.54 to 2.37 g/kg s) according to Fig. 1b. In addition, an increment in the drying temperature causes an increment in both the critical and equilibrium moisture contents, which change from 54.9% to 70.5% and 0.56% to 1.93% on a wet basis, respectively.

Figure 2a shows the mass and heat transfer coefficients at different drying temperatures. It is observed that the heat and mass transfer coefficients increased at high temperatures [18]. In addition, the moisture diffusivity varied from 1.97×10^{-9} to 9.49×10^{-9} m/s^2 according to Fig. 2b. Similar results were obtained with a tray dryer for SCGs [9] in which the mass and heat transfer coefficients also increased from 2.65×10^{-6} to 3.1×10^{-6} m/s and 19.4 to 21.9 W/(m^2 K) when the inlet air drying temperature increased from 30 to 32 °C. Furthermore, the diffusivity coefficients increased from 4.05×10^{-10} to 9.12×10^{-10} m/s^2.

3.2 Model Validation

The drying process was simulated in Aspen Plus 12.1 using the data obtained from the characterization of the SCGs and the drying kinetic study. The simulated final moisture and the experimental final moisture were compared to validate the model developed for the dryer. Table 3 shows that the simulation results have relative errors between 1.9% and 4.0%. Therefore, the simulated dryer can precisely describe the behavior of the drying process between 60 and 120 °C.

Table 2 Results of SCG characterization

Ultimate analysis (% db)		Proximate analysis (% db)		Particle size distribution (% mm)	
Carbon	62.34	Moisture (wb)	84.28	PS > 2	16.3
Hydrogen	13.09	Fixed carbon	14.22	1.4 < PS < 2	19.8
Nitrogen	2.15	Volatile matter	85.38	1 < PS < 1.4	17.4
Sulfur	0.43	Ashes	0.40	0.6 < PS < 1	23.3
Oxygen	21.59			PS < 0.6	23.2

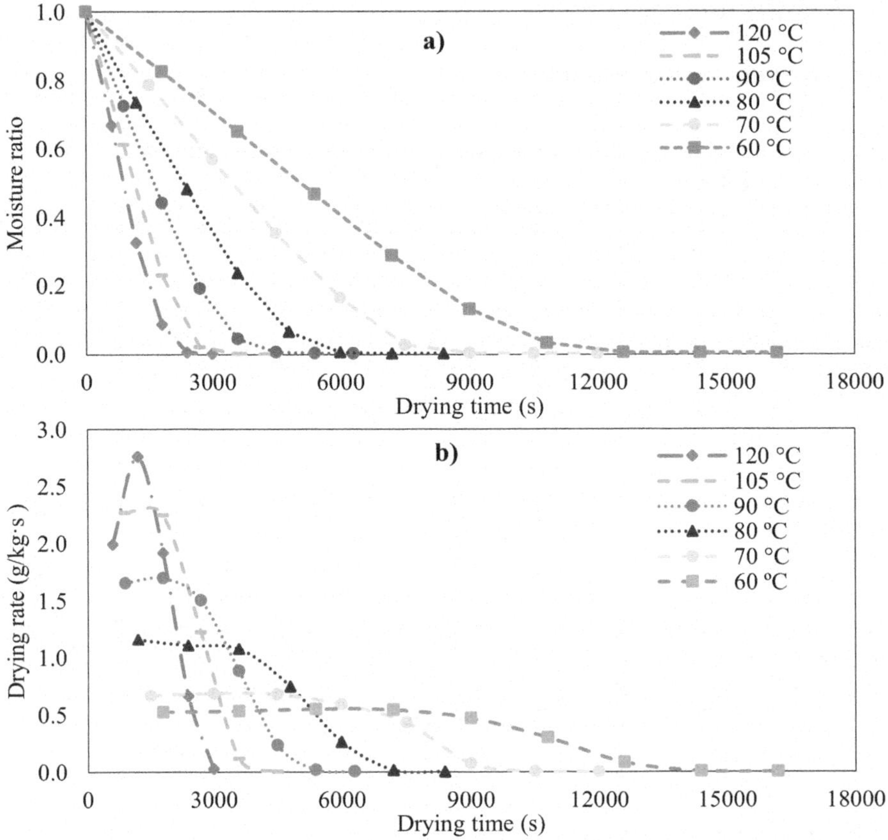

Fig. 1 (**a**) Drying curves and (**b**) drying rate of SCGs vs. drying time at different temperatures

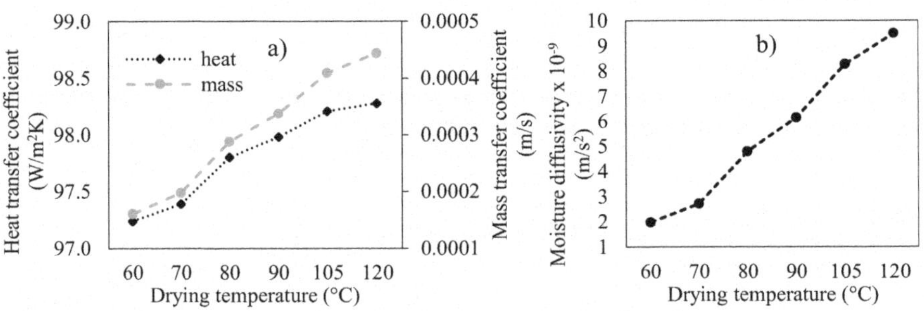

Fig. 2 (**a**) Heat and mass transfer coefficients and (**b**) moisture diffusivity of SCGs at different air temperatures

Table 3 Model validation of the batch-type convective dryer

Temperature (°C)	Moisture (wet basis)		Relative error (%)
	Experiment	Simulation	
60	1.932	1.969	1.90
70	1.024	1.052	2.66
80	1.209	1.258	4.02
90	1.153	1.187	2.91
105	0.556	0.577	3.80

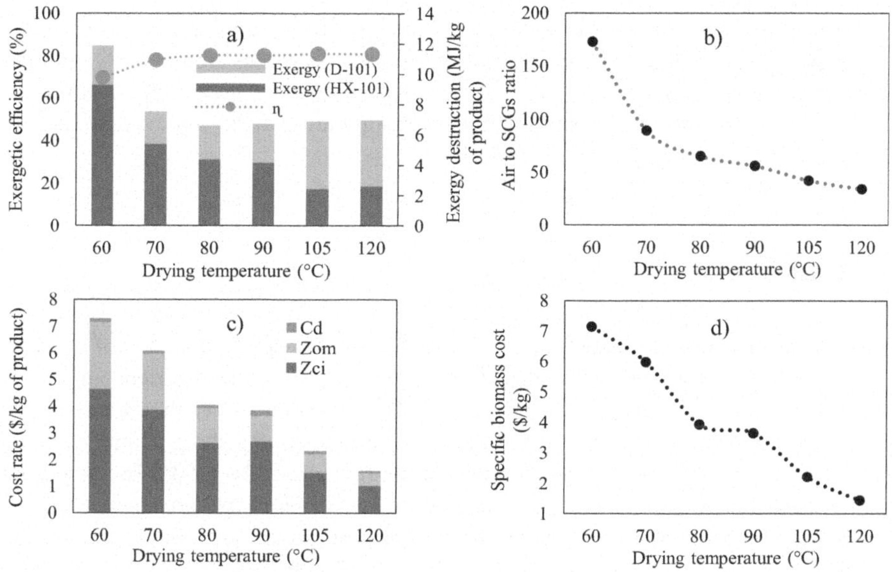

Fig. 3 The effect of drying temperature in (**a**) global exergy destruction and exergetic efficiency per component, (**b**) air to SCGs ratio, (**c**) exergy destruction, investment, operational, and maintenance cost rates of the drying system, and (**d**) specific biomass cost

3.3 Exergoeconomic Analysis

The exergoeconomic indicators of the simulated SCGs convective drying process at different drying air temperatures were determined.

Figure 3a shows that the exergetic efficiency increases significantly between 60 and 70 °C, while it varies insignificantly at higher temperatures. This behavior is similar to the overall exergy destruction rate. The greatest reduction of destroyed exergy (36.7%) occurs between 60 and 70 °C because the required drying airflow decreases by 52.3% as shown in Fig. 3b. Moreover, the air flow rate is reduced up to 82% when the temperature increases from 60 to 120 °C. Therefore, the drying airflow is directly proportional to the destroyed exergy rate. Khanlari et al. (2021) obtained similar results in a solar collector assisted dryer, where the exergetic efficiency decreased from 77.4% to 66.5% when the airflow rate changed from 0.013 to 0.007 kg/s [19].

Figure 3c shows that at 120 °C, the operating costs and the exergy destruction cost rate are the lowest. This change occurs because, at 120 °C, a lower air/flue gas ratio is required. Previous studies have shown other alternatives to reduce operating costs by increasing drying rates, such as microwave-assisted dryers [20].

Furthermore, the operating cost of the HX-101 at 90 °C represents 54% of the overall operating cost of the drying process. To reduce the operating cost of the heat exchanger, previous works have studied the use of renewable energy sources such as solar energy [21] and solar-assisted heat pumps [7], which have found that the investment and operating costs could be reduced by 20.3% and 34.5%, respectively.

Figure 3d shows that the specific cost of biomass decreases significantly with the temperature increment. At 120 °C, it is possible to reduce the cost of dry biomass by 79.9%. This cost reduction responds to the low operating costs achieved at 120 °C. The minimum cost of SCGs was 1.44$/kg.

4 Conclusion

This study aimed to simulate and perform an exergoeconomic analysis of a batch-type convective dryer of SCGs based on experimental studies carried out for air drying temperatures between 60 and 120 °C. This research shows that the exergetic efficiency varies between 69.6% and 80.8% when increasing the temperature from 60 to 120 °C. Furthermore, production costs can be reduced by 69.3% by increasing the temperature from 60 to 120 °C. The specific cost of biomass varies between 7.15 and 1.44$/kg and is at 120 °C when the lowest cost is reached. Finally, the results suggest that convective drying at a drying air temperature of 120 °C benefits exergetic efficiency and operating costs. Future experimental studies should focus on alternative drying technologies that incorporate renewable energies for heating. In addition, exergetic optimization and advanced exergoeconomic analysis should be performed to determine the avoidable exergy destruction cost rate for the process.

References

1. Ladha-Sabur, A., Bakalis, S., Fryer, P.J., Lopez-Quiroga, E.: Mapping energy consumption in food manufacturing. Trends Food Sci. Technol. **86**, 270–280 (2019)
2. Massaya, J., Prates Pereira, A., Mills-Lamptey, B., Benjamin, J., Chuck, C.J.: Conceptualization of a spent coffee grounds biorefinery: a review of existing valorisation approaches. Food Bioprod. Process. **118**, 149–166 (2019)
3. Gómez-De La Cruz, F.J., Cruz-Peragón, F., Casanova-Peláez, P.J., Palomar-Carnicero, J.M.: A vital stage in the large-scale production of biofuels from spent coffee grounds: the drying kinetics. Fuel Process. Technol. **130**, 188–196 (2015)
4. Bejan, A., Tsatsaronis, G.: Purpose in thermodynamics. Energies. **14**, 408 (2021)

5. Zeng, Z., Li, B., Han, C., Wu, W., Chen, T., Dong, C., Gao, C., He, Z., Zhang, F.: Performance of exergetic, energetic and techno-economic analyses on a gas-type industrial drying system of black tea. Foods. **11**, 3281 (2022)
6. Singh, G., Tyagi, V.V., Pandey, A.K., Goel, V., Sari, A.: Comparative exergoeconomic analysis of single, two and three stage spray drying systems. J. Therm. Anal. Calorim. **147**, 8947–8968 (2022)
7. Atalay, H.: Comparative assessment of solar and heat pump dryers with regards to exergy and exergoeconomic performance. Energy. **189**, 116180 (2019)
8. Gómez-de la Cruz, F.J., Palomar-Carnicero, J.M., Hernández-Escobedo, Q., Cruz-Peragón, F.: Experimental studies on mass transfer during convective drying of spent coffee grounds generated in the soluble coffee industry. J. Therm. Anal. Calorim. **145**, 97 (2020)
9. Martinez, E.N.: The drying of spent coffee grounds in a tray drier. Am. J. Biomed. Sci. Res. **6**, 416–425 (2019)
10. Osorio-Arias, J., Delgado-Arias, S., Cano, L., Zapata, S., Quintero, M., Nuñez, H., Ramírez, C., Simpson, R., Vega-Castro, O.: Sustainable management and valorization of spent coffee grounds through the optimization of thin layer hot air-drying process. Waste Biomass Valoriz. **11**, 5015–5026 (2020)
11. Jang, H., Ocon, J.D., Lee, S., Lee, J.K., Lee, J.: Direct power generation from waste coffee grounds in a biomass fuel cell. J. Power Sources. **296**, 433–439 (2015)
12. Cuevas, M., Martínez-Cartas, M.L., Pérez-Villarejo, L., Hernández, L., García-Martín, J.F., Sánchez, S.: Drying kinetics and effective water diffusivities in olive stone and olive-tree pruning. Renew. Energy. **132**, 911–920 (2019)
13. Srikiatden, J., Roberts, J.S.: Moisture transfer in solid food materials: a review of mechanisms, models, and measurements. Int. J. Food Prop. **10**, 739–777 (2007)
14. Incropera, F.P., Dewitt, D.P., Bergman, T.L., Lavine, A.S.: Fundamentals of Heat and Mass Transfer. John Wiley & Sons, Inc., Hoboken, NJ (2007)
15. Afolabi, O.O.D., Sohail, M., Cheng, Y.L.: Optimisation and characterisation of hydrochar production from spent coffee grounds by hydrothermal carbonisation. Renew. Energy. **147**, 1380–1391 (2020)
16. Caicedo, D.L.T., Torres, M.S., Mero-Benavides, M., Lopez, O.P., Medina, A.L., Marigorta, A. M.B.: Simulation and exergoeconomic analysis of a trigeneration system based on biofuels from spent coffee grounds. Energies. **16**, 1816 (2023)
17. Towler, G., Sinnott, R.: Chemical Engineering Design—Principles, Practice and Economics of Plant and Process Design. Elsevier Inc. (2012)
18. Beigi, M., Tohidi, M., Torki-Harchegani, M.: Exergetic analysis of deep-bed drying of rough rice in a convective dryer. Energy. **140**, 374 (2017)
19. Khanlari, A., Sözen, A., Afshari, F., Tuncer, A.D.: Energy-exergy and sustainability analysis of a PV-driven quadruple-flow solar drying system. Renew. Energy. **175**, 1151–1166 (2021)
20. Ranjbaran, M., Zare, D.: Simulation of energetic- and exergetic performance of microwave-assisted fluidized bed drying of soybeans. Energy. **59**, 484–493 (2013)
21. Agbede, O.O., Oke, E.O., Akinfenwa, S.I., Wahab, K.T., Ogundipe, S., Aworanti, O.A., Olatunde Arinkoola, A., Agarry, S.E., Ogunleye, O., Nihinlola Osuolale, F., Babatunde, A.: Thin layer drying of green microalgae (Chlorella sp.) paste biomass: drying characteristics, energy requirement and mathematical modeling. Bioresour. Technol. Rep. J. **11**, 100467 (2020)

Sustainable Energy Transformation: Design of a Municipal Solid Waste Incineration System for Electricity Generation—Energy Efficiency Assessment

Aidaelena Smith-Perera and Álvarez Bermúdez María Eugenia

1 Introduction

1.1 Global Context

Despite the progress in reusing and recycling, landfills continue to be protagonists in waste management at a global level [1]. According to [2] for the year 2016, 2.01 billion tons of solid waste were produced worldwide, and it is expected that by 2050 this figure will increase to more than 3.4 billion tons.

Treatment of municipal waste in developing countries is precarious. More than 90% of waste is disposed incorrectly and is commonly exposed to the open air [2]. In the specific case of Venezuela, approximately 85–90% of solid waste does not receive adequate treatment [3].

It is estimated that the current world population exceeds 7.8 billion people [4] and it is projected to increase to 9.7 billion by 2050 and 11.2 billion by 2100 [5]. This population increase suggests not only a greater production of municipal waste, but also a greater energy demand.

Waste-to-energy (WTE) technology, through incineration is a process in which municipal solid waste (MSW) is burned at high temperatures to reduce it to ash and recover its calorific value for electricity generation [6]. Incineration is one of the MSW burning processes that, according to information from [5], is the most widely used technology for WTE plants.

Recognizing the advantages of treating MSW to harness its calorific value for electricity generation and knowing the characteristics of the MSW of the AMC, the different technologies proposed in the literature are studied to identify the best

A. Smith-Perera (✉) · Á. B. María Eugenia
Universidad Metropolitana, Caracas, Venezuela
e-mail: asmith@unimet.edu.ve

M. Espinoza-Andaluz et al. (eds.), *Congress on Research, Development and Innovation in Renewable Energies*, Green Energy and Technology,
https://doi.org/10.1007/978-3-031-88995-0_7

solution and to determine the power and electrical energy that could be generated in an environmentally appropriate WTE system. According to [7], the most common WTE system in the United States is controlled combustion, in which MSW enters a large furnace without prior processing.

This study aims to evaluate the utilization of MSW from a specific sector in Caracas for electricity production. This investigation includes an analysis of the waste separation process, controlled combustion, electricity generation through a thermodynamic cycle, and post-treatment of gasses. By designing this process based on the quantity of municipal solid waste, the electric energy that can be generated can be estimated.

1.2 Stages of Conversion of MSW into Electrical Energy

WTE controlled incineration plants have four basic stages to produce electric power from MSW: separation and pretreatment, combustion and post-combustion, power generation, and environmental control, as can be seen in Fig. 1 [8].

In the separation and pretreatment stage, the waste is selected and separated to make the selected combustion process the most efficient.

The combustion and post-combustion stage converts waste into thermal energy through a combustion process that takes place inside a furnace. This stage establishes the criteria for the selection of MSW in the pretreatment stage and for the gas filtering mechanisms in the environmental control stage, while in post-combustion, hot air rises through the inside of the furnace and directs the heat flow into the boiler.

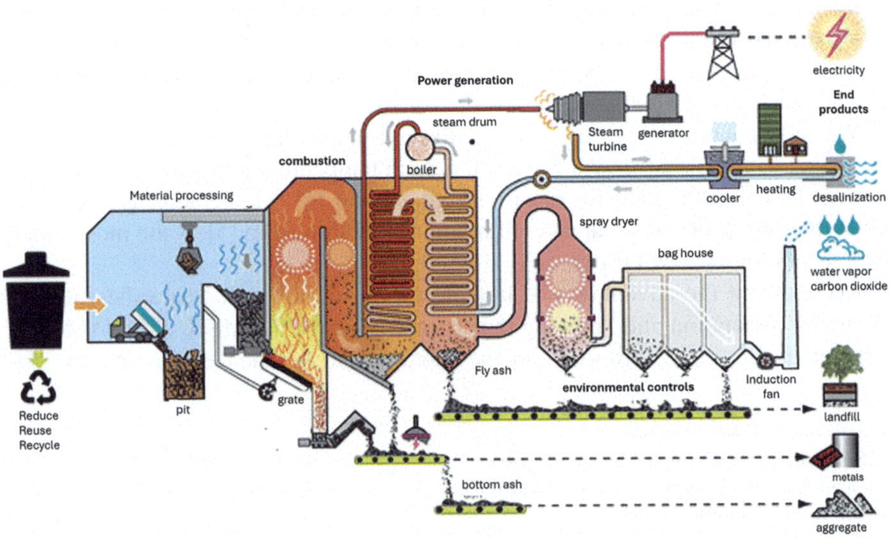

Fig. 1 Diagram of the operation of a controlled combustion WTE plant [7]

In power generation stage, mechanical energy is transformed into electrical energy using a thermodynamic cycle of steam power (Rankine cycle) that allows the energy contained in the steam to be transformed into electrical energy by means of a turbine and an electric generator. And, finally, in the environmental control stage, the gasses produced during the incineration of MSW are treated.

Controlled combustion of solid waste is an exothermic process in which MSW is exposed to high temperatures ranging between 850 and 1000 °C, according to [7], in the presence of oxygen to achieve the release of gasses that contain a high calorific value.

Many authors [1, 8–13] have presented solutions for WTE for different countries and necessities, highlighting the importance of reducing the volume of MSW by taking advantage of it as electric energy. The objective of this investigation is the identification of the most appropriate components of a WTE plant and the study of performance of the solution in terms of the final product which is the power and electrical energy given the MSW and the calorific analysis.

1.3 Elements of the Controlled Combustion Process

It is the objective of this investigation to identify the most appropriate components of the WTE plant to be selected in order to take advantage of MSW in AMC.

The elements of the combustion process are:

- A combustion furnace
- Steam boiler
- Heat exchanger or condenser

The furnace is the element that receives the waste and carries out controlled combustion and produces hot gasses that allow the water in the boiler to evaporate. Ash and slag are also produced and must be filtered.

In the steam boiler, the chemical energy stored in the fuel is transformed into thermal energy through reheated water steam and hot gasses.

In the heat exchanger, a thermal transformation occurs in which the input fluid is gas or water steam (as occurs at the exit of the steam boiler from the combustion of MSW). It is converted into a liquid state so that it is recirculated to the boiler again, the heat exchanger is called a "condenser."

2 Proposed Solution for Converting AMC'S MSW into Electrical Energy

An analysis of the findings in the AMC municipal solid waste literature is presented to then propose a technically valid WTE solution for the environment.

2.1 AMC Municipal Waste Analysis

The reality presented [14] is still in force in Venezuela in 2024 about the lack of systematic information and indicators of solid waste. According to [15], in Venezuela, between 0.8 and 1.3 kg of MSW per day per inhabitant was generated in 2008, totaling at least 28,000 tons of MSW per day, of which about 90% ended up exposed without proper treatment and safe final disposal.

Nowadays, most of the products consumed become solid waste, according to the National Institute of Statistics of the Bolivarian Republic of Venezuela (INE, 2013). Of the 335 municipalities in Venezuela, 128 (38.32%) use sanitary landfills, 138 (41.24%) use dump sites, and 69 (20.44%) use open dumps for the final disposal of solid waste.

According to the data presented by [15], the AMC, which includes the municipalities of Libertador, Sucre, Baruta, Chacao, and El Hatillo (these last four municipalities, territorially accounted for in the state of Miranda), produces 5500 tons of waste daily, but only 60% of the waste is collected and transferred to the La Bonanza landfill, the only one authorized. The other 40% of the waste that remains in the streets is discarded in rivers, streams, and in illegal dumps scattered throughout the area.

Villalba [16] states that of the five municipalities that make up the Metropolitan Area of Caracas (AMC), the one that generates the most MSW is the Libertador Municipality with 63.3%, in second place is the Sucre Municipality with 21.6%, and in third place the Baruta Municipality with 9.4%, the remaining 5.7% is produced between the municipalities of El Hatillo and Chacao.

A per capita generation of 1.11 kg of waste per person per day is taken as the basis for the calculation of MSW generation, an experimental result obtained by the students [10]. The population of the AMC is 4,854,098 inhabitants as stated [11], resulting in an extrapolated generation of 5388 tons of MSW per day in the AMC considering only the collected MSW in AMC municipalities.

Next, the MSW generation obtained in the previous paragraph is multiplied by each of the percentages presented in Table 1 as presented in [12].

Given the percentage distribution presented in Table 1 and considering the usable waste for a WTE plant (organics, cardboard and paper, plastics, textiles, and wood), it is known that there would be 4418.70 tons of MSW/day that could be used

Table 1 Physical composition of MSW in the Caracas Metropolitan Area [12]

Type of waste	% of weight
Organic (OR)	41.32
Cardboard and paper (CA)	22.26
Plastic (PL)	11.70
Other	8.63
Metals (ME)	4.84
Glass (VI)	4.52
Textile (TX)	4.05
Wood (PA)	2.68

energetically through controlled combustion in AMC, which corresponds to 82% of the total MSW of AMC.

2.2 *WTE Plant Solution Design for AMC*

The WTE plant proposed for the AMC, given the characteristics of the MSW, follows the configuration shown in Fig. 2 using the acronyms shown in Table 1.

The more detailed stages of the elements and parts of the WTE plant at AMC are shown in Fig. 3.

Glass and metals are separated and sent for recycling at the separation stage. The remaining waste after the separation stage is pretreated and dried. Controlled combustion is carried out with the rest of the waste. The waste to be treated can be unprocessed MSW or refuse-derived fuel (RDF).

In the combustion stage, the ovens available are: grill (H-P), fluid bed (H-L), and rotary (H-R) [13].

On the other hand, boilers can be water-tube (C-A) or pyrotubular (C-Pi). It is proposed to use the water-tube boilers common in industry due to their high efficiency of 90%.

Different MSW treatments at combustion stage have different feasible technologies, so a set of models are shown on Table 1.

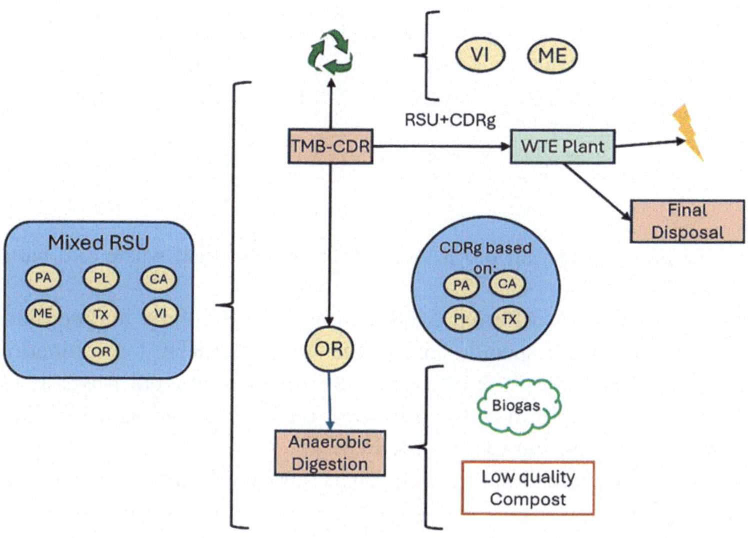

Fig. 2 Configuration of the controlled combustion plant [12]

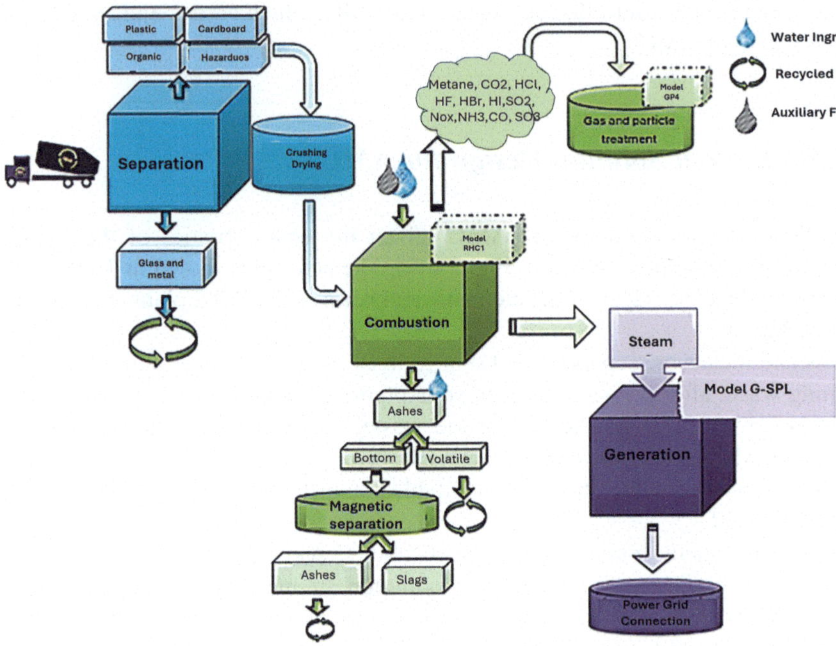

Fig. 3 Equipment layout in the controlled combustion plant [12]

Table 2 Combinations of combustion stage technologies

Element	Feasible models for combustion stage					
	RHC1	RHC2	RHC3	RHC4	RHC5	RHC6
Waste to be treated	MSW	MSW	MSW	RDF	RDF	RDF
Oven type	H-P	H-R	H-L	H-P	H-R	H-L
Boiler type	C-A	C-A	C-A	C-A	C-A	C-A

In the separation stage, all hazardous waste, like medical waste and batteries, is taken out from the waste.

As shown in Table 2, different combinations of technology and equipment are feasible; however, for the specific case of the AMC, the RHC1 combination is the most appropriate because of the characteristics of the municipal waste available to the AMC, which, in addition to the grill oven (H-P), uses pretreated MSW as fuel through crushing and drying to raise the PCI.

Waste from the combustion process such as inert slag is taken to final disposal in the existing landfill.

The gas and particle cleaning system, as well as the reagents supplied in each of them, such as the lime slurry injection reactor or $Ca(OH)_2$ for the wet method of acid gas treatment, bag filters with activated carbon for the retention of particles, as well as dioxins, furans, and heavy metals, and lime or CaO, is also supplied to the baghouse filter as a dry method for the treatment of acid gasses. Finally, the CSR treatment for the reduction of nitrogen oxides (NOx) with ammonia injection can be appreciated. This scheme is identified as GP4.

The turbine can be action (T-A) or reaction (T-R), while electric generators can be synchronous smooth pole (G-SPL) or induction (G-I). The G-SPL generator is selected because it is synchronous, and the coupling to the existing electricity grid is direct.

3 Energy Analysis of the Solution

According to data from the Venezuelan electricity company (CORPOELEC) in 2016, per capita energy consumption in Venezuela is 4179 kWh per person per year, so if the controlled combustion plant uses four processing lines for MSW, it generates 771,986.52 MWh/year for sale and could provide electricity service to 184,730 people or about 61,577 households of three members. The calculated energy generated per ton of MSW is 0.683 MWh/ton MSW and the plant's self-consumption is 15% (125,644.68 MWh/year), resulting in an energy use of 771,986.52 MWh/year for sale.

In the case of using three lines for MSW and one for waste-derived fuel or RDF, the benefit increases as 812,966.69 MWh/year is generated, which is equivalent to the supply of 194,536 individuals or 64,845 families. Finally, the highest generation occurs in the case of assigning two lines for each type of fuel, 913,462.87 MWh/year that supplies 219,056 people or 73,018 homes.

The conversion of the total MSW into the AMC, the number of people who produce it, and the ratio of electrical energy allow us to calculate that the MSW produced per year by 18.4 people would allow the generation of sufficient electricity for the annual consumption of one person, as shown in Fig. 4.

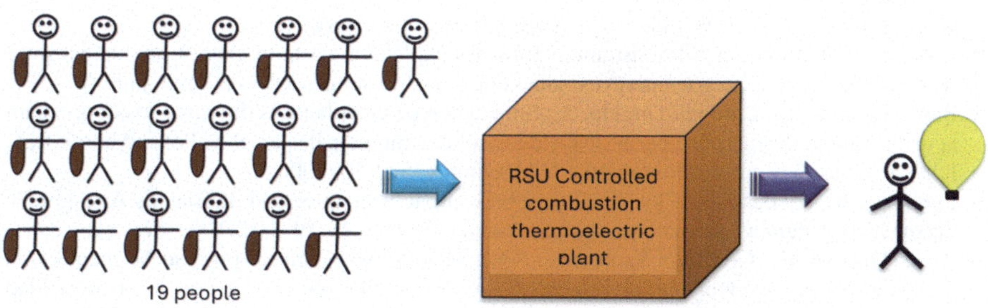

Fig. 4 Schematic of the actual performance of the proposed WTE plant for AMC [13]

4 Conclusion

The conceptual model of a controlled combustion plant with 96 MW of installed capacity was obtained to expand the national electricity grid, which produces annually between 771,986.52 and 913,462.87 MWh, thus benefiting approximately 200,000 people.

The annual waste production of 19 people generates enough energy to supply one person's annual electricity consumption.

Properly conditioned MSW can replace fossil fuels such as coal.

Waste-derived fuel (RDF) is homogeneous in composition and its lower calorific value (PCI) is higher than that of municipal solid waste (MSW), so the combustion of these generates a greater amount of energy and a lower production of slag, ash, and gasses for the same amount of incinerated mass; however, when using RDF, the combustion of organic waste is lost, which represents approximately 50% of MSW.

The creation of thermal MSW combustion plants for the generation of electricity could become a good solution to counteract the difficulties of having MSW in landfills, since it not only decreases the volume and weight of waste but also contributes electricity to the community.

References

1. Kannankai, M.P., Devipriya, S.P.: An introduction to cost-effective technologies for solid waste and wastewater treatment. In: Kathi, S., Thamaraiselvi, K., Devipriya, S. (eds.) Cost Effective Technologies for Solid Waste and Wastewater Treatment, vol. 1, pp. 1–8. Elsevier (2022). https://doi.org/10.1016/B978-0-12-822933-0.00013-9
2. World Bank: Solid waste management. https://www.worldbank.org/en/topic/urbandevelopment/brief/solid-waste-management
3. Vitalis: Situación Ambiental de Venezuela 2015, Caracas (2016)
4. World Bank: Population, total. https://data.worldbank.org/indicator/SP.POP.TOTL
5. Naciones Unidas: Desafíos Globales: Población. https://www.un.org/es/global-issues/population
6. World Energy Council: World energy resources: waste to energy (2013)
7. US Energy Information Administration: Biomass explained waste-to-energy (municipal solid waste). https://www.eia.gov/energyexplained/biomass/waste-to-energy-in-depth.php
8. Barrera-Zapata, R., Cardona-Giraldo, A.: Simulación de una planta WtE (waste to energy) para la recuperación de energía a partir de residuos sólidos municipales en el Valle de Aburrá. Rev. ION. **35**(2) (2022). https://doi.org/10.18273/revion.v35n2-2022006
9. Zubillaga, M.: El destino de los residuos sólidos urbanos de la ciudad de Buenos Aires: breve diagnóstico y algunas alternativas. Agron. Ambiente. **33**(1–2), 79 (2013)
10. Ayala Díaz, A.M., González Sarmiento, A.C.: Metodología para un proceso de recolección desagregado de residuos en el origen. Caso de estudio terrazas del ávila. Universidad Metropolitana (2016)
11. Urdaneta, T.: La gestión urbana del Área Metropolitana de Caracas, Jun 2013
12. Marrero López, E., Pérez Morales, M.: Estudio de factibilidad técnica para la generación de electricidad a partir de la combustión controlada de RSU. Universidad Metropolitana (2018)

13. Rua-Orozco, D.J., Escobar Palacio, J.C., Melo Henríquez, A.I., Martinez Reyes, A.M., Mambeli Barros, R., Lore, E.: Generación de energía a partir de los residuos sólidos urbanos. In: Ricon Martínez, J.M., Silva Lora, E.E. (eds.) Bioenergía: Fuentes, conversión y sustentabilidad, 1st edn, pp. 275–296. Red Iberoamericana de Aprovechamiento de Residuos Orgánicos en Producción de Energía (2015)
14. Valera, V., Acuña, S.: Indicadores de generación de residuos y desechos sólidos en Venezuela. In: I Simposio Iberoamericano de Ingeniería de Residuos (2008)
15. Instituto Nacional de Estadística: Total de residuos y desechos sólidos recolectados según Entidad Federal, 2008–2011. http://www.ine.gob.ve/documentos/Ambiental/PrincIndicadores/html/ambien_medioAmbiente_3.html
16. Villalba, L.: La Gestión de residuos y desechos sólidos en el Área Metropolitana de Caracas, Caracas, Jun 2013

Development of Multifunctional Equipment for Cocoa Drying and Roasting with Energy Consumption Reduction

Ernesto Martínez, Cristhian A. Macay, Edgar B. Vera, Kleber F. Barcia, Emérita Delgado-Plaza, and Guido Abril

1 Introduction

Cocoa is a vital export product for Ecuador, and its commercial significance has led to increased production in recent years. This rise in production has created challenges for cocoa farmers in the Las Paulinas sector of Bolívar Province [1]. The harvest surge has led to market saturation, as many farmers delay selling their cocoa, waiting for optimal trading conditions [2]. A critical issue in this process is drying, which results in approximately 20% loss of the 10 quintals of raw cocoa due to inadequate thermal insulation and an empirically developed transmission system. These inefficiencies impact energy use and complicate the transition to roasting operations.

The Las Paulinas farm in Bolívar Province near Los Ríos and Guayas, two of Ecuador's largest cocoa-producing regions, faces additional challenges. Climatic conditions aggravated the slow commercialization of cocoa because they affect natural drying, necessitating drying technologies to improve the product's marketability. Recently, the sector has been focusing on adding value to cocoa beans to avoid devaluation due to high production volumes. Currently, an artisanal dryer is used in the area. However, it suffers from significant heat losses and poor grain handling, making it unsuitable for roasting without risking substantial economic losses for producers [3, 4]. This study designed a cocoa roaster based on the existing rotary drying machine. The design's improvements and modifications aimed to reduce heat losses and provide optimal conditions for roasting, enhancing the

E. Martínez (✉) · C. A. Macay · E. B. Vera · K. F. Barcia · E. Delgado-Plaza · G. Abril
Polytechnic University (ESPOL), Escuela Superior Politécnica del Litoral, Centro de Desarrollo Tecnológico Sostenible—FIMCP, Gustavo Galindo Campus, Guayaquil, Ecuador
e-mail: emartine@espol.edu.ec

M. Espinoza-Andaluz et al. (eds.), *Congress on Research, Development and Innovation in Renewable Energies*, Green Energy and Technology,
https://doi.org/10.1007/978-3-031-88995-0_8

efficiency of the drying process and increasing the profitability of cocoa production for local farmers.

2 Methodology

The adaptation of the rotary dryer in cocoa bean roasting was based on two main systems: thermal and structural. A decision matrix helped to determine the best configuration.

2.1 Design Requirements

The primary design requirements are the customer's needs and the dryer's operating characteristics.

The equipment must guarantee the user's operation and safety.

It must maintain the product's safety; use materials, supplies, and components designed for food grade; and avoid contamination.

The equipment must be installed on a solid foundation and have a robust structure capable of withstanding operational conditions.

It must be installed in an open environment to avoid gas accumulation.

It must work with electrical connections ranging from 120 to 220 V, a frequency of 60 Hz, and three domestic LPG cylinders (45 kg) to supply the burner.

Withstand more than 500 kg of cocoa for over 6 h and work at a maximum temperature of 200 °C.

Heat insulators' installation should allow operators to manipulate the equipment housing.

It must have autonomy of functionality in the sector where it is located.

Spare parts must be easily acquired, preferably in Ecuadorian markets.

2.2 Design Criteria

Safety: Comply with the safety conditions for operating the equipment adapted as a roaster.

Maintenance: Ease and efficiency so that the equipment can be maintained and repaired, as well as ease of availability of refractions within the region.

Ergonomics: The adaptations in the machine must include the user's needs, skills, and limitations so that the machine can be operated without difficulty.

Adaptability: Ability to admit changes between drying and roasting without requiring assembly or disassembly.

Operability: The adapted equipment must efficiently dry and roast with lower fuel or electricity consumption.
Autonomy: The equipment can transition from drying to roasting automatically.
Lifespan: Ensure that components are durable, reliable, and can function for 10–15 years.

2.3 *LPG Burner System*

The problem with the burner is that it does not distribute heat correctly; it has an entirely yellow flame and strong smells of LPG. The following solutions are proposed:

Alternative A: Install a hood on the burner and metal vents with a 60° opening.
Alternative B: Install stove-type burners with a steerable cannon.
Alternative C: Install ducts for heat construction.

2.4 *Thermal Insulation System*

The problem with the dryer is that it does not have a thermal insulator, and the oven's temperature varies rapidly, increasing gas consumption. The following solutions are proposed:

Alternative A: Install directional ducts (stove type) under the bed support.
Alternative B: Use an insulating layer of rock wool that covers all the equipment externally.
Alternative C: Use refractory bricks that cover the outside of the equipment.

2.5 *Structural Support System*

In the drying operation, it was observed that the handling of cocoa beans on the steel plate was compromised due to the unevenness of the underlying soil, which caused the plate to be unlevel. The following solutions are proposed:

Alternative A: Install structural supports under the perforated plate.
Alternative B: Swap the fixed perforated type for a rotary smooth plate.
Alternative C: Redesign the structure for the installation of another thicker plate.

2.6 Cocoa Beans Drag System

The dryer distributes very little grain; there are no paddles between the plates to guarantee a required mixing at the top. The following solutions are proposed:

Alternative A: Switch to up-flow double helical belt agitators.
Alternative B: Install drag pallets to improve grain movement.
Alternative C: Add two new extensions with palettes like the current ones.

2.7 Top Coating

When switching to the roasting operation, the convection heat escapes directly into the environment through the top. The following alternatives are proposed:

Alternative A: Four removable steel caps covering the structural hexagon.
Alternative B: Steel lid welded to the rotary shaft, completely covering the structure with a gate for the entry of cocoa beans.
Alternative C: A detached stainless-steel dome encloses the entire top, including the stirring transmission mechanism.

2.8 Transmission System

A problem with the dryer's drive mechanism is inadequate revolution ratios.

Alternative A: Installation of a gearbox from 1 to 20 rpm.
Alternative B: Install a gear motor with a variable frequency drive.
Alternative C: Replace the belt system and implement a new gear ratio.

2.9 Current Equipment Status

For the redesign and technological upgrade of the equipment, temperature measurements were taken at various parts of the dryer, and hot air speed was measured. This data was crucial for input into the computer program used. Table 1 presents the recorded values. Twenty-seven points were measured: 12 points in the pre-chamber, located 15 cm from the perforated plate where the cocoa is placed; 12 points at the hot air outlet of the perforated plate; and 3 points in the hot air duct leading from the LPG heating system.

Table 1 Parameters measured in the dryer in operation

Parameters	Valor	Unidad
Air inlet velocity	4–8	m/s
Flame temperature at the burner	700–800	°C
Air temperature at the bed perforations	177–280	°C
Perforated bed temperature	38–41	°C
Tank surface temperature	50–60	°C

2.10 *LPG Burner Operation*

Gas consumption is critical for operating the equipment used for drying and roasting cocoa beans. One way to correctly describe the combustion behavior inside the burner is with Eq. (1):

$$Q_T = W_c \times \text{PCS} \tag{1}$$

where Q_T is the total heat for burner input, W_c is the mass flow rate of LPG entering (kg/h), and PCS is the calorific value of LPG (kJ/kg) [4].

2.11 *Heat and Mass Transfer*

It was necessary to calculate the velocity of the hot air entering the system to select the fan for the chamber, which was designed for dual functions: drying and roasting. The heat and mass transfer equations were applied. The equations utilized are presented in Table 2.

Product Inflow and Outflow Analysis

An analysis of the inflow and output of the wet product of the cocoa beans was conducted. The equipment is designed to operate with a maximum capacity of 10 quintals within 6 h. Equation (2) calculated the wet product rate at the dryer inlet. The product has an initial moisture content of 50% inside the equipment. Equations (3) and (4) were used to calculate the mass flow of dry solid and the mass flow of water, respectively [4, 5].

Equation (5) was used to calculate the mass flow of water at the dryer outlet, while Eq. (6) was used to determine the wet product mass flow at the dryer outlet, which corresponds to the sum of the dry solid mass flow and the water mass flow [4, 5].

Table 2 Heat and mass transfer equations table [3–5]

Description		Equation number
Product inflow and outflow analysis	$\dot{m}_{\text{p in}} = \frac{\text{mass}_{\text{cocoa}}}{t_{\text{drying}}}$	(2)
Mass flow of dry solid	$\dot{m}_{\text{s in}} = (1 - X_{\text{in}}) * \dot{m}_{\text{p in}}$	(3)
Mass flow of water	$\dot{m}_{\text{w in}} = X_{\text{in}} * \dot{m}_{\text{p in}}$	(4)
Mass flow of water at the dryer outlet	$\dot{m}_{\text{w out}} = \frac{\dot{m}_{\text{s out}} * X_{\text{out}}}{1 - X_{\text{out}}}$	(5)
Product mass flow at the dryer outlet	$\dot{m}_{\text{p out}} = \dot{m}_{\text{s out}} + \dot{m}_{\text{w out}}$	(6)
Air heating	$Q_T = \dot{m}_a (h_{a2} - h_{a1})$	(7)
Water evaporation rate	$\Delta \dot{m}_w = \dot{m}_{\text{w in}} - \dot{m}_{\text{w out}}$	(8)
Total heat of vaporization	$\dot{Q}_2 = \Delta \dot{m}_w * h_{fg} + \dot{m}_{\text{cocoa}} * Cp_{\text{cocoa}} * (T_{\text{out}} - T_{\text{in}})$	(9)
Air velocity in the drying	$\dot{V}_{\text{air}} = A_{\text{transverse}} \times v_{\text{air}}$	(10)
Coefficient of natural convection	$h = \frac{k}{L} \text{Nu}$	(11)
Nusselt number	$\text{Nu} = \left\{ 0.825 + \frac{0.387\ \text{Ra}_L{}^{1/16}}{\left[1 + (0.492/\text{Pr})^{9/16}\right]^{8/27}} \right\}^2$	(12)
Rayleigh number	$\text{Ra}_L = \frac{g\beta (T_{\text{Sup}} - T_{\text{Ext}})\ L^3}{\nu \alpha}$	(13)
Heat transfer ratio	$\dot{Q}_{\text{perdido}} = \frac{T_{\text{ext}} - T_{\text{int}}}{\frac{1}{h_{\text{conv1}} \times 2 \times \pi \times R_1} + \frac{\ln(R_2/R_1)}{2 \times \pi \times L \times k_1} + \frac{1}{h_{\text{conv2}} \times 2 \times \pi \times R_2}}$	(14)
Heat loss by radiation	$\dot{Q}_{\text{rad}} = \varepsilon \times A_1 \times \sigma \times \left(T_{s4}^4 - T_{\text{sup}}^4\right)$	(15)

Air Heating

Equation (7) was used to calculate the heat required to heat the air, where $\dot{m}_a$ is the mass flow of the air (kg/s), h_{a2} is the enthalpy of the air at the inlet of the exchanger (kJ/kg), and h_{a1} is the enthalpy at the exit of the exchanger.

Drying Cocoa Beans

To calculate the heat required to dry the cocoa beans, it is first necessary to determine the water evaporation rate $\Delta \dot{m}_w$ using Eq. (8), which accounts for the difference in the evaporation rate over the drying time $\dot{m}_{\text{w in}} - \dot{m}_{\text{w out}}$, and Eq. (9) [4, 5], where $\dot{Q}_2$ is the total heat of vaporization (kJ/s), $\Delta \dot{m}_w$ is the water evaporation rate (kg/s), h_{fg} is the latent heat of vaporization of water (kJ/kg), $\dot{m}_{\text{cocoa}}$ is the mass flow rate of wet cocoa (kg/s), Cp_{cocoa} is the specific heat capacity of cocoa (kJ/kg °C), T_{out} is the temperature of cocoa beans at the outlet, and T_{in} is the temperature of cocoa beans at the inlet to the equipment.

In this project, the air velocity value in the drying chamber was already available, so Eq. (10) was directly used to calculate the required airflow (m^3/s), where

$A_{transverse}$ is the cross-sectional area of the equipment and v_{air} is the air velocity in the drying chamber.

Internal Convection Coefficient

Within the equipment, forced convection was considered, and it was assumed that the area where the hot air enters the burner functions as a flat plate. In these calculations, the properties of the air corresponded to the average surface temperature and the ambient temperature. Equation (12) was used to determine the flow behavior, where V is the airflow velocity (m/s), L is the length of the plate (m), and υ is the kinematic viscosity (m^2/s) of the flowing air and the surface of the plate. The flow is considered laminar if $Re_L < 2000$ and turbulent if $Re_L > 4000$ [3].

External Convection Coefficient

It was established that there were no air currents due to external forcing, and it was also assumed that the air was at rest due to its low velocity in an isolated area. Equation (11) was used to calculate the value of the external medium's natural convection coefficient (W/m^2 °C). Nu corresponds to the Nusselt number of natural convection and is obtained from Eq. (12), where Ra refers to the Rayleigh number. Pr is the Prandtl number obtained from the properties of air at average temperature between surface and room temperature. In Eq. (13), g corresponds to gravity (m/s^2), T_s is the surface temperature and text temperature of the medium (°C), L is the characteristic length (m), ν is the kinematic viscosity (m^2/s), and α is the thermal diffusivity (m^2/s^2).

Heat Transfer in Non-insulated Equipment

Equation (14) calculates the heat transfer ratio through the cylinder without thermal insulation [5].

Heat Loss by Radiation

To determine the heat losses due to radiation to the outside of the equipment, Eq. (15), known as Stefan Boltzmann's law, is used, where $\dot{Q}_{rad}$ is the heat loss by radiation, ε is the emissivity of the resistance material, σ is the Boltzmann constant, A is the radiant area of the cylinder, T_{sup} is the surface temperature, and T_s is the ambient temperature.

2.12 Structural Support Design

For the design of the structural support, the base structure of the dryer was modified by creating a new design with support points distributed along the body to withstand the loads involved in the dryer's movement and the hot air.

Mechanical transmission system: The equipment employs a set of pulleys, pinions, and gears to reduce the speed of an electric motor. Equation (16) shows the ratio between the pulleys, Eq. (17) presents the gear ratio, and Eq. (18) indicates the gear ratio i applied, where D_1 is the diameter of the driving pulley, D_2 is the diameter of the driven pulley, N_1 is the rotational speed of the driving pulley, N_2 is the rotational speed of the driven pulley, Z_1 is the number of teeth on the driven gear, and Z_2 is the number of teeth on the driving gear [6].

$$N_1 \times D_1 = N_2 \times D_2 \tag{16}$$

$$N_1 \times Z_1 = N_2 \times Z_2 N_1 \times D_1 = N_2 \times D_2 \tag{17}$$

$$i = \frac{N_1}{N_2} = \frac{D_2}{D_1} \tag{18}$$

2.13 Equipment Transmission System

The design of the pulleys depends on the type and number of belts to be used, according to the equations by Richard G. Budynas, as detailed in Eq. (19) [6]:

$$n_{\text{belts}} = \frac{H_{\text{design}}}{H_{\text{admitted}}} = \frac{H_{\text{nominal}} * K_s * n_d}{K_1 * K_2 * H_{\text{tab}}} \tag{19}$$

Moment of Inertia of Cocoa Mass

The inertia of the cocoa mass within the cylindrical equipment to be agitated is obtained using Eq. (20), which corresponds to the moment of inertia of a solid cylindrical block. M_{cocoa} is the mass of cocoa (kg), and R_{cocoa} is the base radius [6]:

$$I_{\text{cocoa}} = \frac{1}{12} \times M_{\text{cocoa}} \times R^2_{\text{cocoa}} \tag{20}$$

2.14 Torque of the Cocoa Bean Turner

The torque of the removal system (cocoa bean turner) was obtained from Eq. (21), which represents the sum of the moments of inertia of the cocoa and the turner. The angular acceleration α (rad/s^2) was calculated using Eq. (22), where the time difference corresponds to the startup time of the remover, ω_f is the final angular velocity (rad/s), ω_o is the initial angular velocity (rad/s), t_f is the final time (s), and t_o is the initial time (s) [6]:

$$T_{\text{turner}} = I_{\text{total}} \times \alpha \tag{21}$$

$$\alpha = \frac{\omega_f - \omega_o}{t_f - t_o} \tag{22}$$

2.15 Torque Applied to the Removing Elements

Global torque of the turning system: Equation (23) was used to calculate the total torque of the turning system, corresponding to the sum of the previous torques [6]:

$$T_{\text{global}} = T_{\text{turner}} + T_{\text{component}} \tag{23}$$

Sizing of the main axis: The axis of the equipment is subjected to torsion, resulting in uniform shear stress. Additionally, the concentration factor is not required to calculate the shear stress. Equation (24) allows you to calculate the diameter of the De, where n is the safety factor, S_y is the yield strength of the material, and T_{global} is the maximum torsion:

$$\text{De} = \left(\frac{32n}{\pi}\sqrt{\frac{3}{4}\left(\frac{T_{\text{global}}}{S_y}\right)^2}\right)^{\frac{1}{3}} \tag{24}$$

Superior coating: We worked as a pressure tank for the top coating to determine if the current thickness was adequate. Equation (25) made it possible to calculate the pressure of the cocoa mass, where the cocoa beans exert force on a given area.

The force exerted by the cocoa beans is expressed in Eq. (26), which corresponds to the total mass of beans inside the tank, 10 quintals, and g, the acceleration due to gravity, 9.81 m/s^2. The total pressure was determined using Eq. (27), where P_{atm} is the atmospheric pressure. An additional 10% of pressure was considered as an adjustment for containers. The tank thickness was sized according to the ASME Code Sect. VIII, as detailed in Eq. (28), where t is the minimum required thickness (inches); P is the design internal pressure (psi); R is the internal radius of the shell (inches); S is the maximum allowable stress of stainless steel (psi); and E is the joint efficiency:

$$P_{mass} = \frac{F_{mass}}{A_{base}} \tag{25}$$

$$F_{mass} = m_{cocoa} \times g \tag{26}$$

$$P_t = P_{atm} + P_{mass} \tag{27}$$

$$t = \frac{PR}{SE - 0.6P} \tag{28}$$

Electric motor power: It is essential to know if the proposed modifications require a change of the current electric motor. The engine's mechanical power is obtained by Eq. (29), where $P_{mechanical}$ is the mechanical power of the motor (W); T_{total} is the torque (Nm); ω is the angular velocity (rpm). The motor's electric power is obtained by dividing the mechanical power by the efficiency η, as seen in Eq. (30):

$$P_{mechanical} = T_{total} \times \omega \tag{29}$$

$$P_{electric\ power} = \frac{P_{mechanical}}{\eta} \tag{30}$$

3 Results and Analysis

After analyzing the designs, the following alternatives were selected for the LPG burner system:

Burner system: Alternative A: Install a hood on the burner and metal louvers with a 60° opening. The equations analysis determined that the burning process is adequate; it is required to improve hot air distribution within the pre-chamber to achieve uniform temperatures during drying.

Thermal insulation: Alternative B: Use a rock wool insulation layer to cover the entire equipment externally. After analyzing the temperature of the roasting process to 120 °C and the costs, it is determined that a thickness of 10 cm of insulation will prevent operators from getting burned and reduce heat losses.

Structural support: Alternative A: Install structural supports under the perforated plate. The simulation analysis shows that the perforated plate is adequate. The structural support is being redesigned with new support points distributed along the body.

Cocoa bean conveying system: Alternative B: Install drag paddles to improve the movement of the beans. The option is selected for its better cost, equipment maintenance, and the availability of parts in the market.

Upper covering: Alternative A: Four detachable steel covers covering the structural hexagon. This alternative considers worker handling for cocoa inspection and ease of removing the product.

Transmission system: Alternative C: Replace the belt system and implement a new transmission ratio. Based on the force analysis, moving more than 10 quintals of cocoa is required; therefore, updating the transmission design is necessary.

3.1 Thermodynamic Analysis

Heat required for the equipment operation: It was obtained that the equipment theoretically requires 56.26 kW for grain drying. This value allowed for comparing heat losses due to conduction, convection, and radiation [7]. Table 3 shows the variables involved in the operation of the equipment.

Heat loss: The surface temperature of the equipment initially reached 60 °C with a heat loss of 4.25 kW, resulting in a 7.55% loss in the equipment, which demanded higher fuel consumption. The selected insulating material's surface temperature was 40 °C, with heat losses of 1.77 kW, representing 3.15% in losses.

Table 3 Energy flow for the operation of the machine

Variable	Quantity	Unit	Description
ϕ_{a1}	85	%	Relative humidity of the location
P	101.5	kPa	Atmospheric pressure
$P_{sat\ @\ 25\ ^\circ C}$	3.17	kPa	Saturation vapor pressure
P_{v1}	2.69	kPa	Partial pressure of water vapor
$w_{1,\ 2}$	0.01693	$kg_{water}/kg_{dry\ air}$	Absolute humidity
$T_{dry\ air}$	250	°C	Air temperature in the drying chamber
ρ_{air}	0.6746	kg/m^3	Air density
v_{air}	8	m/s	Air velocity in the drying chamber
$\dot{V}_{air}$	4.02	m^3/s	Volumetric flow rate of air
$\dot{m}_{air}$	2.71	kg/s	Mass flow rate of air
h_{a1}	68.27	kJ/kg	Enthalpy of air at burner inlet
h_{a2}	89.03	kJ/kg	Enthalpy of air at the burner outlet
$\dot{Q}_1$	56.26	kW	Heat required for grain drying

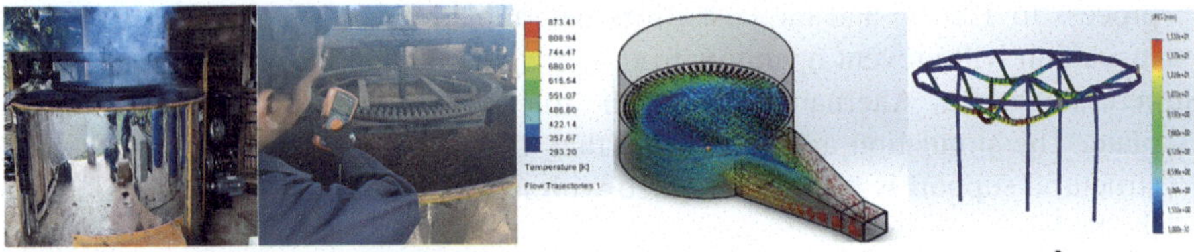

a b c d

Fig. 1 (**a**) Initial dryer, (**b**) temperature measurements, (**c**) fluid dynamics simulation of the equipment, and (**d**) simulation of the metal support structures

Implementation of the new burner: The burner selection was based on existing industrial models in the market, considering the operating temperature of the roaster.

Analysis of internal fin arrangements: For the design, simulation programs were used to analyze the heat flow lines throughout the dryer chamber, including during roasting. Temperatures reached 60–80 °C during drying and 110–210 °C during roasting (Fig. 1). The measurement process of the camera is shown, whose data were entered for the analysis of the simulation, determining the speed and maximum operating temperature of the equipment.

3.2 Mechanical System Analysis

The structure of the perforated plate was reinforced to prevent buckling of the thin plate. A feasible solution involved adjusting the mixing speed from drying (2 rpm) to roasting (5 rpm) using staggered pulleys. New agitators were tested, which reduced grain fractures as they passed through areas with increased buckling.

Structural analysis of the bed: The design software analysis demonstrated favorable results with the proposed reinforcement solutions. With a distributed load of 7500 N, a temperature of 60 °C in the surrounding area, 80% fine meshing, and 51 nodes throughout the structure, maximum shear forces of 7 N were observed, producing moments of 0.5 Nm. The maximum Von Mises stress resulted in a safety factor of 1.7. Deformation was less than one-thousandth of a millimeter, and the maximum displacement was one-tenth of a millimeter.

Top coating of the tank: The minimum required thickness for the tank is 0.724 mm. Since the current equipment thickness is 1.1 mm, implementing the top coating will not pose any issues.

Transmission system for roasting: Table 4 details the characteristics of the transmission system for roasting. The initial stirrer speed is inadequate for roasting cocoa beans, as a speed between 5 and 8 rpm is recommended.

Table 4 Mechanical transmission for roasting

Variable	Quantity	Unit	Description
n_{motor}	1730	rpm	Motor input speed
$n_{pulley\ 2}$	576	rpm	Output speed for roasting
$n_{gearbox}$	52	rpm	Output speed through the gearbox
n_{chain}	52	rpm	Output speed of sprocket chain
n_{pinion}	52	rpm	Input speed of bevel pinion
n_{final}	5.24	rpm	Output speed of remover or crown

Table 5 Main shaft design parameters

Variable	Quantity	Unit	Description
T_{global}	44.71	Nm	Global torque of the removal system
$n_{roasting}$	2		Factor of safety
S_y	250	MPa	Yield strength factor of SAE 36 steel
d	0.015	m	Shaft diameter

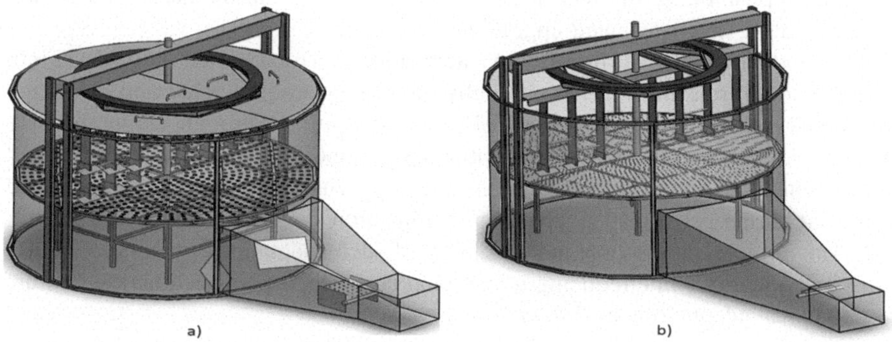

Fig. 2 (**a**) Final cocoa roaster and (**b**) initial cocoa dryer

Moments of inertia of the cocoa and system: The system operates with a total moment of inertia of 36.89 kg m^2, an angular velocity of 0.52 rad/s, and a torque of 9.59 Nm.

Sizing of the main removal shaft: The design parameters provided in Table 5 indicate that the minimum diameter for the shaft should be 15 mm. Given that the current diameter is 38 mm, it remains acceptable to use the existing shaft with the new vane modification, and no change in the shaft is necessary due to the low operating speed of the equipment.

Selection of the electric motor: It was determined that the required power for the electric motor is less than 2 HP. The equipment uses a 2 HP motor operating at 1730 rpm, so no motor replacement is necessary.

Figure 2 presents images comparing the final design of the drying-roasting model with the initial state of the dryer.

4 Conclusion

A cocoa bean roaster was designed based on the operation of a rotary drying machine to enhance value addition. This adaptation enables roasting approximately 454 kg of cocoa with an investment of around $1200.00.

The proposed modifications to the drying machine for cocoa roasting were derived from an analysis of similar equipment. These adaptations included increasing the temperature by 39% below the drying surface, improving the mixing mechanism to reduce residual cocoa by 57%, and utilizing two 30 kg gas cylinders for the entire process. These changes aim to ensure a uniform distribution of heat.

The elements required for the cocoa roasting operation were analyzed using design software through mechanical and heat transfer simulations, yielding an error margin of 14%. This analysis provided critical insights into the phenomena, aiding the development of efficient solutions and consistent results.

The drying procedure's qualities were optimized for successful roasting, increasing machine operation efficiency. Key improvements included reinforcing the plate structure, repositioning the thermocouple, enhancing the mechanical transmission system, sealing heat leaks, and upgrading the burner.

This study demonstrates the effective adaptation of a rotary grain cocoa dryer into a dual-function roasting machine, thereby increasing product value and achieving substantial savings in propane gas (LPG) consumption. Through comprehensive analysis and iterative prototyping, essential design modifications were identified and implemented to optimize cocoa bean roasting. The final equipment exhibited improved burner performance, consistent temperatures, and efficient energy utilization. This project underscores the feasibility of such adaptations to enhance equipment functionality while reducing operational costs, setting a benchmark for future technological adaptations in similar contexts.

References

1. Amador Sacoto, C., Alvarado Barzallo, A., Farah Asang, S.E., Martillo Garcia, J.J.: Caracterización morfológica del cacao nacional "Theobroma cacao L." del cantón Naranjal, Ecuador. RTE. **34**(4), 80–97 (2022). https://doi.org/10.37815/rte.v34n4.978
2. Caballero Cava, J.A.: Diseño de una secadora de cacao para almacenaje con capacidad de 2 T/Día. Tesis de Grado, Pontificia Universidad Católica del Perú. http://hdl.handle.net/20.500.12404/6199 (2015)
3. Mora Caranqui, A.J.: Diseño de un secador rotatorio para granos de cacao con capacidad de 3 toneladas diarias para la producción nacional. Bachelor thesis, Espol. http://www.dspace.espol.edu.ec/handle/123456789/35067 (2015). Accedido 9 de diciembre de 2023
4. Delgado Plaza, E.: Diseño, modelo matemático y construcción de un secadero optimizado con energías limpias. Tesis de Maestria, Universidade de Santiago de Compostela. http://hdl.handle.net/10347/12299 (2015)
5. Guiterrez, M.: Tecnología de tostado del grano de cacao. https://www.academia.edu/51229574/tecnolog%c3%8da_de_tostado_del_grano_de_cacao (2021)
6. Budynas, R., Nisbett, K.: Shigley's Mechanical Engineering Design. McGraw-Hill, New York (2015)

Part III
Alternative Fuels and Renewables

Assessment of Pollutant Emissions in an M2 Vehicle Fueled with Oxyhydrogen-Gasoline Blends Under Static Conditions

M. Estrella-Guayasamín, M. Gómez Berrezueta, J. Cáceres Moreira, and M. Cofre Yanchaguano

1 Introduction

The emission of air pollutants in urban environments constitutes a global environmental challenge, with its rise primarily linked to industrial activities and vehicular traffic [1]. The automotive industry is tasked with the critical objective of reducing emissions of harmful gases, given its significant contribution to global carbon dioxide (CO_2) emissions [2]. The urgency to address climate change has brought the reduction of mobile emissions, particularly CO_2 and carbon monoxide (CO), to the forefront. These pollutants not only degrade environmental quality but also pose serious risks to human health. According to the International Energy Agency, the transportation sector is responsible for approximately one-quarter of global CO_2 emission [3], a key driver of the greenhouse effect and climate change [4].

Limiting the global temperature increase to 1.5 °C is critical to mitigating the most severe climate impacts, as emphasized by the Intergovernmental Panel on Climate Change (IPCC) [5]. In Latin America, the sustained growth of the vehicle fleet, with a 3% increase in demand since 2014, remains heavily reliant on fossil fuels, exacerbating air quality degradation in the region's major metropolitan areas [6]. This issue is further compounded by the global decline in fossil fuel reserves, driven by rising demand linked to current consumption patterns [7]. In response, research into alternative fuels has intensified, aiming to reduce both pollutant emissions and dependence on petroleum [8].

M. Estrella-Guayasamín · J. Cáceres Moreira · M. Cofre Yanchaguano
Universidad Politécnica Salesiana - GMovInt, Guayaquil, Ecuador

M. Gómez Berrezueta (✉)
International University of Ecuador, Guayaquil, Ecuador
e-mail: magomezbe@uide.edu.ec

M. Espinoza-Andaluz et al. (eds.), *Congress on Research, Development and Innovation in Renewable Energies*, Green Energy and Technology,
https://doi.org/10.1007/978-3-031-88995-0_9

In Ecuador, the transportation sector accounts for approximately 25% of the nation's greenhouse gas emissions [9]. To address this, Ecuador has begun implementing measures aligned with its environmental and sustainable development goals, including the promotion of cleaner fuels, the adoption of hybrid vehicles, and the tightening of vehicular emission regulations. Emerging technologies, such as vehicle electrification and the use of biofuels, including oxyhydrogen-gasoline blends, have demonstrated significant potential in reducing pollutant emissions, making them crucial for sustainable mobility. However, it is essential to conduct a comprehensive life cycle assessment of these technologies, considering fuel production, distribution, and usage, to ensure a net positive environmental impact [10].

Over the past decade, hydrogen has emerged as a promising alternative for reducing vehicular pollution due to its ability to eliminate carbon emissions, such as CO and CO_2 [11]. However, the excessive costs associated with hydrogen infrastructure have limited its widespread adoption as a primary fuel for personal and urban vehicles [12]. A potential solution lies in the use of a gasoline and oxyhydrogen (HHO) mixture, which enhances combustion efficiency and reduces both fuel consumption and pollutant emissions. These advantages are attributed to hydrogen's complete combustion and higher flame speed, leading to more thorough fuel oxidation [13]. HHO gas, also known as Brown gas, is obtained from electrolyzers. Electrolysis is the process by which electricity is used to split water into hydrogen and oxygen, this process is facilitated by the use of dry and wet HHO-generating cells.

The dry cell is a device that features a reservoir supplying an alkaline electrolyte to the electrolyzer, which serves as the conductive medium. Upon applying voltage to the electrodes, each plate within the electrolyzer becomes oppositely charged, inducing water dissociation, and generating HHO gas. This gas is then directed through the system via inlet and outlet valves. As depicted in Fig. 1, the produced gas is channeled into a bubbling device containing a slightly acidic water solution, which neutralizes residual impurities, ensuring that only clean gas is delivered to the engine [14].

Conversely, the wet cell is distinguished by its higher gas production and ease of manufacturing. Sodium or potassium hydroxides can be used as electrolytes in this

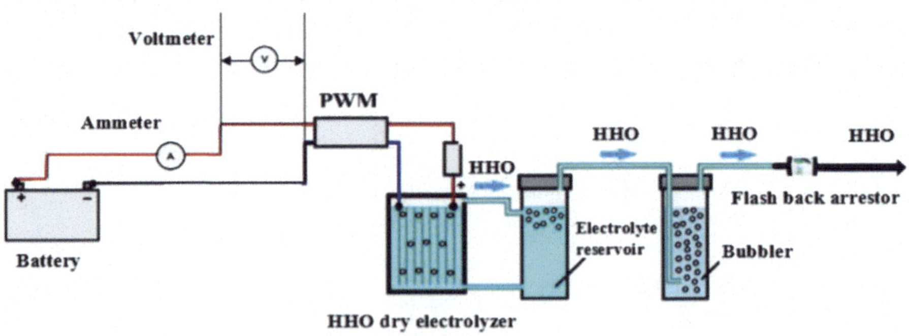

Fig. 1 Schematic drawing of HHO dry electrolyzer [14]

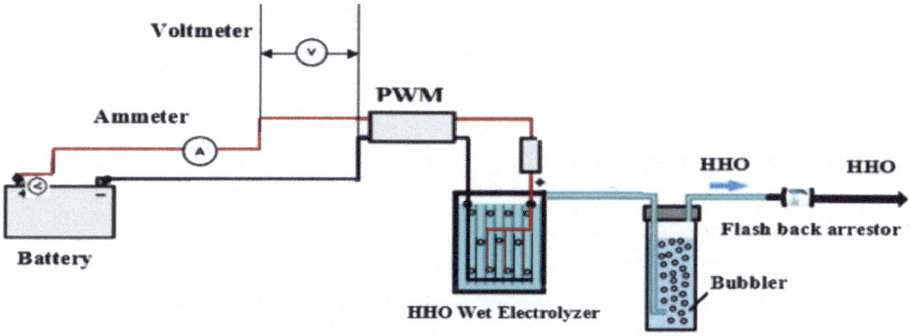

Fig. 2 Schematic drawing of HHO wet electrolyzer [14]

configuration. The design of the wet cell closely resembles that of a lead-acid battery, as illustrated in Fig. 2.

The optimal mechanical and environmental performance of internal combustion engines (ICEs) occurs when the air-fuel mixture reaches a stoichiometric ratio of 14.7:1, ensuring complete combustion with CO_2, H_2O, and N_2 as the primary products. Conversely, when ICEs operate with air-deficient mixtures (rich mixtures), the combustion products include CO_2, CO, HC, H_2O, and N_2, among others. This not only reduces mechanical efficiency but also increases fuel consumption and the emission of harmful gases that pose risks to both health and the environment. While CO_2 emissions, a key contributor to the greenhouse effect, can be mitigated through photosynthesis, unburned hydrocarbons (HC) and carbon monoxide (CO) are highly toxic to human health and are not readily processed by the environment [15, 16].

The implementation of oxyhydrogen gas (HHO) in internal combustion engines has demonstrated significant reductions in pollutant emissions. Studies have shown that CO_2 emissions can be decreased by 10–14%, while hydrocarbon (HC) and carbon monoxide (CO) emissions can be reduced by around 20% [17–20]. Additionally, in diesel engines, the opacity of exhaust emissions has been observed to decline by 8–25%, depending on engine speed, further highlighting the potential of HHO integration to improve environmental performance in both gasoline and diesel engines [21]. These findings collectively underscore the significant potential of HHO gas to enhance the efficiency and environmental performance of both gasoline and diesel internal combustion engines.

Category M2 vehicles are designed for the transportation of passengers and feature more than nine seats, including the driver, with a maximum mass not exceeding 5 tons [22]. This category includes minibuses and cargo vans, which play a vital role in both urban and rural transportation. Given the extensive distances these vehicles often cover, their emission rates are significantly impactful. Consequently, any enhancement in the engine's combustion process could lead to notable reductions in fuel consumption and, by extension, emissions [23]. In response to these findings and building on previous research, this study designed a series of experiments to assess the impact of oxyhydrogen-gasoline mixtures on the generation of CO_2, HC, and CO emissions in an M2 vehicle. The experiments were conducted in Guayaquil, Ecuador.

2 Materials and Methods

2.1 Experimental Setup

The equipment and instruments utilized in the experimental setup of this study are depicted in Fig. 3. The M2-type vehicle selected for the experiment was a 2023 Karry van model Q22L, featuring a 1200 cc engine, four cylinders, 4X2 MT configuration, and an 11-passenger capacity. This model gained prominence in the Ecuadorian market due to its strong sales performance in 2022 and was recognized by the Ecuadorian Association of Automotive Companies (AEADE) as one of the most popular vehicles in its category [24]. Prior to the experiment, comprehensive corrective maintenance was performed on the fuel supply and ignition systems to ensure optimal engine operability throughout the study.

For this study, a wet-cell HHO generator was selected due to its superior efficiency in HHO gas production compared to the dry cell configuration [14]. The generator was powered by an Instek PSW-80-27 variable DC power supply, which can regulate voltages ranging from 0 to 80 V and currents from 0 to 27 A. The cell structure comprised two 316 stainless steel plates, each measuring 100 mm by 100 mm and 1.5 mm thick, with a capacity of 250 cm^3 for the electrolyte. The electrolyte solution consisted of 1 L of distilled water mixed with 1.5% KOH. The HHO generator provided three distinct HHO gas flow rates to the vehicle's intake

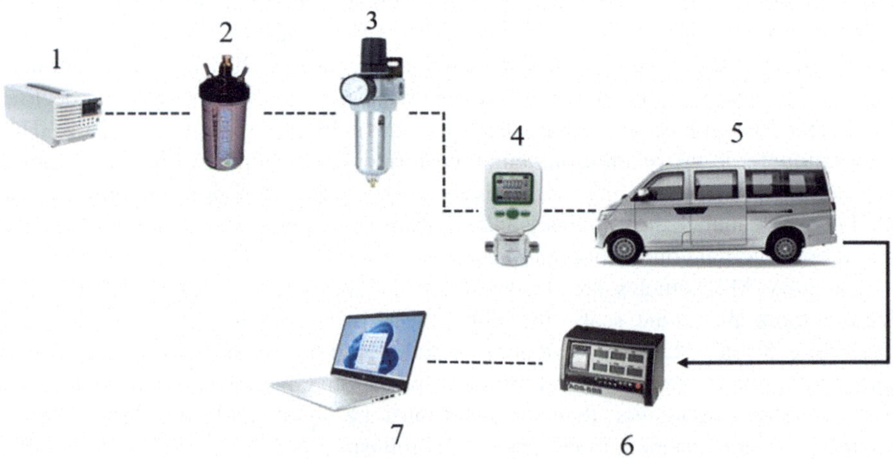

Fig. 3 An experimental setup: (1) variable direct current power supply PSW-80-27 Instek, (2) HHO generator, (3) filter drier, (4) gas flow meter (model MF571), (5) M2-type vehicle, (6) brain Bee AGS 688 gas analyzer, (7) computer

manifold. These flows were directed through a filter regulator and analyzed using an MF5712 gas mass flow meter, with an operational range of 0–250 L/min. Subsequently, a Brain Bee AGS 688 gas analyzer was employed for further analysis.

2.2 *Experimental Methodology*

The experimental methodology employed in this study was designed to evaluate the effect of HHO-gasoline blends on the pollutant emissions (CO, CO_2, and HC) of an internal combustion engine running on regular 87-octane gasoline in static conditions in Guayaquil City. The HHO flow rates were regulated by adjusting the electrical current supplied to the electrolytic cell. Specifically, currents of 2, 7, and 17 A were applied to achieve HHO flow rates (Factor A) of 1.03 standard liters per minute (slpm) for HHO_1, 1.31 slpm for HHO_2, and 1.69 slpm for HHO_3, respectively. These current values were selected based on the operational limits of the HHO wet cell, as currents exceeding 17 A could lead to elevated temperatures, resulting in electrolyte evaporation and water vapor generation. Evaluations were performed under two specific conditions (Factor B): the vehicle at idle (780 rpm) and at 2500 rpm.

Eight treatments were conducted, derived from the combination of the experimental factors. Two of these treatments involved the operation of the engine using only regular gasoline, serving as a baseline for assessing the effect of HHO-gasoline mixtures on the engine's pollutant emissions. In accordance with the NTE INEN 2204 standard, which mandates a minimum of three repetitions for each test [25], six repetitions were performed for each treatment, ensuring the statistical robustness of the obtained results.

The protocol for conducting the experimental tests was as follows: (1) calibrate the gas analyzer, (2) start the vehicle and allow it to reach its operating temperature, (3) ensure that the vehicle's transmission is in neutral and that all auxiliary systems, including air conditioning and lights, remain off, (4) set the test conditions as specified in Table 1, (5) insert the probe into the exhaust pipe, allowing a measurement time of 30 seconds for each test.

3 Results

Figure 4 shows the installation of the wet cell HHO generator in the test vehicle. The results of the six replicates of the completely randomized set of experiments for HC and CO_2 emissions are shown in Tables 1 and 2. A normality analysis was performed on the data for each treatment, indicating that the data follows a normal distribution. This conclusion is further supported by a p-value greater than 0.1, which exceeds the significance threshold of 0.05.

Table 1 HC emissions generated under different load conditions

	Factor		HC (ppm)						
Treatment	A	B	R1	R2	R3	R4	R5	R6	Mean
1	0	1	14.00	16.00	17.00	16.00	16.00	15.00	15.67
2	1	1	9.00	9.00	8.00	8.00	10.00	9.00	8.83
3	2	1	8.00	9.00	5.00	6.00	7.00	8.00	7.17
4	3	1	4.00	5.00	2.00	3.00	4.00	4.00	3.67
5	0	2	10.00	12.00	10.00	9.00	11.00	10.00	10.33
6	1	2	8.00	9.00	7.00	8.00	9.00	8.00	8.17
7	2	2	7.00	8.00	7.00	6.00	6.00	7.00	6.83
8	3	2	3.00	4.00	1.00	2.00	2.00	2.00	2.33

Factor A (Level: 0 = gasoline, 1 = gasoline + HHO_1, 2 = gasoline + HHO_2, 3 = gasoline + HHO_3). Factor B (Level: 1 = 780 rpm and 2 = 2500 rpm)

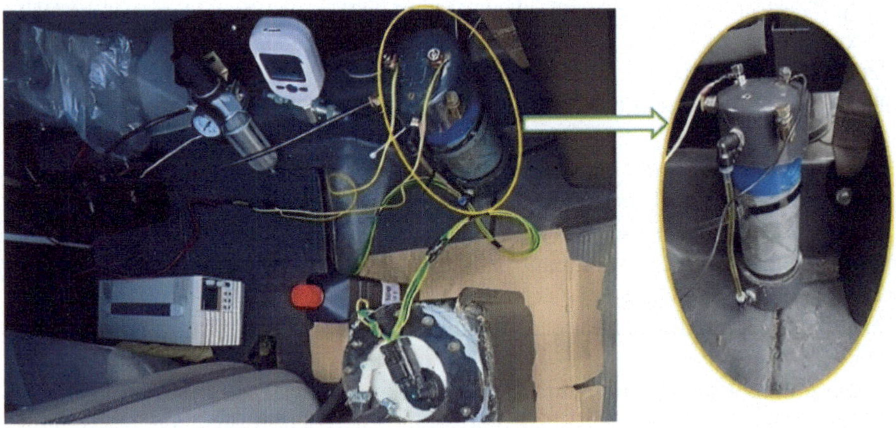

Fig. 4 Installing wet-cell HHO generator on the test vehicle engine

Table 2 CO_2 emissions generated under different load conditions

	Factor		CO_2 (% vol)						
Treatment	A	B	R1	R2	R3	R4	R5	R6	Mean
1	0	1	13.90	14.10	14.30	13.80	14.10	14.10	14.05
2	1	1	14.40	14.30	14.20	14.20	14.50	14.40	14.33
3	2	1	14.50	14.80	14.30	14.20	14.40	14.50	14.45
4	3	1	14.90	15.00	14.90	14.60	14.70	14.80	14.82
5	0	2	14.10	14.30	14.20	14.10	14.40	14.30	14.23
6	1	2	14.70	14.80	14.40	14.50	14.70	14.70	14.63
7	2	2	15.20	14.80	14.80	14.70	14.50	14.70	14.78
8	3	2	15.60	15.70	15.10	14.80	15.00	15.00	15.20

Factor A (Level: 0 = gasoline, 1 = gasoline + HHO_1, 2 = gasoline + HHO_2, 3 = gasoline + HHO_3). Factor B (Level: 1 = 780 rpm and 2 = 2500 rpm)

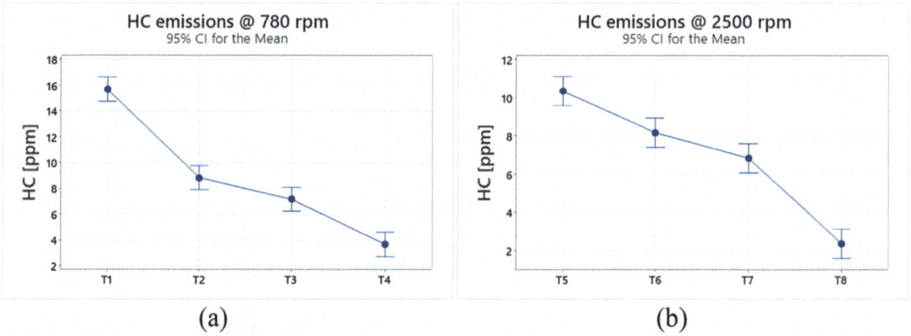

Fig. 5 HC emissions vs. HHO flow rate supplied: (**a**) idling condition and (**b**) 2500 rpm

Multiple ANOVA analyses were conducted to determine whether significant differences exist in the mean HC emissions as the HHO flow rate supplied to the engine as a secondary fuel, alongside gasoline, was varied. Analysis revealed that variations in HHO flow rate significantly affect HC generation, as indicated by a *p*-value below the significance threshold. The obtained data show a decreasing trend in HC emissions as the HHO flow rate increases. Specifically, the mean HC emissions decreased from 15.67 ppm (T1—gasoline only) to 3.67 ppm (T4—gasoline + HHO_3) when the engine operated at idle, and from 15.67 ppm (T5—gasoline only) to 3.67 ppm (T8—gasoline + HHO_3) when the engine ran at 2500 rpm, as shown in Fig. 5a, b. The comparison of the different treatments using Tukey's method indicates that the HC emissions from the control treatments, T1 and T5, under idle and 2500 rpm conditions, respectively, differ significantly from those of the other treatments.

Similar analyses were conducted to determine whether significant differences existed in the mean CO_2 emissions generated by the engine as the HHO flow rate supplied alongside gasoline was varied. The analysis revealed that variations in HHO flow rate significantly affect CO_2 generation, as indicated by a *p*-value below the significance threshold.

Figure 6 illustrates the increasing trend in CO_2 emissions as the HHO flow rate increases. Specifically, the mean CO_2 concentration increased from 14.07% vol (T1—gasoline only) to 14.82% vol (T4—gasoline + HHO_3) when the engine operated at idle, and from 14.23% vol (T5—gasoline only) to 15.20% vol (T8—gasoline + HHO_3) when the engine ran at 2500 rpm, as shown in Fig. 6a, b. The comparison of the different treatments using Tukey's method shows that CO_2 emissions from the control treatments, T1 and T5, differ significantly from those of the other treatments.

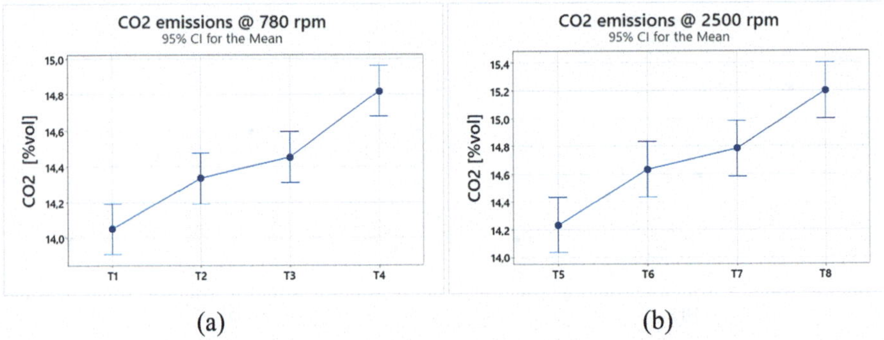

Fig. 6 CO_2 emissions vs. HHO flow rate supplied: (**a**) idling condition and (**b**) 2500 rpm

4 Discussion

Based on the values obtained from the experimental tests, it is observed that increasing the HHO flow rate supplied as a secondary fuel to the M2 category vehicle's engine reduces the generation of unburned hydrocarbons. At idle, HC reductions ranged from 43.61% to 76.63%, with HHO contributions to the engine of 1.03 and 1.7 slpm. Additionally, at 2500 rpm, the reductions achieved ranged from 20.96% to 77.42%. Several studies [17, 19–21] demonstrate that this behavior is due to the addition of HHO gas improving the combustion process, thereby reducing HC generation, and consequently enhancing fuel efficiency.

Conversely, CO_2 levels generated under idle conditions increased by 1.99–5.45% as the HHO flow rate varied from 1 to 1.7 slpm. At 2500 rpm, the CO_2 levels showed a maximum increase of 5.48% under idle conditions, and an increase between 2.81% and 6.82% when operating at 2500 rpm. This rise in CO_2 concentration is attributed to the enhanced combustion efficiency due to the use of HHO, leading to more complete fuel utilization in each engine cycle, resulting in almost complete combustion and, consequently, higher CO_2 production. These findings are consistent with the study conducted by González et al. [26].

Additionally, carbon monoxide (CO) generation was negligible under all conditions, likely due to the high catalytic efficiency of the vehicle used in this study. These results align with those obtained by Jakliński [27], who evaluated the impact of HHO gas on emissions in five vehicles through static tests, with similar CO results observed in three of the five vehicles.

5 Conclusion

A wet cell oxyhydrogen gas supply system was implemented as a secondary fuel, combined with gasoline, in an M2 category vehicle to evaluate its effect on pollutant emissions (CO, CO_2, and HC) through static tests. The assessments were conducted under two specific load conditions: at engine idle and 2500 rpm. The results revealed that increasing the oxyhydrogen gas flow rates led to a reduction in HC emissions, indicating an improvement in combustion, which consequently resulted in an increase in CO_2 emissions.

Maximum HC reductions of approximately 77% were observed across different operating regimes, underscoring the effectiveness of using oxyhydrogen gas as a secondary fuel in reducing unburned hydrocarbon emissions in M2 category vehicles. Regarding CO_2 emissions, peak values of 5.48% at idle and 6.82% at 2500 rpm were recorded. Overall, the benefits of employing HHO-gasoline blends in M2 vehicle engines to mitigate harmful emissions are evident.

Carbon monoxide (CO) emissions were negligible across all tested conditions, likely attributable to the high catalytic efficiency of the vehicle utilized in this study. To further validate this observation, future research should focus on evaluating data collected upstream of the catalytic converter.

Additionally, a detailed investigation into the power consumption of the wet cell and the overall energy balance of the system is recommended.

References

1. Hernández Cerda, C.N., Ávila Galarza, A., Cerda Alonso, D.G.: Impacto de la movilidad urbana en la calidad del aire de la zona metropolitana de San Luis Potosí, México. Rev. Cienc. Ambient. **57**, 1–27 (2022). https://doi.org/10.15359/rca.57-1.8
2. Morocho, V., Paz, P., Ulloa, C., Valencia, M.: Evaluación del consumo y emisiones de un vehículo implementado una válvula MARCLAIS a 2754 msnm. AlfaPublicaciones. **4**, 89–105 (2022). https://doi.org/10.33262/ap.v4i2.2.218
3. Pérez, S.M., Reátegui, J.S., Mendoza, A.S.: Contaminación ambiental producida por el tránsito vehicular y sus efectos en la salud humana: revisión de literatura. INVENTUM. **16**, 20–30 (2021). https://doi.org/10.26620/UNIMINUTO.INVENTUM.16.30.2021.20-30
4. Sancan, D.: Vehículos híbridos, una solución interina para bajar los niveles de contaminación del medio ambiente causados por las emisiones provenientes de los motores de combustión interna. INNOVA Res. J. **2**, 1–10 (2017). https://doi.org/10.33890/innova.v2.n12.2017.527
5. Masson-Delmotte, V., Zhai, P., Pörtner, H.-O., Roberts, D., Skea, J., Shukla, P.R., Pirani, A., Moufouma-Okia, W., Péan, C., Pidcock, R., Connors, S., Robin, J.B., Chen, Y., Zhou, X., Gomis, M.I., Lonnoy, E., Maycock, T., Tignor, M., Waterfield, T.: Calentamiento Global de 1,5°C (2019)
6. Merchan, V., Edwin, P., Julio, L., Santiago, C.: Estudio de emisiones contaminantes producidas por un motor MEP con transmisión automática y transmisión manual. INNOVA Res. J. **3**, 120–128 (2018). https://doi.org/10.33890/innova.v3.n4.2018.644
7. Ortiz, M.: Vehicular Contamination: A Rising Problem with Immediate Needs (2020)
8. Quinchuela, J., Solís, J., Martillo, G., Cali, D.: Estudio del Comportamiento de un vehículo monoplaza con el uso de hidrógeno como combustible suplementario. AlfaPublicaciones. **5**, 6–31 (2023). https://doi.org/10.33262/ap.v5i3.1.382
9. Pérez, D., Diego, G.: Estudio de emisiones contaminantes utilizando combustibles locales. INNOVA Res. J., 23–34 (2018). https://doi.org/10.33890/innova.v3.n3.2018.635
10. Velepucha, J., Sabando, L.: Emisiones de gases contaminantes en vehículos livianos a gasolina. Rev. Cient. INGENIAR Ing. Tecnol. Investig. **4**, 78–95 (2021). https://doi.org/10.46296/ig.v4i8.0024

11. Shadidi, B., Najafi, G., Yusaf, T.: A review of hydrogen as a fuel in internal combustion engines. Energies (Basel). **14** (2021). https://doi.org/10.3390/en14196209
12. Tamayo, E., Rosales, C., Guzman, A., Pasmiño, P.: Efecto del uso de hidrógeno en la potencia y rendimiento de un motor de combustión interna. Enfoque UTE. **7**, 43–54 (2016)
13. Rosero Espín, M., González, H., Ochoa, P., Díaz, M., Jiménez, E., Castillo, C., Chulde, D., Stahl, U.: Producción de oxihidrógeno como alternativa energética. Rev. Perspect. **3**, 62–67 (2021). https://doi.org/10.47187/perspectivas.vol3iss2.pp62-67.2021
14. El Soly, A.K., El Kady, M.A., Farrag, A.E.F., Gad, M.S.: Comparative experimental investigation of oxyhydrogen (HHO) production rate using dry and wet cells. Int. J. Hydrog. Energy. **46**, 12639–12653 (2021). https://doi.org/10.1016/j.ijhydene.2021.01.110
15. Rivera, N., Chica, J., Zambrano, I., García, C.: Study of the behavior of an otto engine of electronic injection in relation to the stoichiometry of the mixture and the advance to the ignition for Cuenca City. Revista Politécnica, Octubre, vol. 40 (2017)
16. Pulkrabek, W.W.: Engineering Fundamentals of the Internal Combustion Engine, 2nd edn, (2013)
17. Dewangan, A., Mallick, A., Yadav, A.K., Islam, S., Saleel, C.A., Shaik, S., Ağbulut, Ü.: Production of oxy-hydrogen gas and the impact of its usability on CI engine combustion, performance, and emission behaviors. Energy. **278** (2023). https://doi.org/10.1016/j.energy.2023.127937
18. Mendoza-Casseres, D., Valencia-Ochoa, G., Duarte-Forero, J.: Evaluación experimental del rendimiento de combustión en motores estacionarios de baja cilindrada que funcionan con mezclas de biodiesel e hidroxi. https://www.sciencedirect.com/science/article/abs/pii/S2451904921000457 (2021). https://doi.org/10.1016/j.tsep.2021.100883
19. Yilmaz, A.C., Uludamar, E., Aydin, K.: Effect of hydroxy (HHO) gas addition on performance and exhaust emissions in compression ignition engines. Int. J. Hydrog. Energy. **35**, 11366–11372 (2010). https://doi.org/10.1016/j.ijhydene.2010.07.040
20. Nabil, T.: Efficient use of oxy-hydrogen gas (HHO) in vehicle engines. J. Eur. Syst. Automat. **52**, 87–96 (2019). https://doi.org/10.18280/JESA.520112
21. Al-Dawody, M.F., Al-Farhany, K.A., Allami, S., Al-Chlaihawi, K.K.I., Jamshed, W., Eid, M. R., Raezah, A.A., Amjad, A., El, S.M.: Using oxy-hydrogen gas to enhance efficacy and reduce emissions of diesel engine. Ain Shams Eng. J. (2023). https://doi.org/10.1016/j.asej.2023.102217
22. NTE INEN 2656: Primera Revisión—Clasificación Vehicular
23. Rojas, E., Romero, V., Pancha, J.: Analysis of behavior of CO_2 emissions, CO and the lambda factor of a vehicle with a conventional injection system with catalyst and without catalyst. Ingenius. **2020**, 23–29 (2020). https://doi.org/10.17163/INGS.N23.2020.02
24. El Universo: Vehículos más vendidos en Ecuador en el año. https://www.eluniverso.com/noticias/economia/cuales-son-los-vehiculos-mas-vendidos-en-ecuador-en-el-2022-nota/ (2022). Last accessed 29 Feb 2024
25. NTE INEN 2204: Gestión ambiental. Aire. Vehículos automotores. Límites permitidos de emisiones producidas por fuentes móviles terrestres de gasolina
26. González Gil, J.E., Díaz Rey, Á.O., González-Estrada, O.A.: Análisis de un generador de HHO de celda seca para su aplicación en motores de combustión interna. Rev. UIS Ing. **17**, 143–154 (2018). https://doi.org/10.18273/revuin.v17n1-2018013
27. Jakliński, P., Czarnigowski, J.: An experimental investigation of the impact of added HHO gas on automotive emissions under idle conditions. Int. J. Hydrog. Energy. **45**, 13119–13128 (2020). https://doi.org/10.1016/j.ijhydene.2020.02.225

Toward Understanding the Role of Chemical Looping Water Splitting in Clean Hydrogen Production: A Scientometric Analysis

Esteban Vanegas, Nini Johana Marín-Rodríguez, Mario Luna-delRisco, Carlos E. Arrieta, Jorge Sierra, Hernando A. Yepes, and Juan Pablo Gómez Montoya

1 Introduction

Global warming and greenhouse gas (GHG) emissions are critical challenges facing humanity in the twenty-first century [1]. Burning fossil fuels is a significant source of GHGs, significantly contributing to climate change and environmental degradation [2]. The harmful effects of these emissions include rising global temperatures, ocean acidification, and disrupted weather patterns, leading to extreme weather events and biodiversity loss [3]. Mitigating these problems is essential for a sustainable future. Clean hydrogen production has become a key goal in the search for sustainable solutions to the energy crisis and climate change. As a versatile and clean energy carrier, hydrogen has the potential to decarbonize various sectors, from power generation to transportation and industry [4]. However, most hydrogen production

E. Vanegas (✉) · M. Luna-delRisco · C. E. Arrieta
Grupo de Investigación en Energía—GRINEN, Universidad de Medellin, Medellín, Colombia
e-mail: evanegas901@soyudemedellin.edu.co

N. J. Marín-Rodríguez
Grupo de Investigación en Ingeniería Financiera GINIF, Programa de Ingeniería Financiera, Facultad de Ingenierías, Universidad de Medellín, Medellin, Colombia

J. Sierra
Department of Mechatronics Engineering—MATyER, Instituto Tecnológico Metropolitano, Medellin, Colombia

H. A. Yepes
Aplicaciones en Termofluidos, Energía y Nanomateriales Avanzados—ATENA, Departamento de Ingeniería Mecánica, Universidad Francisco de Paula Santander Ocaña, Ocaña, Colombia

J. P. Gómez Montoya
Departamento de Ingeniería Mecánica, Universidad Tecnológica del Perú, Lima, Peru

M. Espinoza-Andaluz et al. (eds.), *Congress on Research, Development and Innovation in Renewable Energies*, Green Energy and Technology,
https://doi.org/10.1007/978-3-031-88995-0_10

today relies on carbon-release methods, like reforming natural gas. This process accounts for three-quarters of all hydrogen produced globally each year (about 70 million tons) [5]. It highlights the urgent need for hydrogen production technologies that are both efficient and environmentally friendly.

The transition to cleaner, more sustainable energy sources is imperative. Hydrogen, as an energy carrier, stands out for its ability to efficiently store and release energy without carbon emissions when used, producing only water vapor during combustion, providing more energy per mass than fuels like methane, gasoline, or coal [6]. However, hydrogen production from hydrocarbons, currently dominant, generates GHGs and contradicts the decarbonization goal. Therefore, exploring and developing environmentally benign hydrogen production methods are crucial. Several methodologies exist for hydrogen generation, including water electrolysis, natural gas reforming, and biomass utilization [7].

However, chemical looping water splitting (CLWS) has emerged as a promising technology due to its efficiency and ability to significantly reduce GHG emissions. Being the alternative that best addresses the challenge of reducing greenhouse gas emissions, as it not only generates hydrogen but also offers a more efficient CO_2 capture compared to oxy-combustion and physical and chemical absorption, with lower energy and economic costs [8].

CLWS employs a material known as an oxygen carrier, which consists of a metal oxide such as hematite (Fe_2O_3), to split water into hydrogen and oxygen without the need for polluting agents, offering a cleaner and more sustainable alternative [9]. This process occurs between two fluidized bed reactors: an oxidizer and a reducer, where fuel combustion is carried out indirectly, facilitating hydrogen production without releasing CO_2 into the atmosphere [10]. CLWS generally includes an intermediate reactor where H_2 is produced by breaking water molecules as they react with the oxygen carrier before entering the air reactor to be oxidized. [11].

CLWS's potential to revolutionize clean hydrogen production is significant. However, understanding its role and evolution within the field is not just important, but it is crucial. This requires a thorough analysis of scientific literature. This is where scientometric analysis becomes an invaluable tool. This approach allows us to map research trends, identify key players and collaborations, and discover emerging themes that are driving the field forward.

This research provides a comprehensive overview of the current state and trends in CLWS research using scientometric techniques, offering a complete understanding of the academic landscape in this domain, catering for diverse audiences including researchers, policymakers, and energy sector professionals. This study identifies emerging trajectories in research, providing a foundational framework to drive future research and innovation. Using advanced tools like VOSviewer and Bibliometrix, scientific database data is analyzed to answer crucial questions about the evolution of knowledge, collaboration networks, and emerging themes within the CLWS field.

In this chapter, we present the results of a scientometric analysis focused on understanding the research dynamics in hydrogen production using CLWS. This analysis aims to enhance our understanding of the field and guide for future research and innovation in advancing toward a cleaner and sustainable energy future. This study contributes to academic discourse and practical implementation through a comprehensive approach, facilitating the identification and exploration of emerging themes in literature. This comprehensive view, rich in analytical depth and interpretative nuances, deepens our understanding of the multifaceted aspects of clean hydrogen production through CLWS. As a result, it is poised to inform future research directions, support evidence-based decision-making, and develop holistic strategies to address the challenges associated with clean hydrogen production.

2 Materials and Methods

To conduct this study, various scientometric techniques were employed, including diagram analysis and information mapping, with the aim of evaluating the role and significance of water splitting in hydrogen production. The applied techniques were: co-authorship analysis, co-word analysis, citation analysis, and cluster analysis [12]. Thirty-four duplicate documents were removed from the Scopus and Web of Science databases, resulting in a final database of 78 research studies. The tools VOSviewer version 1.6.20 and the Bibliometrix package in RStudio, developed by [13, 14], respectively, were used in these analyses. This methodology allows for the evaluation of the impact of scientific contributions, monitoring of advancements in the field, identification of emerging trends, and mapping of scientific patterns to reveal the internal structure of the studied topic.

The scientometric techniques were used to track the evolution of the research field through key documents, identify leading researchers and institutions in the area, analyze research keywords and co-citation groups, and identify emerging research topics. The Scopus and Web of Science (WoS) databases were selected as the most suitable for this study. Although [15] mentions that the use of additional databases could increase the robustness of the findings in future studies. In June 2024, 78 documents were obtained using the following search equation: (TITLE-ABS-KEY ("hydrogen production" OR "hydrogen generation") AND TITLE-ABS-KEY ("chemical loops" OR "chemical looping combustion") AND TITLE-ABS-KEY ("water splitting")). Subsequently, all studies were downloaded and indexed in the reference manager Zotero for content reading and analysis. From the 78 documents analyzed, 253 distinct authors were identified as collaborators. The average number of citations per publication, which is 27.32, highlights the academic relevance within the scientific community. Additionally, an average of 5.4 co-authors per document underscores the collaborative nature of the knowledge generation process in this field. The most common type of document in this collection is the journal article, with 45 examples. A more detailed analysis reveals the presence of 716 keywords plus and 188 author keywords, each adding nuanced dimensions to the semantic scope of the discourse. It also highlights an annual growth rate of 9.25%, in addition to the total use of 3562 references.

3 Findings

3.1 Annual Production

The research on CLWS has steadily grown between 2007 and 2018, with an average of 1.6 articles published per year. However, from 2019 to the present, we see a significant increase (Fig. 1), with an average of eight articles per year, reaching a peak of 17 articles in 2023. This growth trend is promising, as we already have nine articles published in 2024. This number is expected to increase further, considering the annual growth rate of 9.25%. This surge in publications points to a rising interest in the academic community, likely driven by the increasing focus on developing efficient and environmentally friendly fuels by companies, alongside the development of modeling software like Aspen Plus and CFD, which simplify academic research in this area.

3.2 Scientific Production Distribution

Global Distribution

This analysis reveals the geographic distribution of countries leading in scientific output on CLWS, highlighting the contribution of China, which leads with a significant number of 35 research publications, followed by the USA and Korea with 17 and 5 publications, respectively. Other countries like Italy and Spain have lower production with three publications each. These findings align with the assessment of corresponding author's countries, as seen in Fig. 2. China again dominates with 44.9%, followed by the USA and Korea with 14.1% and 6.4%, respectively. Interestingly, the United Kingdom comes in fourth with 5.1%, surpassing Italy and Spain, each with 3.8% and 2.6% respectively.

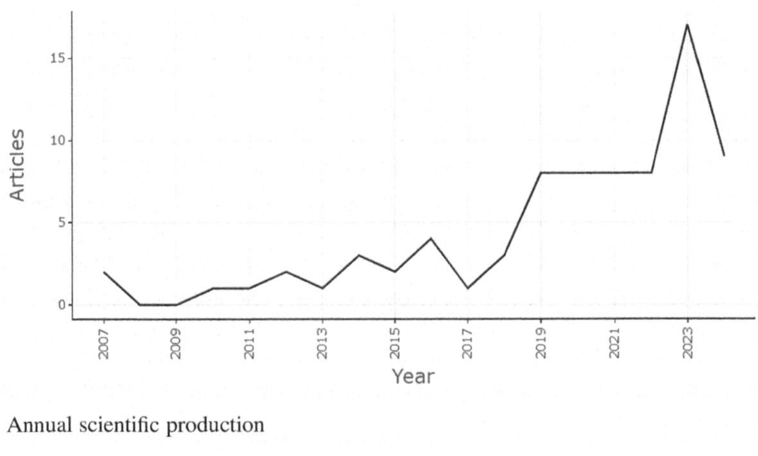

Fig. 1 Annual scientific production

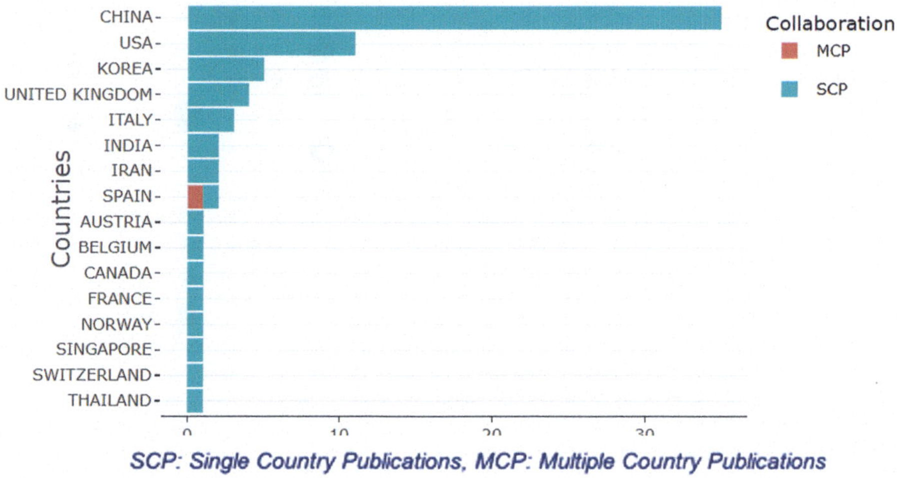

Fig. 2 Corresponding author's countries

Most Relevant Affiliations and Sources

Building upon the analysis by country, the research delves deeper into the institutions most actively involved in CLWS development. As expected, the institutions with the highest levels of participation are located in China. Topping the list is Kunming University of Science and Technology with 22 published articles. Huazhong University of Science and Technology in China follows closely with 14 articles, and Nanjing University of Science and Technology has published six articles. Other prominent institutions include Southeast University in Bangladesh, Imperial College London, Stanford University, and the University of Alabama, both in the USA, each of them with seven, five, five, and five articles, respectively. Mentioning the most relevant sources, it is noted that the International Journal of Hydrogen Energy stands out with the highest number of repositories, totaling 14. It is followed by Industrial and Engineering Chemistry Research and Energy Conversion and Management with six and four articles, respectively. Additionally, the series *Applied Energy, Chemical Engineering Journal, and Fuel* show a similar alignment, each with three notable documents. Finally, *Energy, Energy and Fuels, Energy Technology, and Green Energy and Environment* are mentioned, each with two documents.

3.3 *Author Analysis*

Most Relevant Authors

The study of the most relevant authors is led by Wang H with 11 documents, followed by Li K with 10, Zhu X with 9, and Wang X and Wang Y both with 8. Notably, Wei Y, Xiao B, and Xu T each have 6 publications.

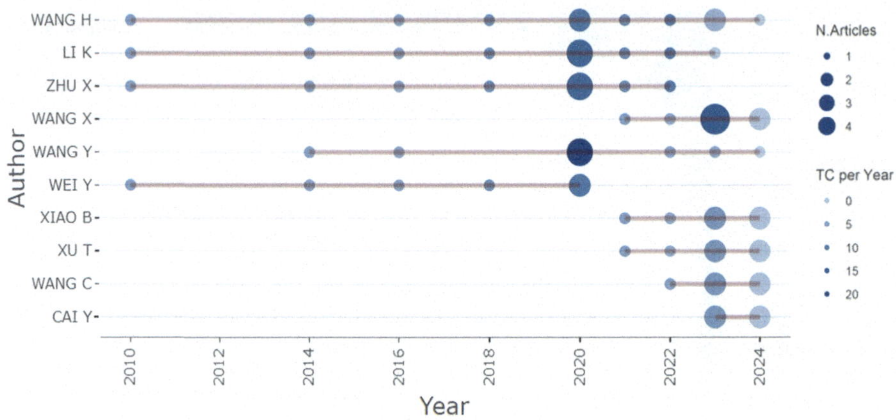

Fig. 3 Authors' production over time

On the other hand, Fig. 3 shows how the research output of the most prominent authors has changed over time. The color of the circles represents the total number of citations each year, and the size of the circles indicates the number of articles published. We see a significant increase in research output starting in 2020, moving from an average of 4–5 articles per year to 13.

Production has remained steady in the following years, with another increase in 2023, reaching a total of 16 articles. This trend seems to continue to the present day. It is important to highlight the work of Wang H, Li K, and Zhu X, who have been researching CLWS since 2010. They have made advancements in coupling water splitting with a steam methane reforming (SMR) system to generate both hydrogen and syngas [16]. These authors continue to develop this field by incorporating the use of a macroporous solid solution of CeO_2–ZrO_2 into the SMR process coupled with CLWS, along with tuning the macroporous structure [17, 18]. Since 2020, there has been a special focus on oxygen carriers. Researchers are evaluating different compounds, support elements, and various treatments, aiming to diversify and optimize the development of this technique.

Co-authorship Network

In the analysis of co-authorship, it is essential to consider the frequency of collaborations among researchers, as indicated by the number of documents they jointly contribute to. This network visualization is achieved through a diagram where connections between researchers are depicted as color-coded lines linking various elements. The significance of this network is portrayed through weighted attributes that determine the size of each element. The quantity of connections among co-authors in bibliographic data analysis indicates their relevance and prominence in the collaborative framework [15, 19]. Figure 4 illustrates connections between researchers based on jointly authored documents. This graphical representation

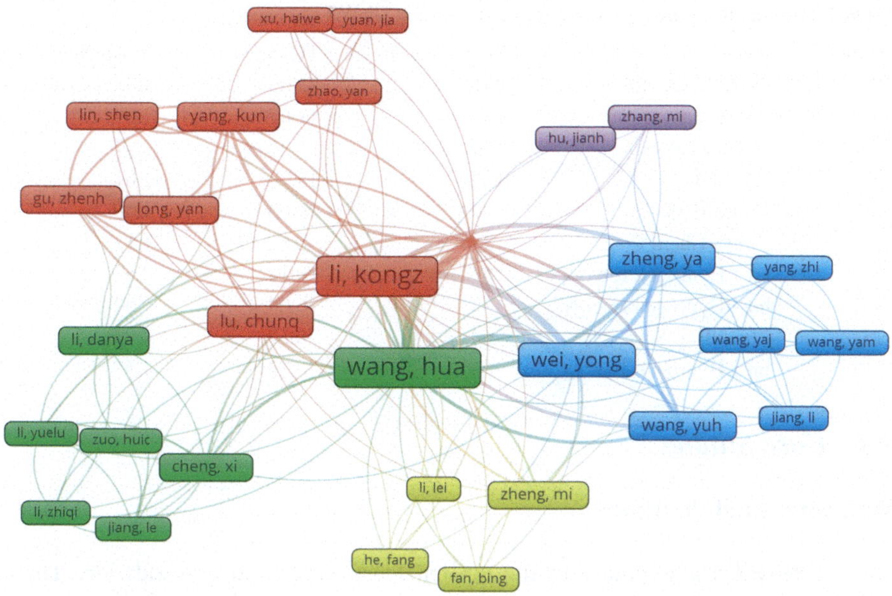

Fig. 4 Co-authorship network

allows for examination of the existence and characteristics of collaboration networks and potential author groups focused on CLWS. The obtained network reveals the presence of 30 authors and 165 links organized into five clusters. Despite their proximity, these clusters reveal multiple research collaborations. For instance, the red cluster comprises Kongzhai Li and Chunqiang Lu [20], in addition to Yanhui Long, Shen Lin, Kun Yang, and Zhenhua Gu [21]. Cluster analysis has identified key focus areas within these research groups, highlighting significant advancements in hydrogen production technologies such as gas reforming. The studies reviewed demonstrate the use of Mg–Al and Ce–Fe–Zr–O-based oxygen carriers, the substitution of Fe in perovskites like $LaMnO_3$ for high-purity hydrogen production, and the enhancement of CeO_2–ZrO_2 solid solutions for methane reforming. Additionally, improvements in magnetite oxygen carriers through ultrasonic treatment and water splitting with ceria-modified iron oxide have been investigated. These findings underscore a clear trend toward optimizing oxygen carriers and exploring innovative methods, indicating that future research could further enhance the efficiency and sustainability of these processes [16–18, 20–25].

Table 1 Top ten most globally cited studies exploring CLWS

Author	Source	Total citations	TC per year	Normalized TC
[26]	Environ Sci Nano	253	23.00	1.96
[29]	Chem Mater	155	11.07	1.00
[30]	J Mater Chem A	149	21.29	1.97
[9]	RSC Adv	142	15.78	2.01
[16]	J Rare Earth	90	6.00	1.00
[27]	J Phys Chem C	78	8.67	1.11
[31]	Int J Hydrogen Energy	68	6.18	0.53
[17]	Int J Hydrogen Energy	67	6.09	0.52
[25]	Chem Eng Sci	60	8.57	0.79
[28]	Int J Hydrogen Energy	50	8.33	1.57

3.4 *Core Studies*

Most Influential Publications

Table 1 provides a summary of the top ten most cited studies worldwide. These studies focus primarily on developing alternatives to optimize H_2 production through CLWS, with special emphasis on researching various materials, alloys, and techniques. Among the recurring themes, the use of cerium compounds stands out, especially their implementation in the form of ceria particles (CeO_2). Their application in CLWS is specifically mentioned in studies on redox kinetics, their combination with iron and zirconium oxides, and exploration of novel technologies such as photolytic and thermolytic water splitting, as well as the simultaneous generation of syngas and H_2 in two-stage chemical looping processes [16, 17, 25–27].

Secondly, various oxygen carrier (OC) materials and their applications are analyzed, including iron oxides and cobalt ferrites synthesized by atomic layer deposition and production performance through controllable redox exsolution within CoFeAlO spinel oxygen carrier materials. Sustainable iron oxide uses are also considered [28–30]. Finally, thirdly, studies are presented that comprehensively evaluate CLWS, including cost-benefit analyses and various techniques for H_2 generation. A bibliographic review of advancements in CLWS covers diverse alloys, kinetic parameters, operating conditions, sizing, among other aspects [9, 31]. Collectively, these studies contribute to refining CLWS, aiming for it to become a more competitive methodology for H_2 production with enhancements in energy efficiency, productivity, and sustainability.

Most Frequent Keywords

Co-occurring keyword analysis is employed to identify the predominant terms in the examined bibliographic records. This process aids in determining the most relevant research categories within the study field. This analytical approach proves valuable as it enables researchers to focus on the most significant terms highlighted in the research findings. Author keywords such as "hydrogen," "chemical looping," "water splitting," "hydrogen production," and "oxygen carrier" stand out as the primary focal points in the context of CLWS, reported with frequencies of 28, 26, 19, 18, and 17 times respectively. On the other hand, the KeyWords Plus section emphasizes terms like "hydrogen production," "chemical looping," "oxygen," "water splitting," and "oxygen carrier," with frequencies of 68, 47, 41, 37, and 28 times, respectively.

Trend Topics

Figure 5 presents the themes identified with KeyWords Plus, considering at least five words per article. This analysis highlights the most relevant themes that have emerged between 2018 and 2023. Among these topics, hydrogen (H_2) production stands out, along with chemical stability, greenhouse gases, chemical looping, water splitting, oxygen carriers, and hydrogen production. It is no surprise that hydrogen production is the central theme of both the article and the CLWS. Similarly, "chemical looping," "water splitting," "oxygen carriers," and "hydrogen production"are essential topics for CLWS development, though they do not refer to specific research areas and can be considered common themes.

Additionally, "chemical stability" emerges as highly relevant. Throughout this chapter, a recurring theme is the search for materials that optimize processes, with particular emphasis on literature highlighting the components of oxygen carriers. Properties such as stability are crucial for efficiency, making this field continuously

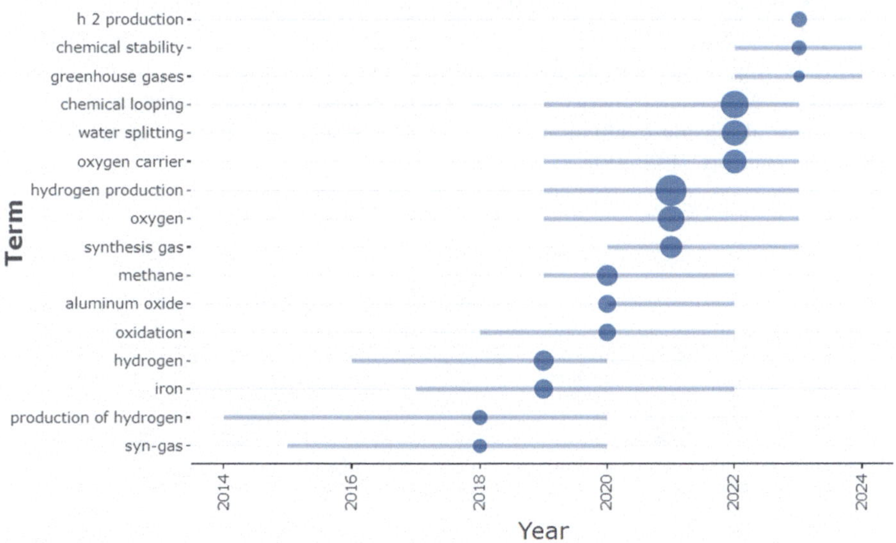

Fig. 5 Trend topics

evolving. Research focuses not only on the most stable compounds but also on their economic competitiveness and sustainability. Other notable topics include "synthesis gas" and "methane." These refer to alternative techniques and products that can integrate into CLWS processes, their frequent appearance indicating their high relevance.

Thematic Evolution

Thematic evolution analysis allows the study of a topic from a historical viewpoint, facilitating the observation of its development over time. This technique highlights the most relevant research themes that have evolved, providing an insight into the future dynamics of the topic [12]. Figure 6 illustrates the evolution of CLWS themes from 2007 to 2024. The analysis period is divided into three phases: 2007–2019, 2020–2022, and 2023–2024. This division clarifies the development of the topic from its inception in 2007 to the present. Initially (2007–2019), the terminology was quite broad, encompassing terms like "carbon monoxide," "carbon dioxide," "iron oxides," "hydrogen production," and "steam reforming." These terms suggest an early stage of knowledge establishment in CLWS, where concepts such as "carbon monoxide and dioxide" highlight environmental benefits like greenhouse gas capture. "Iron oxides" refer to historically significant and widely studied oxygen carriers (OC), and "high resolution transmission" denotes pioneering studies in OC analysis and process efficiency enhancement through material modification [30–32]. During 2020–2022, despite the shorter timeframe, there was a noticeable increase in focus on "hydrogen production"and the emergence of "aluminum oxide" and "redox cycles." This indicates progress in understanding redox processes in CLWS, the use of OC materials, and the application of support materials like aluminum [22, 24, 33, 34]. From 2023 onward, there has been a significant rise in "hydrogen production," reflecting its growing importance due to the energy benefits of hydrogen and the shift toward cleaner energy sources. There is also a modest increase in "steam reforming," alongside evaluations of various OCs such as "iron compounds" and "chemical stability" [35–38].

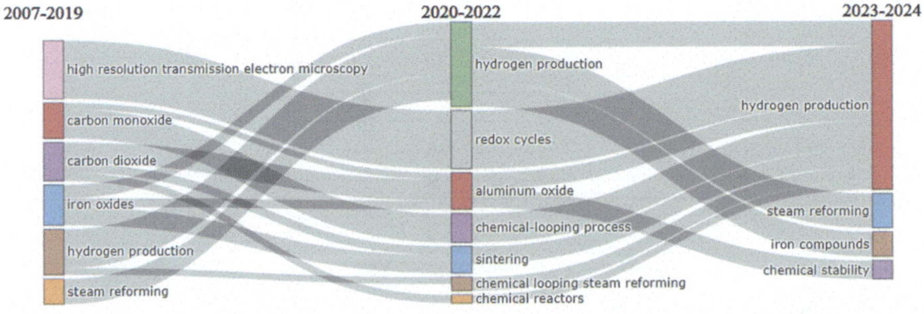

Fig. 6 Thematic evolution

4 Discussion

A scientometric analysis of research on chemical looping water splitting (CLWS) for clean hydrogen production reveals a constantly evolving research landscape. This evolution is evident in the development of research over the years, starting in 2007, where CLWS emerged as a novel alternative for hydrogen generation and greenhouse gas capture. Today, the concept has matured, and the focus has shifted toward optimization through the identification of alternative materials and processes that maximize efficiency. These shifts in the research landscape are driven by the dynamic nature of technological fields, where research in CLWS is subject to changing trends, the development of new technologies, and the emergence of innovative methods. These dynamics can influence the relevance and significance of certain topics, potentially impacting the conclusions drawn from the analysis. However, the growing need for clean energy production presents a promising future for this field. Similar conclusions are obtained from the work of Nandiyanto et al. [39], who performed a bibliometric analysis on iron-based oxygen carriers in hydrogen production, highlighting the value of bibliometric analysis for tracking scientific trends in metal oxide processes for energy storage. Their study offers insights into global research trends and theoretical advancements in iron-based catalysis from 2014 to 2023. Demonstrating tools like VOS viewer facilitate visual representations of network structures and density maps, enhancing the understanding of research patterns and guiding future research directions. As part of their study, the relevance and significant opportunities offered by topics such as "iron" and "hydrogen" up to the present are highlighted. The high degree of interest in developing compounds that promise greater efficiency since 2014 is emphasized, with notable terms including "electron microscopy," "electrocatalyst," and "water splitting." These terms primarily refer to study techniques for material characteristics and the implementation of catalysts to improve processes. The importance of water splitting, in particular, is highlighted, reaffirming its relevance in the field. Furthermore, the study identifies China as the principal source of bibliographic material on this topic, followed by the USA and Korea, supporting our findings and reinforcing the significance of these regions in the development of chemical looping water splitting (CLWS).

5 Conclusion

This research has been valuable for understanding the thematic development, identifying leading authors and countries, and highlighting promising research areas. However, it is important to recognize certain limitations. The analysis relies solely on two databases, which may introduce some bias into the results. Additionally, the study did not account for grey literature, which could contain relevant information that might affect the perspectives obtained. Therefore, future research should consider a broader range of sources to achieve a more comprehensive and balanced view of the field. To expand the scope of this analysis, incorporating additional databases and grey literature can provide a more complete understanding and potentially reveal new insights for future investigations.

Acknowledgments The authors would like to thank Ecopetrol, Ministerio de Ciencia Tecnología e Innovación and Ministerio de Minas y Energía for their support in the development of the project "*Proceso integral para la obtención de hidrógeno de bajas emisiones a partir de biogás producido en plantas de tratamiento de aguas residuales usando transportadores sólidos de oxígeno de bajo costo*" Code 1115-929-93266 through the financing contract 274-2022 derived from the special cooperation agreement 753-2021 and to Universidades G8 Antioquia—Gobernación de Antioquia through the "Segunda convocatoria conjunta de proyectos de I+D+i en el marco de la agenda regional de I+D→i" program.

References

1. Shakoor, A., Ashraf, F., Shakoor, S., Mustafa, A., Rehman, A., Altaf, M.M.: Biogeochemical transformation of greenhouse gas emissions from terrestrial to atmospheric environment and potential feedback to climate forcing. Environ. Sci. Pollut. Res. **27**, 38513–38536 (2020). https://doi.org/10.1007/s11356-020-10151-1
2. Gani, A.: Fossil fuel energy and environmental performance in an extended STIRPAT model. J. Clean. Prod. **297**, 126526 (2021). https://doi.org/10.1016/j.jclepro.2021.126526
3. Gattuso, J.-P., Magnan, A., Billé, R., Cheung, W.W.L., Howes, E.L., Joos, F., et al.: Contrasting futures for ocean and society from different anthropogenic CO_2 emissions scenarios. Science. **349**, aac4722 (2015). https://doi.org/10.1126/science.aac4722
4. Das, A., Peu, S.D.: A comprehensive review on recent advancements in thermochemical processes for clean hydrogen production to decarbonize the energy sector. Sustainability. **14**, 11206 (2022). https://doi.org/10.3390/su141811206
5. Chen, L., Qi, Z., Zhang, S., Su, J., Somorjai, G.A.: Catalytic hydrogen production from methane: a review on recent progress and prospect. Catalysts. **10**, 858 (2020). https://doi.org/10.3390/catal10080858
6. Luo, M., Yi, Y., Wang, S., Wang, Z., Du, M., Pan, J., et al.: Review of hydrogen production using chemical-looping technology. Renew. Sustain. Energy Rev. **81**, 3186–3214 (2018). https://doi.org/10.1016/j.rser.2017.07.007
7. Ju, H., Badwal, S., Giddey, S.: A comprehensive review of carbon and hydrocarbon assisted water electrolysis for hydrogen production. Appl. Energy. **231**, 502–533 (2018). https://doi.org/10.1016/j.apenergy.2018.09.125
8. Zhou, L., Deshpande, K., Zhang, X., Agarwal, R.K.: Process simulation of chemical looping combustion using ASPEN plus for a mixture of biomass and coal with various oxygen carriers. Energy. **195**, 116955 (2020). https://doi.org/10.1016/j.energy.2020.116955
9. Voitic, G., Hacker, V.: Recent advancements in chemical looping water splitting for the production of hydrogen. RSC Adv. **6**, 98267–98296 (2016). https://doi.org/10.1039/C6RA21180A
10. Das, S., Biswas, A., Tiwary, C.S., Paliwal, M.: Hydrogen production using chemical looping technology: a review with emphasis on H2 yield of various oxygen carriers. Int. J. Hydrog. Energy. **47**, 28322–28352 (2022). https://doi.org/10.1016/j.ijhydene.2022.06.170
11. Wang, I., Liu, L., Yu, S., Lai, N.-C., Gao, Y., Li, Z., et al.: Highly sintering-resistant iron oxide with a hetero-oxide shell for chemical looping water splitting. Int. J. Hydrog. Energy. **57**, 438–449 (2024). https://doi.org/10.1016/j.ijhydene.2024.01.051

12. Marín-Rodríguez, N.J., Vega, J., Zanabria, O.B., González-Ruiz, J.D., Botero, S.: Towards an understanding of landslide risk assessment and its economic losses: a scientometric analysis. Landslides. **21**, 1865 (2024). https://doi.org/10.1007/s10346-024-02272-2
13. van Eck, N.J., Waltman, L.: Citation-based clustering of publications using CitNetExplorer and VOSviewer. Scientometrics. **111**, 1053–1070 (2017). https://doi.org/10.1007/s11192-017-2300-7
14. Aria, M., Cuccurullo, C.: *bibliometrix*: an R-tool for comprehensive science mapping analysis. J. Informet. **11**, 959–975 (2017). https://doi.org/10.1016/j.joi.2017.08.007
15. Marín-Rodríguez, N.J., González-Ruiz, J.D., Botero Botero, S.: Dynamic co-movements among oil prices and financial assets: a scientometric analysis. Sustainability. **14**, 12796 (2022). https://doi.org/10.3390/su141912796
16. Zhu, X., Wang, H., Wei, Y., Li, K., Cheng, X.: J. Rare Earth. **28**, 907–913 (2010). https://doi.org/10.1016/S1002-0721(09)60225-8
17. Zheng, Y., Wei, Y., Li, K., Zhu, X., Wang, H., Wang, Y.: Chemical-looping steam methane reforming over macroporous CeO2-ZrO2 solid solution: effect of calcination temperature. Int. J. Hydrog. Energy. **39**, 13361–13368 (2014). https://doi.org/10.1016/j.ijhydene.2014.04.116
18. Zheng, Y., Li, K., Wang, H., Zhu, X., Wei, Y., Zheng, M., et al.: Enhanced activity of CeO2-ZrO2 solid solutions for chemical-looping reforming of methane via tuning the macroporous structure. Energy Fuel. **30**, 638–647 (2016). https://doi.org/10.1021/acs.energyfuels.5b02151
19. Lin, B., Su, T.: Mapping the oil price-stock market nexus researches: a scientometric review. Int. Rev. Econ. Finance. **67**, 133–147 (2020). https://doi.org/10.1016/j.iref.2020.01.007
20. Zuo, H., Lu, C., Jiang, L., Cheng, X., Li, Z., Li, Y., et al.: Chem. Eng. J. **477** (2023). https://doi.org/10.1016/j.cej.2023.146870
21. Long, Y., Yang, K., Gu, Z., Lin, S., Li, D., Zhu, X., et al.: Hydrogen generation from water splitting over polyfunctional perovskite oxygen carriers by using coke oven gas as reducing agent. Appl. Catal. B Environ. **301**, 301 (2022). https://doi.org/10.1016/j.apcatb.2021.120778
22. Yuan, J., Zhao, Y., Xu, H., Lu, C., Yang, K., Zhu, X., et al.: Layered Mg-Al spinel supported Ce-Fe-Zr-O oxygen carriers for chemical looping reforming. Chin. J. Chem. Eng. **28**, 2668–2676 (2020). https://doi.org/10.1016/j.cjche.2020.06.035
23. Wang, Y., Zheng, Y., Wang, Y., Wang, H., Zhu, X., Wei, Y., et al.: Evaluation of Fe substitution in perovskite LaMnO3 for the production of high purity syngas and hydrogen. J. Power Sources. **449**, 227505 (2020). https://doi.org/10.1016/j.jpowsour.2019.227505
24. Lu, C., Li, K., Zhu, X., Wei, Y., Li, L., Zheng, M., et al.: Improved activity of magnetite oxygen carrier for chemical looping steam reforming by ultrasonic treatment. Appl. Energy. **261**, 261 (2020). https://doi.org/10.1016/j.apenergy.2019.114437
25. Zhu, X., Zhang, M., Li, K., Wei, Y., Zheng, Y., Hu, J., et al.: Chem. Eng. Sci. **179**, 92–103 (2018). https://doi.org/10.1016/j.ces.2018.01.015
26. Reed, K., Cormack, A., Kulkarni, A., Mayton, M., Sayle, D., Klaessig, F., et al.: Exploring the properties and applications of nanoceria: is there still plenty of room at the bottom? Environ. Sci. Nano. **1**, 390–405 (2014). https://doi.org/10.1039/c4en00079j
27. Zhao, Z., Uddi, M., Tsvetkov, N., Yildiz, B., Ghoniem, A.F.: Redox kinetics study of fuel reduced ceria for chemical-looping water splitting. J. Phys. Chem. C. **120**, 16271–16289 (2016). https://doi.org/10.1021/acs.jpcc.6b01847
28. De Vos, Y., Jacobs, M., Van Der Voort, P., Van Driessche, I., Snijkers, F., Verberckmoes, A.: Sustainable iron-based oxygen carriers for chemical looping for hydrogen generation. Int. J. Hydrog. Energy. **44**, 1374–1391 (2019). https://doi.org/10.1016/j.ijhydene.2018.11.099
29. Scheffe, J.R., Allendorf, M.D., Coker, E.N., Jacobs, B.W., McDaniel, A.H., Weimer, A.W.: Hydrogen production via chemical looping redox cycles using atomic layer deposition-synthesized iron oxide and cobalt ferrites. Chem. Mater. **23**, 2030–2038 (2011). https://doi.org/10.1021/cm103622e

30. Zeng, D., Qiu, Y., Peng, S., Chen, C., Zeng, J., Zhang, S., et al.: Enhanced hydrogen production performance through controllable redox exsolution within CoFeAlO: X spinel oxygen carrier materials. J. Mater. Chem. A. **6**, 11306–11316 (2018). https://doi.org/10.1039/c8ta02477d
31. Thengane, S.K., Hoadley, A., Bhattacharya, S., Mitra, S., Bandyopadhyay, S.: Cost-benefit analysis of different hydrogen production technologies using AHP and fuzzy AHP. Int. J. Hydrog. Energy. **39**, 15293–15306 (2014). https://doi.org/10.1016/j.ijhydene.2014.07.107
32. Bahzad, H., Shah, N., Dowell, N.M., Boot-Handford, M., Soltani, S.M., Ho, M., et al.: Development and techno-economic analyses of a novel hydrogen production process via chemical looping. Int. J. Hydrog. Energy. **44**, 21251–21263 (2019). https://doi.org/10.1016/j.ijhydene.2019.05.202
33. Wang, X., Fu, G., Xiao, B., Xu, T.: Optimization of nickel-iron bimetallic oxides for coproduction of hydrogen and syngas in chemical looping reforming with water splitting process. Energy. **246**, 246 (2022). https://doi.org/10.1016/j.energy.2022.123410
34. Damizia, M., de Caprariis, B., Bracciale, M.P., Anania, F., D'Alvia, L., Del Prete, Z., et al.: Utilization of Al2O3 and MgO as structural promoters of Fe into 2 and 3 steps chemical looping hydrogen process: pure and green H2 production. In: Dincer, I., Colpan, C.C., Ezan, M.A. (eds.) Proc. WHEC—World Hydrogen Energy Conference: Bridging Continents by H2, pp. 74–76. International Association for Hydrogen Energy, IAHE (2022)
35. Ji, J., Shen, L.: Enhanced morphological maintenance and redox stability by dispersing nickel ferrite into silica matrix for chemical looping hydrogen production via water splitting. Fuel Process. Technol. **251**, 107946 (2023). https://doi.org/10.1016/j.fuproc.2023.107946
36. Cai, Y., Wang, C., Zhang, Z., Zhong, M., Wu, Q., Xiao, B., et al.: Performance optimization of Ca2Fe2O5 oxygen carrier by doping different metals for coproduction syngas and hydrogen with chemical looping gasification and water splitting. J. Energy Inst. **111**, 101391 (2023). https://doi.org/10.1016/j.joei.2023.101391
37. He, J., Yang, Q., Song, Z., Chang, W., Huang, C., Zhu, Y., et al.: Improving the carbon resistance of iron-based oxygen carrier for hydrogen production via chemical looping steam methane reforming: a review. Fuel. **351** (2023). https://doi.org/10.1016/j.fuel.2023.128864
38. Liu, M., Wu, H., Wang, H., Chen, T., Wang, Z., Zhang, J., et al.: Enhancing redox stability through metal substitution in nickel ferrite for chemical looping hydrogen production via water splitting. Int. J. Hydrog. Energy. **73**, 221–230 (2024). https://doi.org/10.1016/j.ijhydene.2024.06.017
39. Nandiyanto, A.B.D., Ragadhita, R., Fiandini, M., Husaeni, D.N.A., Aziz, M.: The role of iron oxide in hydrogen production: theory and bibliometric analyses. Moroccan J. Chem. **11**(4), 897–916 (2023). https://doi.org/10.48317/IMIST.PRSM/morjchem-v11i04.41591

Foreign Direct Investment and Renewable Energy Consumption: Evidence from Latin American Countries

Segundo F. Vilema-Escudero, José Pérez-Villamar, Jaime M. Samaniego-López, Juan C. Lata-García, and Miguel Ruiz-Martínez

1 Introduction

1.1 *Renewable Energy Consumption*

Global renewable energy consumption has experienced significant growth over the past 20 years, driven primarily by the need to mitigate climate change and reduce greenhouse gas emissions [1]. Renewable sources, including solar, wind, hydroelectric, and biomass energy, have increased their share in the global energy matrix, evolving from a marginal alternative to becoming an essential component of the energy supply [2]. This increase has been reflected in the exponential growth of global investment in renewable energy technologies over the last decade [3]. Despite these advances, the energy transition toward renewables faces significant challenges, including the need for efficient storage and integration into existing power grids [4]. It is projected that renewable energies could meet up to two-thirds of global energy demand by 2050, provided that the pace of growth and technological innovation remains accelerated [5].

S. F. Vilema-Escudero (✉)
Universidad ECOTEC, Samborondón, Guayas, Ecuador
e-mail: svilema@ecotec.edu.ec

J. Pérez-Villamar · J. M. Samaniego-López
Universidad Católica de Santiago de Guayaquil, Guayaquil, Guayas, Ecuador

J. C. Lata-García
Universidad Politécnica Salesiana, GIPI Research Group, Guayaquil, Guayas, Ecuador

M. Ruiz-Martínez
Escuela Superior Politécnica del Litoral—ESPOL, Guayaquil, Guayas, Ecuador

M. Espinoza-Andaluz et al. (eds.), *Congress on Research, Development and Innovation in Renewable Energies*, Green Energy and Technology,
https://doi.org/10.1007/978-3-031-88995-0_11

In Latin America, renewable energy consumption has been driven by the abundance of natural resources and the growing need to diversify the energy matrix. Policies implemented by countries such as Brazil, Chile, and Mexico have facilitated investment in wind, solar, and hydroelectric energy projects [6–8]. Through FDI, the transfer of clean technologies and improvements in energy infrastructure have been promoted, significantly contributing to the increase in renewable energy consumption in the region [9]. However, economic stability and regulatory frameworks play a crucial role in attracting sustainable investments [10]. Despite the growth, the share of renewable energy in total consumption varies considerably among countries in the region, indicating the need for more comprehensive policies and greater international cooperation to maximize the potential of available renewable resources [11].

Renewable energy consumption has been linked to multiple social and environmental benefits [12]. The adoption of clean energy sources has been observed to significantly reduce greenhouse gas emissions, mitigating climate change and improving air quality [13]. It has also been demonstrated that renewable energy generation promotes sustainable economic development by creating local jobs in rural communities, thereby reducing poverty and improving social welfare [14]. Additionally, renewable energies, being more accessible and less volatile in terms of costs, contribute to energy security by reducing dependence on fossil fuels and strengthening economic stability [15]. However, challenges have been identified, such as the need for adequate infrastructure and the management of technological waste associated with these technologies [16].

1.2 Renewable Energy Policy

Global energy policy has increasingly shifted toward the promotion of renewable energy sources due to growing concerns about climate change and sustainability [17]. In the European Union, directives have been implemented to increase the proportion of renewables in final energy consumption, with binding targets for member states [18]. Germany has led this transition through the "Energiewende" program, which has promoted wind and solar generation, resulting in a significant reduction in CO_2 emissions [19]. In the United States, energy policy has varied by administration, though progress has been seen in the adoption of renewables at the state level, particularly in California, where 60% of electricity generation now comes from clean sources [20]. In Asia, China has emerged as the world's largest producer of solar energy, driven by government policies that incentivize investment in renewable technology [21].

In Latin America, various energy policies have been implemented to promote the transition to renewable energy sources [7]. In Brazil, the National Biofuels Policy has been crucial for the development of bioethanol, positioning the country as a world leader in biofuel production [22]. Chile has established a robust regulatory framework, including Law 20.257, which mandates a minimum share of non-conventional renewable energy in the energy matrix [23]. Mexico, with its

2013 Energy Reform, has incentivized FDI in solar and wind energy projects, consolidating the country as a key player in the region [24]. In contrast, countries like Venezuela have shown limited progress in adopting renewable energy, due to their reliance on oil revenues [8]. These policies reflect a trend toward energy diversification and reduced dependence on fossil fuels, though with significant variations in their implementation and effectiveness across the region [25].

1.3 FDI and Renewable Energy Consumption

FDI has played a crucial role in the expansion of renewable energy consumption globally [26]. In countries such as China and Germany, FDI has facilitated the transfer of clean technologies, significantly contributing to the growth of installed renewable energy capacity [27]. In China, FDI has acted as a catalyst for the development of large-scale solar and wind projects, enabling the country to become a global leader in renewable energy [28]. In Germany, FDI has enhanced the integration of advanced technologies into the electrical grid, improving the efficiency and reliability of energy supply [29]. FDI has also been pivotal in India, where foreign investment has driven the rapid expansion of wind and solar infrastructure, helping the country achieve its sustainable energy goals [30]. In sub-Saharan Africa, although to a lesser extent, FDI has begun to support renewable energy projects, contributing to the diversification of the energy matrix in countries like South Africa [9].

In Latin America, FDI has played a crucial role in the development of the renewable energy sector [10]. This region, characterized by its abundance of natural resources, has seen a significant increase in the adoption of clean technologies driven by foreign capital inflows [31]. Countries like Brazil, Chile, and Mexico have led this process by implementing favorable policies that have facilitated the attraction of FDI into wind, solar, and hydroelectric energy projects [7]. In Brazil, foreign investment has been fundamental for the expansion of installed wind energy capacity, while in Chile, regulatory stability has incentivized foreign participation in solar energy development [32]. Mexico, on the other hand, has received significant investments in hydroelectric projects, contributing to the diversification of its energy matrix [33]. FDI has not only enabled the transfer of clean technologies but has also improved energy infrastructure, fostering more sustainable growth in the region [12].

Research on FDI and renewable energy consumption in Latin America has been limited, leaving a significant gap in the economic and energy literature [7, 22]. Although studies have explored how FDI drives economic growth and environmental sustainability, few have examined its specific impact on the adoption of renewable energies in the region [10]. Given the growing concerns about climate change and the transition to low-carbon economies, it is crucial to understand the factors driving renewable energy consumption in Latin American countries [34]. Future prospects indicate significant potential for the region to become a leader in renewable energy, but deeper analysis is needed to guide the formulation of effective policies. Ultimately, the objective of this study is to analyze the impact of FDI on renewable energy consumption in Latin American countries. This study is organized into Sects. 3–6.

2 Hypotheses and Theoretical Model

Based on a review of existing studies, the effect of foreign direct investment (FDI) on renewable energy consumption is identified, along with the moderating effects of economic growth and CO_2 emissions.

2.1 The Effect of FDI on Renewable Energy Consumption

Previous studies suggest that foreign direct investment (FDI) can play a crucial role in promoting the use of renewable energy through the transfer of technology, knowledge, and sustainable practices [9, 35]. In developing countries, FDI can provide the necessary financial resources for the implementation of renewable energy projects, which often require significant investments that may not be locally available [36]. In Latin America, where energy demand is constantly growing, the need to diversify energy sources is critical [31, 33]. Specific research has found a positive correlation between FDI and the development of green infrastructure in several countries in the region, suggesting that FDI not only drives economic growth but also facilitates the transition to cleaner energy sources [11, 37, 38]. It is emphasized that FDI can be an important driver of the energy transition in Latin America, facilitating investment in clean and efficient technologies [33]. A significant positive relationship between FDI and renewable energy consumption in the region has been found [25]. The bidirectional relationship between FDI and renewable energy consumption suggests that foreign investment promotes the use of clean energy and is, in turn, favored by it [6]. It is concluded that FDI is positively linked to the demand for renewable energy, highlighting its impact on economic growth and clean energy consumption [23]. In accordance with the analysis, this study proposes Hypothesis 1. *H1*: FDI increases renewable energy consumption in Latin American countries.

2.2 Moderating Effect of Economic Growth

In developing economies, such as those in Latin America, foreign direct investment (FDI) can increase renewable energy consumption, conditional on the level of economic development [2, 25]. It is suggested that FDI may stimulate economic growth by facilitating the adoption of clean technologies and the implementation of renewable energy projects [11]. In more developed economies, existing infrastructure and policies can maximize the positive impact of FDI in the renewable energy sector [30]. FDI has a significant positive effect on energy consumption, which is amplified in contexts of rapid economic growth [39]. However, the relationship between FDI and economic growth may have mixed effects on renewable energy consumption [23]. In countries with strong economic growth, FDI promotes greater adoption of renewable energy technologies due to increased capacity for investment in clean and sustainable infrastructure [40]. The study suggests that economic growth may moderate the impact of FDI by providing the necessary resources to implement environmental policies and green technologies [41]. In accordance with the analysis, this study proposes Hypothesis 2. *H2*: Economic growth moderates the effect of FDI on renewable energy consumption in Latin American countries.

2.3 Moderating Effect of CO_2 Emissions

Previous research has shown that FDI can have a positive effect on the adoption of renewable energy, particularly when environmental policies are stringent [42]. CO_2 emissions reflect the level of industrialization and fossil fuel usage in a country [9]. Studies suggest that high levels of emissions may drive the adoption of clean technologies in response to environmental pressure and international regulations [43]. In this context, FDI facilitates the transfer of technology and capital necessary for the transition to renewable energies [44]. Therefore, CO_2 emissions not only indicate the need for change but can also enhance the impact of FDI on renewable energy consumption by acting as a catalyst for sustainable policies and projects [39]. The relationship between renewable energy consumption, FDI flows, and economic growth is examined, highlighting that while economic growth and FDI increase emissions, renewable energy consumption reduces them [42]. It is also found that CO_2 emissions are positively related to economic growth and FDI, suggesting the need for policies that mitigate emissions through sustainable energy [23]. In conclusion, FDI can reduce CO_2 emissions in developing economies if the use of renewable energies is promoted [45]. In accordance with the analysis, this study proposes Hypothesis 3. *H3*: CO_2 emissions moderate the impact of FDI on renewable energy consumption in Latin American countries.

Based on the information gathered from the literature review, a theoretical model is proposed, as illustrated in Fig. 1.

3 Methodology

3.1 Data

Data from 13 Latin American countries were used, covering the period from 2000 to 2021. A balanced panel was constructed to analyze the effect of FDI on

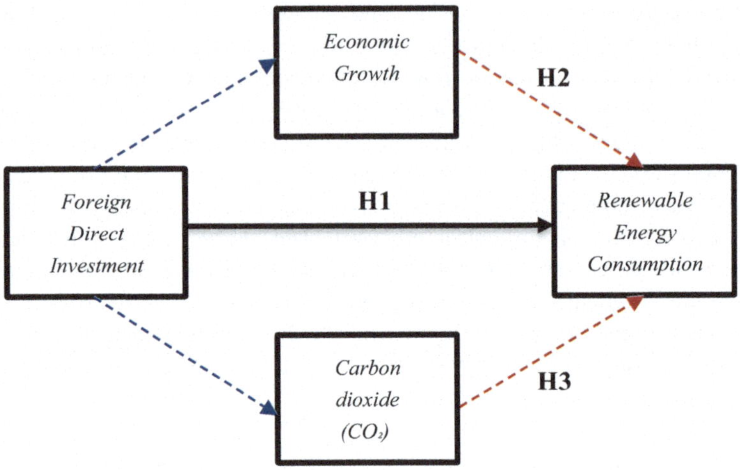

Fig. 1 Theoretical model

renewable energy consumption. The countries selected in the sample include Chile, Costa Rica, Colombia, Mexico, Argentina, Brazil, Peru, Panama, Venezuela, Ecuador, Bolivia, Paraguay, and Uruguay. The data were obtained from the CEPALSTAT statistical platform, developed by the Economic Commission for Latin America and the Caribbean (ECLAC) of the United Nations (UN) [46], which compiles information from the main official sources in Latin American countries.

3.2 Variables

Dependent Variable

The dependent variable, renewable energy consumption, is widely used in studies on energy sustainability due to its ability to capture the degree of penetration of clean energy in a country's energy matrix [47, 48]. By expressing renewable energy consumption as a percentage of total final energy consumption, it allows for more meaningful comparisons between countries and over time, regardless of the absolute size of their energy consumption [18]. This approach has been adopted in previous studies to evaluate the effectiveness of energy policies and the adoption of clean technologies in different national and regional contexts [49]. Furthermore, the proportion of renewable energy in total consumption is considered a key indicator of progress toward a more sustainable energy model, as highlighted by recent research emphasizing the importance of this indicator in energy policy formulation [3, 16].

Independent Variable

Foreign direct investment (FDI) influences economic development and the transition to renewable energies in developing countries. FDI, measured by net foreign direct investment, has been widely recognized in the literature as a key catalyst for the transfer of clean technologies and the improvement of energy infrastructure [2, 7, 10]. In the context of Latin America, FDI facilitates the diversification of the energy matrix, reduces dependence on fossil fuels, and promotes sustainable economic growth [25]. Moreover, recent studies have demonstrated that FDI directly influences the adoption of green technologies, which is crucial for achieving carbon emission reduction goals [28, 45].

Moderating Variables

The moderating variable *Economic Growth* is measured by each country's GDP per capita. GDP per capita serves as a key indicator of a country's level of economic development, influencing both the capacity for clean technology adoption and the existing energy infrastructure [50]. Previous studies have shown that higher GDP per capita can facilitate the implementation of sustainable energy policies and improve resource efficiency [43]. Additionally, in more developed economies, FDI tends to have a more significant impact on the transition to renewable energies due to the availability of advanced infrastructure and a stable political environment [30]. On the other hand, the moderating variable *CO_2 Emissions* reflects the level of industrialization and fossil fuel use, making it a key indicator of the environmental impact of economic activities [41]. Furthermore, recent studies have noted that high emissions can act as a catalyst for the adoption of clean technologies in response to environmental pressure and international regulations [51].

Control Variables

The control variable *Energy Intensity* is included as a key control, influencing the relationship between FDI and energy consumption by indicating the efficiency with which investment is transformed into economic growth and energy sustainability [52]. Additionally, previous research has demonstrated that energy intensity is a significant determinant of energy demand, directly affecting an economy's ability to integrate renewable energies into its energy matrix [53]. Therefore, the inclusion of this variable allows for a more robust and precise analysis of the impact of FDI on renewable energy consumption, controlling for variations in energy efficiency across different countries and periods [54].

The control variable *Sustainable Innovation* is measured by the adoption of ISO 14001 certification. Sustainable innovation has been identified as a critical factor for the adoption of renewable technologies and the improvement of energy efficiency across multiple national and sectoral contexts [6, 15]. It has been demonstrated that innovation in clean technologies facilitates the reduction of CO_2 emissions and promotes more efficient use of energy resources [28, 55]. Moreover, ISO 14001 certification serves as a relevant proxy for measuring sustainable innovation at the corporate level [56]. Recent studies have found that ISO 14001 certification is positively correlated with the adoption of renewable energies and the reduction of CO_2 emissions [57].

The control variable *Trade Openness* directly affects the transfer of clean technologies and sustainable practices, influencing the demand for and adoption of renewable energies [9]. It has been shown that more open economies have greater access to advanced technologies, facilitating the transition to more sustainable energy sources [27]. Additionally, trade openness promotes competition, which incentivizes energy efficiency and technological innovation [58]. Failing to consider trade openness could bias estimators due to the omission of a key determinant of energy development in Latin American countries [44].

Finally, the control variable *Industrial Development* influences the energy structure of an economy, as energy-intensive industrial sectors tend to increase total energy demand, including renewable sources. Moreover, industrial expansion is often linked to an increase in CO_2 emissions, which could condition the transition to cleaner energies [59]. On the other hand, robust industrial growth can increase investments in renewable energy technologies, particularly in countries with strict environmental policies [60].

To improve variance stabilization and mitigate heteroscedasticity in the data, a transformation of some variables using the natural logarithm was applied. This transformation facilitates clearer interpretations and proportional comparisons between economic variables, thereby increasing the robustness of the econometric model [61]. Detailed information on the study variables is presented in Table 1.

3.3 Panel Regression Model

To improve variance stabilization and mitigate heteroscedasticity in the data, a transformation of the variables was applied using the natural logarithm. This transformation facilitates clearer interpretations and proportional comparisons between economic variables, thereby increasing the robustness of the econometric model [61].

Three baseline linear regression models are used to test the study hypotheses. To investigate the effect of foreign direct investment on renewable energy consumption, the following reference model is proposed:

$$\begin{aligned} \mathrm{ERC}_{it} = {} & \beta_0 + \beta_1 \mathrm{Ln}(\mathrm{FDI}_{it}) + \beta_2 \mathrm{EG}_{it} + \beta_3 \mathrm{Ln}(\mathrm{CO}_{2it}) + \beta_4 \mathrm{Ln}(\mathrm{EI}_{it}) + \beta_5 \mathrm{SI}_{it} \\ & + \beta_6 \mathrm{Ln}(\mathrm{TO}_{it}) + \beta_7 \mathrm{ID}_{it} + \varepsilon_{it} \end{aligned} \tag{1}$$

where ERC_{it} is the dependent variable representing renewable energy consumption, with the subscript *it* indicates country *i* at time *t*. $\mathrm{Ln}(\mathrm{FDI}_{it})$ is the independent

Table 1 Variable definitions and explanations

Variable	Code	Description	Measurement	Source
Dependent	ERC	*Renewable Energy Consumption* measured by the proportion of renewable energy in total final energy consumption	Percentage	ECLAC
Independent	LnFDI	*Net Foreign Direct Investment* in the reporting economy	Ln (millions of dollars)	ECLAC
Moderator	EG	*Economic Growth* measured by the gross domestic product per inhabitant generated during a year by a country at market prices for each year	Dollars per inhabitant at current prices	ECLAC
	$LnCO_2$	*Carbon dioxide (CO_2) emissions*, total excluding land use change and deforestation	Ln (millions of tons of carbon dioxide equivalent ($MtCO_2e$))	ECLAC
Control	LnEI	*Energy Intensity* of the gross domestic product (GDP) (final energy consumption/ GDP at constant prices in 2018 dollars)	Ln (primary energy supply (in thousands of barrels of oil equivalent) per million dollars of GDP (at constant 2018 prices))	ECLAC
	SI	*Sustainable Innovation* measured by the number of companies certified with ISO 14001 in relation to the gross domestic product (GDP) in constant 2010 dollars	Number of companies per $1 billion of GDP (at constant 2010 prices)	ECLAC
	LnTO	*Trade Opening* measured by the unit value index for exports	Ln (index 2018 = 100)	ECLAC
	ID	*Industrial Development* measured by the growth rate of the gross domestic product of the industrial sector at constant prices	Percentage	ECLAC

variable representing foreign direct investment. EG_{it} represents economic growth. $Ln(CO_{2it})$ represents CO_2 emissions. $Ln(EI_{it})$ represents energy intensity. SI_{it} represents sustainable innovation. $Ln(TO_{it})$ represents trade opening. ID_{it} represents industrial development. β_0, β_1, ..., β_7 are the regression coefficients that represent the relationship between the variables, and ε_{it} is the error term, which captures the variation not explained by the independent variables and time.

To further investigate the effect of foreign direct investment on renewable energy consumption, the baseline model is modified by including interaction effects of economic growth and CO_2 emissions. The modified models to be estimated are as follows:

$$\mathrm{ERC}_{it} = \beta_0 + \beta_1 \mathrm{Ln}(\mathrm{FDI}_{it}) + \beta_2 \mathrm{EG}_{it} + \beta_3 \mathrm{Ln}(\mathrm{CO}_{2it}) + \beta_4 \mathrm{Ln}(\mathrm{EI}_{it}) + \beta_5 \mathrm{SI}_{it} + \beta_6 \mathrm{Ln}(\mathrm{TO}_{it}) + \beta_7 \mathrm{ID}_{it} + \beta_8 \mathrm{Ln}(\mathrm{FDI}_{it}) * \mathrm{EG}_{it} + \varepsilon_{it} \quad (2)$$

$$\mathrm{ERC}_{it} = \beta_0 + \beta_1 \mathrm{Ln}(\mathrm{FDI}_{it}) + \beta_2 \mathrm{EG}_{it} + \beta_3 \mathrm{Ln}(\mathrm{CO}_{2it}) + \beta_4 \mathrm{Ln}(\mathrm{EI}_{it}) + \beta_5 \mathrm{SI}_{it} + \beta_6 \mathrm{Ln}(\mathrm{TO}_{it}) + \beta_7 \mathrm{ID}_{it} + \beta_8 \mathrm{Ln}(\mathrm{FDI}_{it}) * \mathrm{Ln}(\mathrm{CO}_{2it}) + \varepsilon_{it} \quad (3)$$

The effect of foreign direct investment on renewable energy consumption conditioned on the moderation of economic growth and CO_2 emissions is represented by the following equations:

$$\frac{\partial \mathrm{ERC}_{it}}{\partial \mathrm{Ln}(\mathrm{FDI}_{it})} = \beta_1 + \beta_8 * \mathrm{EG}_{it} \quad (4)$$

$$\frac{\partial \mathrm{ERC}_{it}}{\partial \mathrm{Ln}(\mathrm{FDI}_{it})} = \beta_1 + \beta_8 * \mathrm{Ln}(\mathrm{CO}_{2it}) \quad (5)$$

The additional effect of economic growth and CO_2 emissions on the effect of FDI on renewable energy consumption is assessed in the measure resulting from the coefficients β_1 and β_8.

4 Results

The methodological approach used to perform the analysis of results and hypotheses is shown in Fig. 2.

Table 2 presents a descriptive analysis of the main variables used in the study. Renewable energy consumption (ERC) shows a mean of 29.51%, with a significant

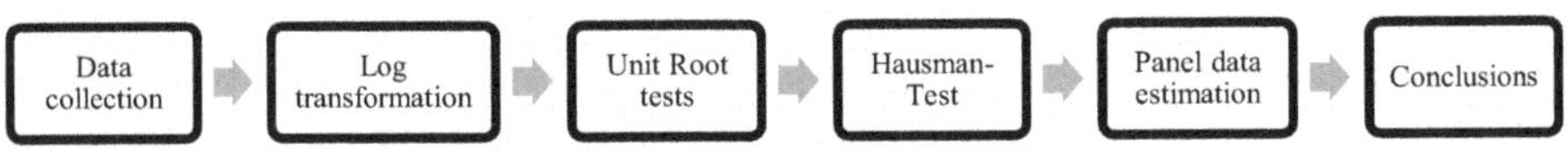

Fig. 2 Methodological approach

Table 2 Descriptive analysis

Variable	Obs.	Mean	Std. dev.	Min	Max
ERC	286	29.51	16.25	7.65	70.41
LnFDI	286	7.44	2.43	0.00	11.41
EG	286	7425.23	4324.59	888.21	19,100.10
$LnCO_2$	286	3.68	1.52	1.26	6.24
LnEI	286	−0.24	0.34	−0.97	0.63
SI	286	1.67	1.68	0.02	9.81
LnTO	286	4.24	0.59	2.54	5.67
ID	286	1.36	8.65	−54.00	21.44

standard deviation of 16.25%, indicating considerable variability in the adoption of clean energy among the countries in the sample. Foreign direct investment (LnFDI) exhibits a wide range, with a minimum value of 0 and a maximum of 11.41, suggesting disparities in the attraction of foreign capital across the region. Economic growth (EG) has a mean of $7425 per capita, with a standard deviation that denotes economic heterogeneity among the countries. CO_2 emissions ($LnCO_2$) also vary widely, reflecting different levels of industrialization and fossil fuel use. Energy intensity (LnEI) and sustainable innovation (SI) show moderate values, but the high dispersion of trade openness (LnTO) suggests an uneven impact on the transfer of clean technologies.

Table 3 presents the correlation results between the key variables of the study: renewable energy consumption (ERC), foreign direct investment (LnFDI), economic growth (EG), CO_2 emissions ($LnCO_2$), energy intensity (LnEI), sustainable innovation (SI), trade openness (LnTO), and industrial development (ID). A significant negative correlation is observed between ERC and LnFDI (-0.139, $p < 0.05$), suggesting that an increase in foreign direct investment may not be directly related to an increase in renewable energy consumption in the countries analyzed. Additionally, a significant negative correlation between ERC and $LnCO_2$ (-0.500, $p < 0.01$) is highlighted, indicating that higher renewable energy consumption is associated with lower carbon emissions. The positive relationship between LnFDI and $LnCO_2$ (0.460, $p < 0.01$) reveals a potentially adverse impact of foreign direct investment on CO_2 emissions, underscoring the need for policies that direct these capital flows toward sustainable projects.

Table 4 presents the results of the unit root tests for the variables included in the analysis. Three methods were employed: Levin-Lin-Chu, Fisher Augmented Dickey Fuller (ADF), and Fisher Phillips Perron (PP). The results indicate that most variables meet the condition of stationarity, which is crucial to ensure the validity of the panel models used. Specifically, foreign direct investment (LnFDI) shows statistical significance across all three tests ($p < 0.01$), suggesting that the variable does not have a unit root and is stationary. On the other hand, CO_2 emissions ($LnCO_2$) and energy intensity (LnEI) exhibit stationarity in the Levin-Lin-Chu test, but not in the Fisher ADF and PP tests, implying potential weaknesses in the robustness of the results for these variables.

Table 5 presents the results of the Hausman test, used to determine the suitability between fixed effects and random effects models in panel analysis. The results indicate a preference for the fixed effects model, as evidenced by a significant chi-square value ($chi^2(6) = 19.34$, $p < 0.05$). This result suggests that unobserved differences between countries significantly affect the relationship between foreign direct investment (FDI) and renewable energy consumption. Specifically, FDI showed a considerable difference between the fixed effects and random effects models (-0.009, $p < 0.05$), implying that unobserved country-specific characteristics influence the impact of FDI on renewable energy consumption.

Figure 3 provides a graphical interpretation of the behavior of the ERC and LnFDI variables over time. Similarly, Fig. 4 graphically illustrates the significance of ERC and LnFDI across Latin American countries.

Table 3 Correlations between variables

Variables	ERC	LnFDI	EG	$LnCO_2$	LnEI	SI	LnTO	ID
ERC	1.000							
LnFDI	−0.139**	1.000						
EG	−0.005	0.301***	1.000					
$LnCO_2$	−0.500***	0.460***	0.153***	1.000				
LnEI	−0.120**	−0.286***	−0.515***	0.300***	1.000			
SI	0.019	0.211***	0.353***	0.057	−0.285***	1.000		
LnTO	−0.217***	0.119**	0.549***	0.251***	0.050	0.345***	1.000	
ID	0.044	0.112*	0.018	−0.113*	−0.241***	0.075	−0.056	1.000

*** $p < 0.01$, ** $p < 0.05$, * $p < 0.1$

Table 4 Results of the panel unit root test

Variable	Levin-Lin-Chu test	Fisher ADF test	Fisher PP test
ERC	−2.142 (0.016)	−0.096 (0.461)	0.621 (0.732)
LnFDI	−3.107 (0.000)	−4.483 (0.000)	−8.183 (0.000)
EG	−1.658 (0.048)	1.444 (0.925)	2.005 (0.977)
$LnCO_2$	−2.635 (0.004)	0.308 (0.621)	0.935 (0.825)
LnEI	−1.972 (0.024)	0.211 (0.583)	1.034 (0.849)
SI	−1.989 (0.023)	0.297 (0.617)	0.128 (0.551)
LnTO	−3.671 (0.000)	−0.614 (0.269)	1.061 (0.855)
ID	−1.438 (0.075)	−5.750 (0.000)	−10.292 (0.000)

Note: p values are shown in parentheses

Table 5 Hausman test

Variable	Fixed	Random	Difference	Std. err.
LnFDI	−0.002	0.007	−0.009	0.002
EG	0.000	0.000	0.000	0.000
$LnCO_2$	−20.350	−18.891	−1.459	0.350
LnEI	15.860	16.196	−0.335	0.191
SI	1.072	1.069	0.002	0.005
LnTO	0.201	−0.199	0.400	0.094
ID	−0.011	−0.013	0.001	0.000
Test	$chi^2(6) = 19.34$			
	$Prob > chi^2 = 0.0036$			

Table 6 presents two regression models that explore the relationship between foreign direct investment (FDI) and renewable energy consumption, controlling for various macroeconomic variables. In the first model, without additional controls, a negative and significant coefficient (−0.297) is observed for the LnFDI variable (foreign direct investment in logarithmic terms), suggesting that, in the absence of controls, FDI might be associated with a reduction in renewable energy consumption. This result, which may initially seem contradictory, could reflect the complexity of the relationship between FDI and the development of renewable technologies in the region. It is possible that FDI in sectors unrelated to clean energy, such as natural resource extraction, predominates in some countries, negatively influencing the proportion of renewable energy in the energy matrix.

However, the second model, which includes a series of control variables (economic growth, CO_2 emissions, energy intensity, sustainable innovation, trade openness, and industrial development), reveals a different picture. When these controls are included, the coefficient for LnFDI becomes nonsignificant (−0.00213), indicating that when considering these factors, the relationship between FDI and renewable energy consumption is not statistically relevant. This suggests that the influence of FDI on the renewable energy sector is mediated by a more complex set of economic and regulatory factors.

The model also highlights the importance of other variables. For example, energy intensity (LnEI) and sustainable innovation (SI) show positive and significant

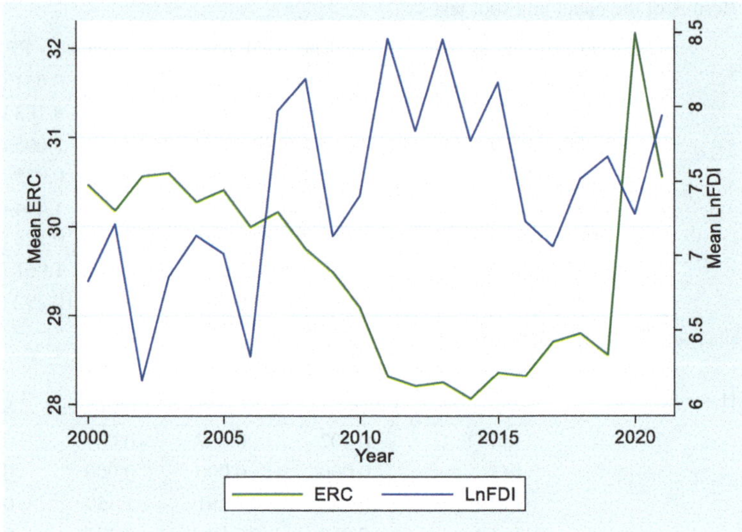

Fig. 3 Mean ERC and mean LnFDI (2000–2021)

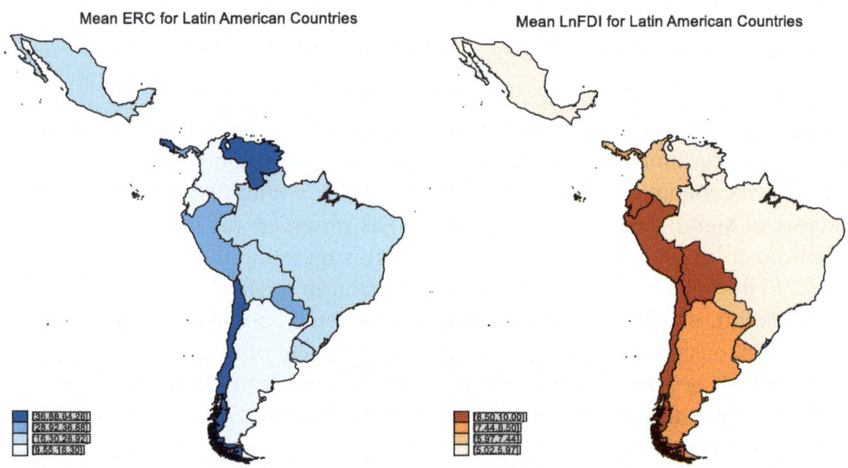

Fig. 4 ERC and LnFDI for Latin American sampled countries

coefficients, suggesting that greater efficiency in energy use and the adoption of sustainable practices promote renewable energy consumption. Additionally, CO_2 emissions have a significant negative impact, reflecting the challenge these countries face in transitioning to a cleaner energy matrix. These results have important implications for Hypothesis 1 of the study, which posits that FDI increases renewable energy consumption in Latin America. Although FDI has the potential to foster

Table 6 Effect of FDI on renewable energy consumption

Variables	(1) ERC	(2) ERC
LnFDI	−0.297**	−0.00213
	(0.138)	(0.0803)
EG		0.000945***
		(7.25e−05)
$LnCO_2$		−20.35***
		(1.056)
LnEI		15.86***
		(1.701)
SI		1.072***
		(0.142)
LnTO		0.201
		(0.517)
ID		−0.0116
		(0.0181)
Constant	31.72***	98.63***
	(1.054)	(3.010)
Observations	286	286
R-squared	0.017	0.705
Number of country	13	13
Controls	No	Yes
Country FE	Yes	Yes

Standard errors in parentheses
*** $p < 0.01$, ** $p < 0.05$, * $p < 0.1$

the adoption of clean technologies, this effect appears to be strongly dependent on the macroeconomic context and the implementation of appropriate energy policies.

Table 7 presents the results for testing Hypothesis 2, which posits that economic growth moderates the influence of foreign direct investment (FDI) on renewable energy consumption. The results indicate that the direct effect of FDI on renewable energy consumption, when considering economic growth, shows significant variability. In Model 3, the coefficient for FDI is negative and not significant (−0.172, $p > 0.1$), suggesting that FDI alone does not drive renewable energy consumption in the region without the consideration of other factors. However, in Model 4, when the interaction between FDI and economic growth is introduced, a drastic change is observed. The coefficient for FDI becomes positive and highly significant (0.400, $p < 0.01$), indicating that the relationship between FDI and renewable energy consumption is significantly strengthened in contexts of higher economic growth.

The interaction coefficient between FDI and economic growth is negative and significant in both Model 3 (−1.13e−05, $p < 0.1$) and Model 4 (−4.71e−05, $p < 0.01$). This result suggests that although economic growth amplifies the positive impact of FDI on renewable energy consumption, this effect slightly diminishes as economic growth increases. This could be interpreted as an indication that in high-growth economies, FDI tends to focus on sectors that do not necessarily promote the

Table 7 Moderating effect of economic growth

Variables	(3) ERC	(4) ERC
LnFDI	−0.172	0.400***
	(0.176)	(0.134)
LnFDI × EG	−1.13e−05	−4.71e−05***
	(9.90e−06)	(1.28e−05)
EG		0.00128***
		(0.000116)
$LnCO_2$		−20.05***
		(1.035)
LnEI		14.62***
		(1.696)
SI		1.055***
		(0.139)
LnTO		−0.0459
		(0.510)
ID		−0.0170
		(0.0178)
Constant	31.46***	95.56***
	(1.078)	(3.057)
Observations	286	286
R-squared	0.022	0.720
Number of country	13	13
Controls	No	Yes
Country FE	Yes	Yes

Standard errors in parentheses
*** $p < 0.01$, ** $p < 0.05$, * $p < 0.1$

use of renewable energies, or that the existing infrastructure has already reached a level where additional FDI yields diminishing returns in the adoption of these technologies.

Table 8 presents the results for evaluating Hypothesis 3, which posits that CO_2 emissions moderate the impact of foreign direct investment (FDI) on renewable energy consumption. In Model 5, the coefficient for FDI is not significant, indicating that in the absence of moderators, FDI alone does not have a significant direct impact on renewable energy consumption. However, in Model 6, a significant negative coefficient for FDI (−0.434) is observed when the interaction with CO_2 emissions is introduced. This suggests that in contexts of higher emissions, FDI tends to reduce its positive impact on renewable energy consumption. This could be attributed to a potential diversion of investment toward less sustainable sectors in countries with high levels of industrialization or reliance on fossil fuels, which contrasts with climate change mitigation goals.

On the other hand, in Model 7, the interaction between FDI and economic growth shows a significant positive coefficient (6.40e−05), indicating that in contexts of economic growth, FDI can have a more pronounced positive effect on renewable

Table 8 Moderating effect of CO_2 emissions

	(5)	(6)	(7)
Variables	ERC	ERC	ERC
LnFDI	−0.00920	−0.434	−0.753***
	(0.363)	(0.271)	(0.255)
LnFDI × $LnCO_2$	−0.0675	0.0878	0.0412
	(0.0789)	(0.0601)	(0.0561)
LnFDI × EG			6.40e−05***
			(9.47e−06)
$LnCO_2$		−18.36***	−19.72***
		(1.350)	(1.265)
LnEI		8.398***	13.73***
		(2.060)	(2.064)
SI		1.025***	1.073***
		(0.184)	(0.171)
LnTO		3.361***	2.152***
		(0.584)	(0.570)
ID		−0.0382*	−0.0179
		(0.0230)	(0.0215)
Constant	31.55***	83.89***	95.20***
	(1.074)	(3.629)	(3.753)
Observations	286	286	286
R-squared	0.019	0.521	0.592
Number of country	13	13	13
Controls	No	Yes	Yes
Country FE	Yes	Yes	Yes

Standard errors in parentheses
*** $p < 0.01$, ** $p < 0.05$, * $p < 0.1$

energy consumption. This suggests that a robust economic environment facilitates the adoption of clean technologies, allowing countries to more effectively leverage FDI to promote renewable energies. The CO_2 emissions variable remains significant and negative, reaffirming that high levels of CO_2 are associated with a reduced positive impact of FDI on renewable energies.

5 Discussion

The results confirm the complexity of the relationship between foreign direct investment (FDI) and renewable energy consumption in Latin America. In particular, the findings indicate that the impact of FDI on renewable energy consumption is significantly moderated by economic growth and CO_2 emissions. These findings are consistent with previous studies suggesting that FDI can play a dual role: while it facilitates the transfer of clean technologies, it can also contribute to increasing emissions in countries with an energy infrastructure dominated by fossil fuels [9–11, 28]. This study significantly contributes to the existing literature by highlighting that the relationship between FDI and renewable energy consumption is neither linear nor direct, but largely depends on the macroeconomic and environmental context of the recipient countries.

The observed negative interaction between FDI and CO_2 emissions suggests that in countries with high emissions, foreign investment tends to shift toward less sustainable sectors, potentially offsetting the benefits of FDI in promoting renewable energy. This finding underscores the need for stricter policies that direct FDI toward clean energy projects, especially in highly industrialized economies. On the other hand, the positive interaction between FDI and economic growth suggests that a robust economic environment amplifies the positive impact of FDI on renewable energy consumption. This result aligns with previous research indicating that growing economies are better positioned to leverage FDI for the adoption of clean technologies due to greater resource availability and a more stable political environment [12, 30, 35, 60].

The limitations of the study include the reliance on aggregated data, which may not capture microeconomic or sectoral differences. Additionally, the lack of detailed data on the specific energy policies of each country prevents a more granular analysis of how these policies might influence the effectiveness of FDI. Future studies should focus on specific case analyses and the interaction between FDI and other factors such as technological innovation and energy efficiency to provide a deeper understanding of the underlying mechanisms.

6 Conclusion

It is confirmed that FDI has the potential to drive the development of clean energy in the region, but this impact is deeply conditioned by economic and environmental factors. The results indicate that while FDI can facilitate the transfer of clean technologies, its effectiveness largely depends on the economic context of the recipient country and the energy policies implemented. Notably, in high-growth economies, FDI tends to have a more significant effect on promoting renewable energy consumption. This finding suggests that a stable and expanding macroeconomic environment can enhance the positive impact of FDI, facilitating the adoption of clean technologies and the diversification of the energy matrix. However, the negative interaction between FDI and CO_2 emissions highlights that in high-emission contexts, foreign investment may be diverted toward less sustainable sectors, limiting its contribution to the energy transition.

This study also underscores the importance of designing public policies that not only attract FDI but also ensure that such investment is directed toward projects that contribute to energy sustainability and carbon emission reduction. The conclusions suggest that the region needs to strengthen its regulatory frameworks to maximize the benefits of FDI in the renewable energy sector. Finally, there is a recognized need for future research to explore in greater detail the interaction of FDI with other factors, such as technological innovation and energy efficiency, and to use more disaggregated approaches to capture the specific dynamics of each country.

References

1. Gielen, D., Boshell, F., Saygin, D., Bazilian, M.D., Wagner, N., Gorini, R.: The role of renewable energy in the global energy transformation. Energy Strat. Rev. **24**, 38–50 (2019). https://doi.org/10.1016/j.esr.2019.01.006
2. Sadorsky, P.: Modeling renewable energy consumption for a greener global economy. In: Planet Earth 2011—Global Warming Challenges and Opportunities for Policy and Practice. IntechOpen (2011)
3. Mehedintu, A., Sterpu, M., Soava, G.: Estimation and forecasts for the share of renewable energy consumption in final energy consumption by 2020 in the European Union. Sustainability. **10**, 1515 (2018). https://doi.org/10.3390/su10051515
4. Holechek, J.L., Geli, H.M.E., Sawalhah, M.N., Valdez, R.: A global assessment: can renewable energy replace fossil fuels by 2050? Sustainability. **14**, 4792 (2022). https://doi.org/10.3390/su14084792
5. Yi, J., Hou, Y., Zhang, Z.Z.: The impact of foreign direct investment (FDI) on China's manufacturing carbon emissions. Innov. Green Dev. **2**, 100086 (2023). https://doi.org/10.1016/j.igd.2023.100086
6. Jebli, M.B., Youssef, S.B., Apergis, N.: The dynamic linkage between renewable energy, tourism, CO_2 emissions, economic growth, foreign direct investment, and trade. Latin Am. Econ. Rev. **1**, 1–19 (2019). https://doi.org/10.1186/s40503-019-0063-7
7. Soto, G.H.: The effects of foreign direct investment on environmentally related technologies in Latin America. Resour. Policy. **90**, 104711 (2024). https://doi.org/10.1016/j.resourpol.2024.104711
8. Soto, G.H.: The role of outward foreign direct investment as a filter for high energy intensity economies in the European Union. Appl. Energy. **361**, 122941 (2024). https://doi.org/10.1016/j.apenergy.2024.122941
9. Al-Mulali, U., Binti Che Sab, C.N.: The impact of energy consumption and CO_2 emission on the economic growth and financial development in the sub Saharan African countries. Energy. **39**, 180–186 (2012). https://doi.org/10.1016/j.energy.2012.01.032
10. Alvarado, R., Iñiguez, M., Ponce, P.: Foreign direct investment and economic growth in Latin America. Econ. Anal. Policy. **56**, 176–187 (2017). https://doi.org/10.1016/j.eap.2017.09.006
11. Omri, A.: CO_2 emissions, energy consumption and economic growth nexus in MENA countries: evidence from simultaneous equations models. Energy Econ. **40**, 657–664 (2013). https://doi.org/10.1016/j.eneco.2013.09.003
12. Ergun, S.J., Rivas, M.F.: Does higher income lead to more renewable energy consumption? Evidence from emerging-Asian countries. Heliyon. **9**, e13049 (2023). https://doi.org/10.1016/j.heliyon.2023.e13049
13. Alola, A.A., Adebayo, T.S., Olanipekun, I.O.: Examining the energy efficiency and economic growth potential in the world energy trilemma countries. Energies. **16**, 2036 (2023). https://doi.org/10.3390/en16042036
14. Friedman, E.: Constructing the adaptation economy: climate resilient development and the economization of vulnerability. Glob. Environ. Change. (2023). https://doi.org/10.1016/j.gloenvcha.2023.102673
15. Marques, A.C., Fuinhas, J.A.: Drivers promoting renewable energy: a dynamic panel approach. Renew. Sustain. Energy Rev. **16**, 1608–1613 (2012)

16. Naz, S., Sultan, R., Zaman, K., Aldakhil, A., Nassani, A., Abro, M.M.Q.: Moderating and mediating role of renewable energy consumption, FDI inflows, and economic growth on carbon dioxide emissions: evidence from robust least square estimator. Environ. Sci. Pollut. Res. **26**, 2806–2819 (2018). https://doi.org/10.1007/s11356-018-3837-6
17. Owusu, P.A., Asumadu-Sarkodie, S.: A review of renewable energy sources, sustainability issues and climate change mitigation. Cogent. Eng. **3**, 1167990 (2016). https://doi.org/10.1080/23311916.2016.1167990
18. Scarlat, N., Dallemand, J.-F., Monforti-Ferrario, F., Banja, M., Motola, V.: Renewable energy policy framework and bioenergy contribution in the European Union—an overview from National Renewable Energy Action Plans and Progress Reports. Renew. Sustain. Energy Rev. **51**, 969–985 (2015). https://doi.org/10.1016/j.rser.2015.06.062
19. Schiffer, H.-W., Trüby, J.: A review of the German energy transition: taking stock, looking ahead, and drawing conclusions for the Middle East and North Africa. Energy Transit. **2**, 1–14 (2018). https://doi.org/10.1007/s41825-018-0010-2
20. Stokes, L.C., Breetz, H.L.: Politics in the U.S. energy transition: case studies of solar, wind, biofuels and electric vehicles policy. Energy Policy. **113**, 76–86 (2018). https://doi.org/10.1016/j.enpol.2017.10.057
21. Zhao, N., Liu, X., Zhang, Z.: Does competition from the informal sector affect firms' energy intensity? Evidence from China. Struct. Change Econ. Dyn. **62**, 130–142 (2022). https://doi.org/10.1016/j.strueco.2022.04.008
22. Alvarado, R., Ponce, P., Alvarado, R., Ponce, K., Huachizaca, V., Toledo, E.: Sustainable and non-sustainable energy and output in Latin America: a cointegration and causality approach with panel data. Energy Strat. Rev. **26**, 100369 (2019). https://doi.org/10.1016/j.esr.2019.100369
23. Ozturk, I.: Measuring the impact of alternative and nuclear energy consumption, carbon dioxide emissions and oil rents on specific growth factors in the panel of Latin American countries. Prog. Nucl. Energy. **100**, 71–81 (2017). https://doi.org/10.1016/j.pnucene.2017.05.030
24. Gil Perez, A.J., Hansen, T.: Technology characteristics and catching-up policies: solar energy technologies in Mexico. Energy Sustain. Dev. **56**, 51–66 (2020). https://doi.org/10.1016/j.esd.2020.03.003
25. Zeeshan, M., Han, J., Rehman, A., Bilal, H., Farooq, N., Waseem, M., Hussain, A., Khan, M., Ahmad, I.: Nexus between foreign direct investment, energy consumption, natural resource, and economic growth in Latin American countries. Int. J. Energy Econ. Policy. (2020). https://doi.org/10.32479/ijeep.10255
26. Le, T.T.H., Nguyen, V.C., Phan, T.H.N.: Foreign direct investment, environmental pollution and economic growth—an insight from non-linear ARDL co-integration approach. Sustainability. **14**, 8146 (2022). https://doi.org/10.3390/su14138146
27. Zhang, M., Merchant, H.: A causal analysis of the role of institutions and organizational proficiencies on the innovation capability of Chinese SMEs. Int. Bus. Rev. **29**, 101638 (2020). https://doi.org/10.1016/j.ibusrev.2019.101638
28. Lee, J.W.: The contribution of foreign direct investment to clean energy use, carbon emissions and economic growth. Energy Policy. **55**, 483–489 (2013). https://doi.org/10.1016/j.enpol.2012.12.039
29. Pata, U.K., Karlilar, S., Kartal, M.T.: On the road to sustainable development: the role of ICT and R&D investments in renewable and nuclear energy on energy transition in Germany. Clean Techn. Environ. Policy. **26**, 2323–2335 (2024). https://doi.org/10.1007/s10098-023-02677-y
30. Tang, C.F., Tan, B.W.: The impact of energy consumption, income and foreign direct investment on carbon dioxide emissions in Vietnam. Energy. **79**, 447–454 (2015). https://doi.org/10.1016/j.energy.2014.11.033
31. Calderón, C., Alvarez, R.: Impact of foreign direct investment on renewable energy consumption in Latin America. Energy Econ. **86**, 104616 (2020). https://doi.org/10.1016/j.eneco.2019.104616

32. Shadman, M., Roldan-Carvajal, M., Pierart, F.G., Haim, P.A., Alonso, R., Silva, C., Osorio, A. F., Almonacid, N., Carreras, G., Maali Amiri, M., Arango-Aramburo, S., Rosas, M.A., Pelissero, M., Tula, R., Estefen, S.F., Pastor, M.L., Saavedra, O.R.: A review of offshore renewable energy in South America: current status and future perspectives. Sustainability. **15**, 1740 (2023). https://doi.org/10.3390/su15021740
33. Sentishcheva, E., Parkhomchuk, M., Kuzmina, V., Golovin, A.: Investments of countries in renewable energy sources in Latin America and the Caribbean: state and prospects. E3S Web Conf. **460**, 07009 (2023). https://doi.org/10.1051/e3sconf/202346007009
34. Alonso Sanabria, J.D., Melo-Velandia, L.F., Parra-Amado, D.: Unveiling the critical role of forest areas amidst climate change: the Latin American case (2023)
35. Bu, M., Li, S., Jiang, L.: Foreign direct investment and energy intensity in China: firm-level evidence. Energy Econ. **80**, 366–376 (2019). https://doi.org/10.1016/j.eneco.2019.01.003
36. Sadorsky, P.: Renewable energy consumption and income in emerging economies. Energy Policy. **37**, 4021–4028 (2009). https://doi.org/10.1016/j.enpol.2009.05.003
37. Kumari, R., Sharma, A.K.: Determinants of foreign direct investment in developing countries: a panel data study. Int. J. Emerg. Mark. **12**, 658–682 (2017). https://doi.org/10.1108/IJoEM-10-2014-0169
38. ur Rahman, S., Bakar, N.A.: A review of foreign direct investment and manufacturing sector of Pakistan. Pak. J. Human. Soc. Sci. **6**, 582–599 (2018). https://doi.org/10.52131/pjhss.2018.0604.0065
39. Shao, Y.: Does FDI affect carbon intensity? New evidence from dynamic panel analysis. Int. J. Clim. Change Strat. Manag. **10**, 27–42 (2017). https://doi.org/10.1108/IJCCSM-03-2017-0062
40. Bastida, R.: Innovation in Latin America: paradigm shift toward sustainable development. Technol. Forecast. Soc. Change. **114**, 314–320 (2017)
41. Nyeadi, J.D., Banyen, K.T., Mbilla, S.A.E.: FDI, energy consumption, and institutional quality: the case of Africa. In: Handbook of Research on Energy and Environmental Finance 4.0, pp. 159–188. IGI Global (2022)
42. Zheng, Y., Sheng, B., Yin, H.: Impact of foreign direct investment (FDI) on the energy intensity of host countries based on panel data estimation. Energy Procedia. **5**, 265–271 (2011)
43. Shahbaz, M., Hye, Q.M.A., Tiwari, A.K., Leitao, N.C.: CO_2 emissions, energy consumption, economic and population growth in Pakistan. Renew. Sustain. Energy Rev. **47**, 29–44 (2015)
44. Jannuzzi, G.: Energy efficiency and renewable energy in Latin America: policies towards sustainable development. J. Int. Affairs. **70**, 47–58 (2017). https://doi.org/10.18327/JIAS.2017.07.21.2.9
45. Paramati, S.R., et al.: The effects of stock market growth and renewable energy use on CO_2 emissions: evidence from G20 countries. Energy Econ. **66**, 360–371 (2017)
46. CEPAL: CEPALSTAT Bases de Datos y Publicaciones Estadísticas. https://statistics.cepal.org/portal/cepalstat/index.html?lang=es (2024)
47. Papież, M., Śmiech, S., Frodyma, K.: Determinants of renewable energy development in the EU countries. A 20-year perspective. Renew. Sustain. Energy Rev. **91**, 918–934 (2018). https://doi.org/10.1016/j.rser.2018.04.075
48. Polcyn, J., Us, Y., Lyulyov, O., Pimonenko, T., Kwilinski, A.: Factors influencing the renewable energy consumption in selected European countries. Energies. **15**, 108 (2022). https://doi.org/10.3390/en15010108
49. Armeanu, D.Ş., Vintilă, G., Gherghina, Ş.C.: Does renewable energy drive sustainable economic growth? Multivariate panel data evidence for EU-28 countries. Energies. **10**, 381 (2017). https://doi.org/10.3390/en10030381
50. Grossman, G.M., Krueger, A.B.: Economic growth and the environment. Q. J. Econ. **110**, 353–377 (1995). https://doi.org/10.2307/2118443
51. Wang, S., Wang, Z., Fang, C.: Evolutionary characteristics and driving factors of carbon emission performance at the city level in China. Sci. China Earth Sci. **65**, 1292–1307 (2022). https://doi.org/10.1007/s11430-021-9928-2

52. Zhang, L., Liu, X.: Impact of energy intensity on economic growth and environmental sustainability. Energy Econ. **85**, 104632 (2020)
53. Wang, X., Xu, X., Liu, Y., Zhao, J., Li, H.: An economic impact evaluation of environmental policy. IOP Conf. Ser. Earth Environ. Sci. **295**, 012075 (2019)
54. Smith, J.: Impact of COVID-19 pandemic in Latin America. J. Latin Am. Stud. **23**, 45–60 (2021)
55. Ahmad, N., Youjin, L., Hdia, M.: The role of innovation and tourism in sustainability: why is environment-friendly tourism necessary for entrepreneurship? J. Clean. Prod. **379**, 134799 (2022). https://doi.org/10.1016/j.jclepro.2022.134799
56. Arocena, P., Orcos, R., Zouaghi, F.: The impact of ISO 14001 on firm environmental and economic performance: the moderating role of size and environmental awareness. Business Strat. Environ. **30**, 955–967 (2021). https://doi.org/10.1002/bse.2663
57. Pesce, M., Shi, C., Critto, A., Wang, X., Marcomini, A.: SWOT analysis of the application of International Standard ISO 14001 in the Chinese context. A case study of Guangdong Province. Sustainability. (2018). https://doi.org/10.3390/SU10093196
58. Ikram, M., Zhang, Q., Sroufe, R., Shah, S.: Towards environmental sustainability: exploring the nexus among ISO 14001, renewable energy consumption, access to electricity, agriculture and CO_2 emissions in SAARC countries. Sustain. Prod. Consump. **22**, 218–230 (2020). https://doi.org/10.1016/j.spc.2020.03.011
59. Ji, Y., Xue, J., Fu, Z.: Sustainable development of economic growth, energy-intensive industries and energy consumption: empirical evidence from China's provinces. Sustainability. **14**, 7009 (2022). https://doi.org/10.3390/su14127009
60. Feng, Z., Chen, W.: Environmental regulation, green innovation, and industrial green development: an empirical analysis based on the spatial Durbin model. Sustainability. **10**, 223 (2018). https://doi.org/10.3390/SU10010223
61. Wooldridge, J.M.: Introductory Econometrics: A Modern Approach. Cengage Learning (2013)

Operating Factors for Hydroelectric Power Plants: A Systematic Literature Review

Raudelki Caraballo Ramos, Miguel Angel Caraballo Núñez, and Juan Carlos Solano

1 Introduction

Hydroelectric power plants are facilities that make it possible to harness the gravitational potential energy contained in the mass of water carried by rivers to convert it into electrical energy, using turbines coupled to alternators. Sandoval Erazo poses in [1], this is a process that allows a high level of energy efficiency since yields of up to 90% can be achieved. At present, hydroelectric power generation accounts for 17% of all electricity generation worldwide [2].

The division between hydroelectric plants is important to determine their constructive and functional characteristics, in [3] it is highlighted that hydroelectric power plants are divided into two main types, run-of-river plants, when the natural flow of a river or stream is used directly, whose water is channeled to the turbines and then returned to the natural flow downstream of the plant; and those called reservoir plants, which are cases in which the water of one or more streams is accumulated in a reservoir, natural or artificial (created by a dam), to drive the turbines.

One of the most recurrent aspects in research related to hydroelectric generation is the relationship between the generating capacity of the plant and what is called the low water or drought period. The impact of droughts in hydroelectric power plants is not only determinant in the generation capacity but also influences the determination

R. Caraballo Ramos (✉)
Universidad Nacional de Loja, Facultad de la Energía, Loja, Ecuador
e-mail: raudelki.caraballo@unl.edu.ec

M. A. Caraballo Núñez · J. C. Solano
Universidad Nacional de Loja, Facultad de la Energía, Loja, Ecuador

Universidad Nacional de Loja, Centro de Investigaciones Tecnológicas y Energéticas - CITE, Loja, Ecuador

M. Espinoza-Andaluz et al. (eds.), *Congress on Research, Development and Innovation in Renewable Energies*, Green Energy and Technology,
https://doi.org/10.1007/978-3-031-88995-0_12

of the final price of energy [4]. According to [5], one of the fundamental parameters to quantify the energy production in a hydroelectric plant is the measurement of flows and their lower limits. In this publication, a model is developed that simulates, from several parameters, the impact of drought in the production of electric energy at even international levels. This variation of the hydroelectric generation capacity emphasizes the need to promote a scheme that guarantees greater stability in the supply of electric power because of the growing need for this service. This aspect is detailed in [6].

At present, it is relevant to relate how hydroelectric power generation influences the surrounding environment, for this purpose, environmental impact analyses associated with the construction and operation of such power plants are carried out. In this regard, [7] indicates that although hydroelectric generation uses a renewable energy source, many specialists consider it as non-renewable due to the high environmental impact they have both in the construction stage of the plant and in the arrangement of the reservoir to store the amount of water needed to cope with low water seasons.

According to this aspect, [8] emphasizes that various construction parameters of power plants should be explored and their impact on the environment should be evaluated when designing them, a variant of which proposes solutions that are not so large and so global but that can be an important resource for small settlements and their energy needs while being environmentally responsible. But the environmental impact caused by hydroelectric power plants should not only be evaluated in their direct influence on the construction area; as illustrated by [9], but these effects are also seen downstream when considering the turbulence caused by each type of turbine and the mass of water they can generate.

In any installation of any kind, another factor to consider when making decisions is the technical-economic feasibility of the installation. For this case, [10] proposes a general evaluation framework that makes the performance of a hydroelectric power plant more efficient and optimizes its performance, and a specific evaluation framework that synthesizes several technical-economic indicators to determine when hydro generation is profitable in terms of its production variables. The unavailability or scarce availability of financial resources makes it necessary to execute the hydroelectric power plant implementation study and from there to develop a model to identify the areas with the greatest generation potential in already built reservoirs [11]. Another vision, according to [12], is that the design of small run-of-river hydroelectric power plants should be based on a levelized cost of energy.

The development of new information and computer technologies has led to an avalanche of modeling and process simulation systems, and hydroelectric generation is no exception. Many developers are working on algorithms that simulate and automate the constructive and operational aspects of hydroelectric power plants. Several models have been developed, among them is the one that introduces the concept of equal reservoir full storage rate (ERFSR) to guide the distribution of cascaded hydroelectric power plants [13].

Among the fundamental objectives of any hydro generation system are the operation and maintenance factors to be implemented when intervening in the power plants. For this purpose, several researches have been carried out; [14] develops a hydroelectric power scheduling model that prioritizes multiple objectives in order to make more effective the operational tasks in a hydroelectric power plant, while [15] analyzes the problems associated with the sedimentation of waste and organic material transported by the river as these cause loss of load that negatively impact the efficiency of the generators, and proposes a technique to optimize maintenance that uses hidden Markov models and Bayesian networks to predict the downtime of the power plant.

A systematic literature review (SLR) was proposed to systematize and group the scientific knowledge of the last 6 years on the operational factors of hydroelectric generation plants. Using the methodology described in [16], an SLR was conducted to establish the state of the art related to these operational factors. Additionally, five research questions were formulated and answered to guide the SLR, aiming to identify and analyze the most relevant research articles or those with the greatest contribution to the search conducted.

1.1 Need for Conducting the SLR

The methodological aspect of this SLR is based on the method used by [16, 17], based on the proposal made in [18] and used by [19] to establish the state of the art of research on self-consumption photovoltaic systems in Ecuador. The basis of the method used is the conception of a conceptual map, the formulation of research questions, and the structuring of a semantic search script to guide the literature search.

Hydroelectric power plants are becoming increasingly important worldwide, attracting a large part of the scientific community in the energy field to investigate the behavior of their operating parameters. The knowledge in the field is diverse, ranging from the influence on the generation capacity of various parameters, with the low water level being the most mentioned, to the consideration of operation and maintenance criteria to be implemented when intervening in the power plants.

As it is considered a renewable energy, the impact of the production and operation of hydroelectric power plants on the nature surrounding the plant is vital in its planning. To what extent is the construction of hydroelectric power plants technically and economically feasible? What are the operating conditions to which a hydroelectric power plant is exposed? What are the various maintenance systems to which the plants are subjected? What automation and control processes are the most effective in hydroelectric power plants? These are some of the questions that concern the scientific community in the field of energy, but which are not addressed systemically in the same publication.

The main objective of this publication is to integrate, in a systematic literature review, the main operational factors that affect the operation of hydroelectric power plants and to establish how far science has come in this field, in the last 6 years, at a global level.

The work presents a specific structure divided into the following sections. Section 1 establishes the background that guided the research, where a general description of the method used, the identification of the need for a systematic literature review (SLR), as well as the main objective of the study are described. Section 2 formulates the research questions. Section 2.2 describes the conceptual framework. Section 2.3 establishes the elaboration of the semantic search structure and the corresponding search script, considering different inclusion and exclusion criteria, as well as the analysis of the possible existence of related systematic reviews. Section 2.4 defines the categories and variables. Finally, Sect. 3 presents and analyzes the results obtained and the corresponding conclusions of the work.

2 Methodology

2.1 Research Questions

Research questions constitute the central question that a study sets out to answer and whose solution must be reached during the research. It is an expression of the scientific problem and, therefore, the first category of the scientific research process. Posing a good research question helps to establish what is to be addressed and to adequately plan the strategy to be followed. This allows greater concreteness in the writing of a proposal, which will later be reflected in the scientific article. Research questions lead to assumptions that facilitate the formulation of operational statements and clear hypotheses.

By focusing on the production of electric power from hydroelectric projects, the following research questions are formulated:

1. To what extent is the generation capacity of a hydroelectric power plant related to low water level?
2. How can the environmental impact of a hydroelectric power plant be evaluated?
3. Under what conditions is hydro generation technically and economically feasible?
4. What automation and control processes are integrated into the operation of hydroelectric power plants?
5. What operation and maintenance criteria determine the operation of hydroelectric power plants?

2.2 Conceptual Map

Conceptual maps are ideograms or visual representations of the internal structure of concepts. Their synthetic and visual properties act as powerful cognitive synthesizers. These conceptual maps graphically organize a network of essential propositions extracted from the original text and categorize them into four classes using the

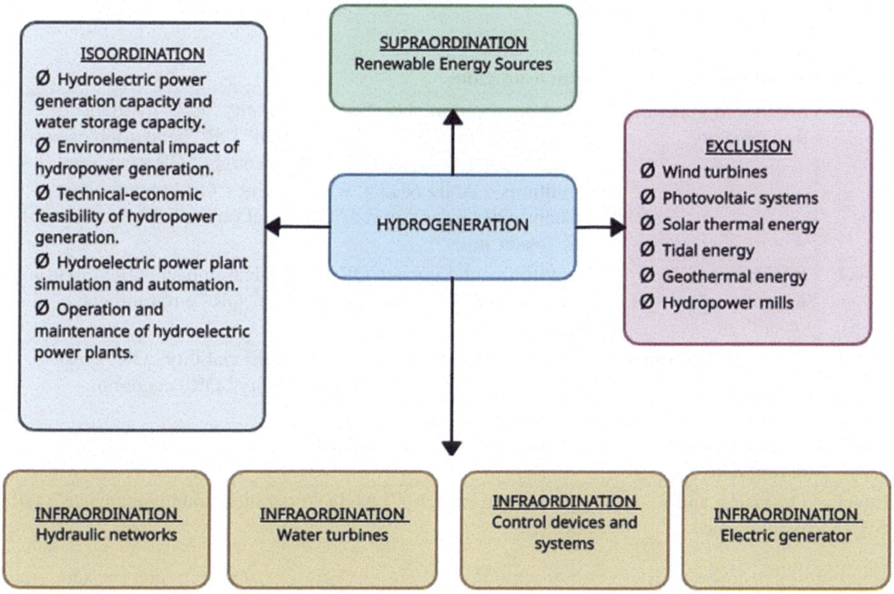

Fig. 1 Conceptual map

four basic operations of conceptual thinking [20]. In Fig. 1, the conceptual map is presented to orient and organize the bibliographic search linked to the research question.

The conceptual map starts from the superior concept or supraordination where the general concept to be developed is shown and fixed. It also presents other elements such as isoordination which establish the qualities and characteristics of the concept that describe them, the exclusion or concepts not directly related to the research and the classes or subtypes of the concept as such that correspond to the infraordination [17].

2.3 Structure of the Bibliographic Search

The first step in establishing the state of the art in any field of research is to carry out a rigorous and meticulous bibliographic search strictly guided to orient the research toward the resolution of the research questions posed.

The search was oriented in the SCOPUS database and was organized based on a search script with a semantic structure in five different levels and with general characteristics where the fundamental idea of the search is developed as shown in Table 1. Each level is directly related to each of the research questions and integrates the qualities that determine and characterize the central concept developed in the conceptual map; that is, the operating parameters of the hydroelectric power plants.

Table 1 Semantic search structure

Level	Research question keywords	Search structure
Level 1	Strain and generation capacity	("hydrogeneration" OR "hydroelectricity" OR "hydroelectric power" OR "hydroelectric generation" OR "generation capacity") AND ("low water flow" OR "drought" OR "low flow conditions") AND NOT ("wind turbines"OR "photovoltaic systems"OR "solar thermal" OR "tidal power" OR "geothermal" OR "water mills")
Level 2	Environmental impact	("environmental impact"OR "ecological impact" OR "environmental effect" OR "ecosystem effect" OR "environmental footprint")
Level 3	Technical-economic feasibility	("technical feasibility" OR "economic viability" OR "cost-effectiveness" OR "financial feasibility" OR "economic assessment")
Level 4	Simulation and automation	("simulation" OR "automation")
Level 5	Operation and maintenance	("generation capacity") AND ("operation and maintenance") OR (O&M)

Table 2 Script entry for SCOPUS search

Script search
(TITLE-ABS-KEY (("low water flow" OR "drought" OR "low flow conditions") AND ("hydrogeneration" OR "hydroelectricity" OR "hydroelectric power" OR "hydroelectric generation") AND NOT ("wind turbines"OR "photovoltaic systems"OR "solar thermal" OR "tidal power" OR "geothermal" OR "water mills") AND generation AND capacity)) AND PUBYEAR > 2018 AND PUBYEAR < 2025 AND (LIMIT-TO (LANGUAGE, "English")) AND (LIMIT-TO (EXACTKEYWORD, "hydroelectric power" OR "drought" OR "hydropower" OR "rivers" OR "water levels" OR "reservoir management" OR "electricity generation" OR "renewable energies"OR "hydropower plants" OR "dry season" OR "water" OR "rain" OR "hydrology" OR "water resources management" OR "seasonal variation" OR "river flow" OR "renewable energy"OR "hydropower reservoirs" OR "hydrological regime" OR "dry seasons" OR "renewable energy source" OR "water level" OR "hydropower potential" OR "hydropower generation" OR "hydrological droughts" OR "hydroelectricity" OR "river water" OR "hydrological change" OR "hydroelectric generators" OR "electricity" OR "drought stress" OR "drought conditions" OR "maintenance" OR "optimization")) AND (LIMIT-TO (SUBJAREA, "ENVI" OR "ENER" OR "ENGI" OR "EART" OR "ECON"))

Based on this structure, the definitive search script was established, which shows in detail the complete search structure with the respective inclusion and exclusion criteria, among them; limiting publications from 2018 to 2025, in addition to limiting only to the pre-established area of interest, ignoring other systems and areas related to the case study (Table 2).

The search script did not yield any other SLR in the results that addressed the general topic of operating parameters in hydroelectric power plants, therefore, it is considered as a novelty in this field of research.

2.4 *Definition of Categories and Variables*

The implementation of the methodology used in this study, considering the experience of other researchers [16–18], implies the definition of categories and variables to address each of the research questions posed. Table 3 presents the categories and variables associated with each of these research questions.

3 Results

The results of the literature search are shown below through a review report related to the operating factors for hydroelectric power plants, using tables with specificity on the research that addresses the topics of analysis defined as variables in each category, and their frequency or number of citations for each one of them; all this in close relation to the research questions.

Table 3 Categories and variables

Research questions	Categories	Variables
1. To what extent is the generating capacity of a hydroelectric power plant related to low water?	Loss of generation	Economic impact Level limits
	Influence by hydroelectric power plant type	Flexibility of the system Changes of schemes Reservoir power plants Run-of-river power plants
2. How to assess the environmental impact of a hydroelectric power plant?	Impact on the construction stage	Environmental impact of reservoir construction Animal habitat
	Impact on surrounding areas	Reincorporation of water to the river flow below the hydroelectric power plants. Flood risks
3. Under what conditions is hydro generation technically and economically feasible?	Technical feasibility	Technical profitability Identification of potentialities Technological additions
	Economic viability	Economic profitability Levelized energy costs
4. What automation and control processes are integrated into the operation of hydroelectric power plants?	Simulation and automation of hydroelectric power plants	Computational modeling Control algorithms
5. What operation and maintenance criteria determine the operation of hydroelectric power plants?	Operation of hydroelectric power plants Maintenance of hydroelectric power plants	Programmatic models Reservoir monitoring Monitoring of parameters Reduction of electrical faults Reduction of mechanical failures

3.1 Question 1: To What Extent Is the Generating Capacity of a Hydroelectric Power Plant Related to Low Water? (Table 4)

The greatest research interest in this field is in two variables: the levels' limit of water storage that guarantees hydroelectric generation and the economic impact caused by periods of drought, with 28.6% and 21.4%, respectively.

The issue of low water levels is vital among the factors that operationalize the operation of hydroelectric power plants; in this sense, the limit levels at which the plants operate are of great importance [2, 5, 25, 27, 28]. In this same line, [25] shows how for hydroelectric power plants with reservoirs it is necessary to introduce efficiency parameters in the management of dam flows to have sufficient water resources in times of drought and to be able to operate without reducing the amount of energy delivered to the system in this type of power plants.

The literature reviews the need to incorporate schemes that make production operational [26, 30–32]. When energy systems depend heavily on hydroelectric power, reliable statistics must be available to help determine how much the generation capacity of hydroelectric power plants increases or decreases with variations in rainy and dry seasons to make decision-making effective [29].

An interesting aspect detailed in the literature search is how, faced with these problems, many power plants are forced to make their operating schemes more flexible to reduce the percentage of loss of generation capacity [25, 28–30]. According to [33], to mitigate drought and the impact on the reduction of generation capacity in hydroelectric power plants, variants can be adopted that integrate other types of energy resources; however, to do so, it is necessary to quantify the amount of energy to be implemented, and to do so, it is necessary to know how much the reduction of energy production is during low water levels. In this sense, [30] proposes a system model that permits to take better advantage of water flows and thus optimize the difference between the actual monthly consumption of the plants and the expected one to make the operation more effective in periods of drought.

To know this influence, [5] shows, through a global hydrological model WaterGAP with a record of more than 8700 hydroelectric power plants, that the impact in loss of hydroelectric generation capacity produced by low water levels

Table 4 Bibliographic references analyzed for Question 1

Categories	Variables	Bibliographic references	Frequency
Loss of generation	Economic impact	[4, 6, 21–24]	6
	Level limits	[2, 5, 25–28]	8
	Flexibility of the system	[8, 25, 29, 30]	5
	Changes of schemes	[26, 30–32]	4
Influence by hydroelectric power plant type	Reservoir power plants	[33–35]	3
	Run-of-river power plants	[26, 36]	2

amounts to 40% in most cases. According to [4], it is estimated that times of drought cause losses of millions of dollars and the proliferation of thermal power plants and their terrible effects such as the emission of greenhouse gases.

The operation feasibility of a hydroelectric power plant depends on the capacity to store water resources, [35] adds a condition called "energy domain-specific drought," which can be reduced if the storage space is properly managed. The literature also highlights that reservoir power plants can better plan the reduction of the impact of low water levels on generation capacity [33–36], but there are insufficient studies that reflect, at quantitative levels, this qualitative advantage of reservoir power plants concerning run-of-river power plants.

3.2 *Question 2: How to Assess the Environmental Impact of a Hydropower Plant? (Table 5)*

In every work built by human beings, there is an impact on nature both during the construction and operation stages, and hydroelectric power plants are no exception. In this aspect, the greatest research interest is evidenced in the impact during the construction of the reservoirs, with 35% of the reviews, and the influence of the reincorporation of the flows downstream of the plant, with 30% of the research in the field in question.

The literature shows that the highest degree of impact during construction is registered in the construction of reservoirs [7, 27, 37–41]. Due to this, the adaptation of reservoirs not intended in principle to supply hydroelectric generation plants causes minimal modifications for the construction of plants from scratch. In these projects, the environmental impact is minimal since the referenced dams or natural reservoirs have already mitigated their effect on nature. According to [39], these projects also have an economic improvement that, in the evaluated sites, was estimated to be close to eight million dollars.

In some studies, as in [32], and considering various construction aspects, it is shown that the environmental impact sometimes makes it necessary to reduce the capacity of the power plants to mitigate the damage that can be done to the environment. This environmental influence has caused the governments of many

Table 5 Bibliographic references analyzed for Question 2

Categories	Variables	Bibliographic references	Frequency
Impact on the construction stage	Environmental impact of reservoir construction	[7, 27, 37–41]	7
	Animal habitat	[38, 42, 43]	3
Impact on surrounding areas	Reincorporation of water to the river flow below the hydroelectric power plants	[9, 44–48]	6
	Flood risks	[32, 49–51]	4

countries to implement regulations regarding the incorporation of flows downstream of the power plants [44]. According to [38], the impact on the environment surrounding hydroelectric power plants is not only reflected in the animal habitat—especially the fish populations in rivers and the terrestrial ecosystem, which are forced to migrate to other areas since their natural habitat is affected [38, 42, 43]—but also in the change in land use caused by the coverage of dams and reservoirs. Quantifying this impact is also very important when developing projects of this type and thus anticipating the expected changes.

The spillway area in the reservoirs of hydroelectric power plants that returns water to the river course largely contains decomposed organic material, which affects the areas downstream of the power plants, influencing the risk that several of these materials will solidify and create a natural barrier that increases the danger of overflows and floods in these areas [9]. For this reason, it is necessary to determine the mitigation thresholds for ecological disasters caused by these plants, as well as the operation and the works to be carried out downstream to reincorporate the natural flow with the least possible environmental impact.

3.3 *Question 3: Under What Conditions Is Hydro Generation Technically and Economically Feasible? (Table 6)*

One of the most practical evaluation criteria when determining the value of a facility is to analyze it from the point of view of its technical-economic viability. In this factor, the literature analyzed for hydroelectric power plants proposes a general evaluation framework that makes the performance of a hydroelectric power plant more efficient and optimizes its performance, and another more specific one that synthesizes several technical-economic indicators to determine, based on its production variables, its level of profitability [52, 54]. The analyses with econometric factors are varied, some propose a technical-financial analysis both for integrating several small hydropower plants into a micro-generation network [52] and the design of small run-of-river hydropower plants based on a levelized cost of energy [12].

Other assessments go more toward an economic approach that determines the profitability of hydrological projects of new reservoirs [61]. What coincides with most of the analyses is that the unavailability or scarce availability of financial resources makes it necessary to carry out the study of the implementation of

Table 6 Bibliographic references analyzed for Question 3

Categories	Variables	Bibliographic references	Frequency
Technical feasibility	Technical profitability	[52–54]	3
	Identification of potentialities	[12, 52, 55]	3
	Technological additions	[56–59]	4
Economic viability	Economic profitability	[10, 11, 57, 60, 61]	5
	Levelized energy costs	[12, 54, 62, 63]	4

hydroelectric power plants, and based on this conclusion it is recommended to develop models to identify the areas with the greatest generation potential and, above all, to have the possibility of doing so in areas with already constructed reservoirs [11].

In terms of technical feasibility, the greatest contributions are made in the field of incorporating various technologies to optimize the performance of power plants. An interesting proposal is the use of hydraulic pumps as PaT turbines [56]; according to this study, a 2 kW installation can generate an annual energy of approximately 17.52 MWh/year. The incorporation of energy storage elements in hydroelectric power plants to operationalize their operation is another proposal that can be incorporated to obtain greater technical feasibility of the facilities.

In this effort to make the facilities more viable, [54] uses a model to minimize the net annual cost (NPC) and the levelized cost of energy (LCOE) in decentralized power plants. A hybrid system is used with greater emphasis on hydraulic systems. The incorporation of energy storage elements in hydroelectric power plants is also used to operationalize their operation, [59] determines that the incorporation of batteries increases the profitability of the power plants by approximately 2%. For reservoir power plants, the dam constitutes a fundamental element of the system.

Finally, considering the viability factor, [57] examines the presence of sediments in these reservoirs and proposes strategies that reduce by up to 88% the risk of overflow due to the accumulation of organic material in the gates and the consequent economic losses due to plant inactivity.

3.4 Question 4: What Automation and Control Processes Are Integrated into the Operation of Hydroelectric Power Plants? (Table 7)

Modeling and simulation are the factors that are gaining more emphasis nowadays. According to the literature analyzed for the case of hydroelectric power plants, information processing is performed in two fundamental ways: computational models ranging from huge global hydropower databases (GHD) and the global hydrological model WaterGAP, which include a total of 44 records encompassing 92% of the world's hydropower production [5], to specific systems to determine the concept of equal reservoir filled storage rate ratio (ERFSR) [24]; which is very useful to guide the distribution of cascading hydropower plants and the use of Soil and Water Assessment Tool (SWAT) to estimate the impact of climate change on

Table 7 Bibliographic references analyzed for Question 4

Categories	Variables	Bibliographic references	Frequency
Simulation and automation of hydroelectric power plants	Computational modeling	[5, 13, 64, 65]	4
	Control algorithms	[65–70]	6

hydropower generation [64]. These computational models allow not only data visualization but also process optimization.

As a hydroelectric power plant is no stranger to the uncertainties inherent to any production process and to get as close as possible to the state of how the operational work will be in these plants, several simulation processes have also been developed that the literature collects and frames in what are called control algorithms [65, 67–69]. Among these are smart grids to obtain output parameters through various methods and systems of optimal algorithms that make it possible to increase the development, and the applications of emerging triboelectric nanogenerators (TENG) in the field of hydroelectric power plants [68], which produce electrical energy from the conversion of all types of mechanical energy for self-powered devices and portable devices with low power consumption.

In [66], references are made to the practical operation and performance of multistage turbine-operated pumps (PaT) in the future. In the same line, [67] develops a simulation method for PaT turbine pump installations based on water distribution networks. Other researchers emphasize the operationalization of the plants with "shared socioeconomics pathways" (SSP) scenarios, and to improve the efficiency of the system, they propose the use of the moderate scenario SSP126 and the medium-high scenario SSP370 [27].

However, there are still insufficient references on the potential of these models to take advantage of big data science and machine learning in hydropower plants.

3.5 *Question 5: What Operation and Maintenance Criteria Determine the Operation of Hydroelectric Power Plants? (Table 8)*

From the point of view of operation and maintenance, the general criteria with which these factors are manifested in hydroelectric power plants are not very different from the same category in other power plants. Of great importance for hydroelectric power plants is the monitoring of parameters and the reduction of both electrical and mechanical failures [74–76]. The analyzed literature includes practices that prioritize the review of oscillations in rotating machines with models and methods of multicriteria decision analysis to determine the main problems at power plant sites

Table 8 Bibliographic references analyzed for Question 5

Categories	Variables	Bibliographic references	Frequency
Operation of hydroelectric power plants	Programmatic models	[14, 71, 72]	3
	Reservoir monitoring	[15, 73]	2
Maintenance of hydroelectric power plants	Monitoring of parameters	[74–76]	3
	Reduction of electrical faults	[74, 77, 78]	3
	Reduction of mechanical failures	[2, 79, 80]	3

[2]. For this, the analysis and monitoring of programmatic models [14, 71], which guarantee the functionality of the hydroelectric power plant system with the best possible operating conditions, are vital [72].

In [2], it is stated that inadequate water flow in the dry season is not the only reason for the reduction in power generation, but mechanical failures due to lack of monitoring and periodic maintenance is the predominant reason for the reduction in power production. This criterion is reinforced in the studied literature where it is shown that mechanical failures have a great incidence in the reduction of energy production, due to the lack of monitoring and periodic maintenance. References to the major maintenance factors to be considered in large hydroelectric generation plants are shown in [80].

Reservoir control is one of the most important sections in such power plants [15, 73]. The problems associated with abrasive erosion, sedimentation of residues, and organic materials transported by the river cause load loss that negatively impacts the efficiency of the generators. For these problems, techniques using hidden Markov models and Bayesian networks are proposed to predict the downtime [15]. Another major problem associated with the operation of reservoir hydropower plants is abrasive erosion, which causes various problems in the efficient operation of these hydropower plants [73]. It follows those factors such as the combination of turbine design and selection, desander capacity, sediment concentration limits that cause turbine shutdown, and preventive measures such as coatings that need to be analyzed for the proper operation of hydropower plants.

4 Conclusion

In the bibliographic research conducted, it was possible to establish the state of the art concerning the operational factors of hydroelectric power plants using the RSL methodology through which the conceptual map and the script with semantic search structure were elaborated, which yielded a total of 122 research articles. Subsequent analysis determined that 74 articles (60.7%) were relevant to the topic in question. Finally, among these research articles and other works consulted, a total of 83 bibliographic citations were selected that comprehensively address the operational factors studied. The most significant points of the analysis can be summarized in the following paragraphs.

Low water levels are directly related to the loss of generation capacity in hydroelectric plants, which amounts to 40%. In turn, this causes the final price of energy to increase, and supply schemes are forced to return to generation systems with thermal power plants with the consequent emission of greenhouse gases such as CO_2. To make the operating system more flexible or directly make changes in schemes that link hydroelectric generation with other renewable energy plants, it is estimated that times of drought generate losses of millions of dollars.

Although hydroelectric generation uses a renewable energy source as its primary energy source, many specialists consider it to be non-renewable due to the high environmental impact during its construction, especially in reservoir plants, and to the change in soil characteristics in surrounding areas during its continuous operation. The evaluation of this impact is vital to determine how these plants are built and to what extent impact mitigation work must be carried out in the river basins that serve as tributaries to these plants.

There are several analysis criteria to determine the technical-economic feasibility of a hydroelectric power plant, but given the low availability of financing, it is the financial analysis that has the most weight when making a final decision.

The various computational models and control algorithms used in this field are still insufficient to deal with the most modern analysis methods such as big data and machine learning systems.

During the operation of hydroelectric power plants, emphasis is placed on the monitoring of parameters that lead to reduced mechanical failures in the facilities, which are the most determining factor in the loss of generation capacity of these plants after the low water level.

However, it would be interesting to establish how these factors are reflected in certain specific regions, and which would be the most influential in each area, evaluating their impact on energy production through hydroelectric generation in a comprehensive manner.

References

1. Sandoval Erazo, W.: Capítulo 6: Conceptos Básicos de Centrales Hidroeléctricas. Washington Sandoval (2018)
2. Pandey, R., Shrestha, R., Bhattarai, N., Dhakal, R.: Problems identification and performance analysis in small hydropower plants in Nepal. Int. J. Low-Carbon Technol. **18**, 561–569 (2023). https://doi.org/10.1093/ijlct/ctad043
3. Centrales_hidroelectricas_BAJO_Azcapotzalco.pdf. Accedido: 6 de abril de 2024. [En línea]. Disponible en: http://zaloamati.azc.uam.mx/bitstream/handle/11191/1821/Centrales_hidroelectricas_BAJO_Azcapotzalco.pdf?sequence=1&isAllowed=y
4. Kern, J.D., Su, Y., Hill, J.: A retrospective study of the 2012-2016 California drought and its impacts on the power sector. Environ. Res. Lett. **15**(9) (2020). https://doi.org/10.1088/1748-9326/ab9db1
5. Wan, W., Zhao, J., Popat, E., Herbert, C., Döll, P.: Analyzing the impact of streamflow drought on hydroelectricity production: a global-scale study. Water Resour. Res. **57**(4) (2021). https://doi.org/10.1029/2020WR028087
6. Qin, P., et al.: Assessing concurrent effects of climate change on hydropower supply, electricity demand, and greenhouse gas emissions in the Upper Yangtze River Basin of China. Appl. Energy. **279** (2020). https://doi.org/10.1016/j.apenergy.2020.115694
7. Koç, C.: A study on solutions and problems of hydroelectric power plants in the operation. Sustain. Water Res. Manag. **8**(3) (2022). https://doi.org/10.1007/s40899-022-00677-2
8. YoosefDoost, A., Lubitz, W.D.: Archimedes screw turbines: a sustainable development solution for green and renewable energy generation-a review of potential and design procedures. Sustainability. **12**(18) (2020). https://doi.org/10.3390/SU12187352
9. Halleraker, J.H., Kenawi, M.S., L'Abée-Lund, J.H., Bakken, T.H., Alfredsen, K.: Assessment of flow ramping in water bodies impacted by hydropower operation in Norway – is hydropower

with environmental restrictions more sustainable? Sci. Total Environ. **832** (2022). https://doi.org/10.1016/j.scitotenv.2022.154776
10. Jiang, J., et al.: A holistic techno-economic evaluation framework for sizing renewable power plant in a hydro-based hybrid generation system. Appl. Energy. **348** (2023). https://doi.org/10.1016/j.apenergy.2023.121537
11. Schroeder, R.E., Loots, I., van Dijk, M., Coetzee, G.L.: Development of a procedure and tool for retrofit hydropower evaluation at South African dams. Water SA. **49**(3), 230–238 (2023). https://doi.org/10.17159/wsa/2023.v49.i3.3980
12. Amougou, C.B., Tsuanyo, D., Fioriti, D., Kenfack, J., Aziz, A., Elé Abiama, P.: LCOE-based optimization for the design of small run-of-river hydropower plants. Energies. **15**(20) (2022). https://doi.org/10.3390/en15207507
13. Lei, K., Chang, J., Wang, Y., Guo, A., Huang, M., Xu, B.: Cascade hydropower stations short-term operation for load distribution considering water level synchronous variation. Renew. Energy. **196**, 683–693 (2022). https://doi.org/10.1016/j.renene.2022.07.039
14. Xie, M., et al.: A hydropower scheduling model to analyze the impacts from integrated wind and solar powers. Sustain. Energy Grids Netw. **27** (2021). https://doi.org/10.1016/j.segan.2021.100499
15. Prado, L., Fonseca, M., Bernardes, J.V., Santos, M.G., Bortoni, E.C., Bastos, G.S.: Forecast of operational downtime of the generating units for sediment cleaning in the water intakes: a case of the Jirau Hydropower Plant. Energies. **16**(17) (2023). https://doi.org/10.3390/en16176354
16. Maldonado-Correa, J., Martín-Martínez, S., Artigao, E., Gómez-Lázaro, E.: Using SCADA data for wind turbine condition monitoring: a systematic literature review. Universidad Nacional de Loja». Accedido: 7 de abril de 2024. [En línea]. Disponible en: https://www.unl.edu.ec/cite/articulos/using-scada-data-wind-turbine-condition-monitoring-systematic-literature-review
17. Torres-Carrion, P., González González, C., Aciar, S., Rodriguez, G.: Methodology for systematic literature review applied to engineering and education. In: 2018 IEEE Global Engineering Education Conference (EDUCON). IEEE (2018). https://doi.org/10.1109/EDUCON.2018.8363388
18. Kitchenham, B.: Procedures for performing systematic reviews. Keele UK Keele Univ. **33** (2004)
19. Núñez, M.Á.C., Jiménez, J.C.S., Mendieta, Á.J.O., Salazar, V.D.R.H., Mendoza, A.A.L., Sigcho, J.R.J.: PV self-consumption in Ecuador: systematic literature review and case studies. In: 2023 IEEE Seventh Ecuador Technical Chapters Meeting (ECTM), pp. 1–8. IEEE (2023). https://doi.org/10.1109/ETCM58927.2023.10309015
20. Und 1_ESQUEMAS_CONCEPTUALES.pdf. Accedido: 17 de abril de 2024. [En línea]. Disponible en: https://tic.uis.edu.co/users/ipred/repositorio/Empresarial/GestionEmpresarial/Nivel3/Analisis_Macroeconomico/Unidad1/Documentos/Und%201_ESQUEMAS_CONCEPTUALES.pdf
21. Paim, M.-A., et al.: Evaluating regulatory strategies for mitigating hydrological risk in Brazil through diversification of its electricity mix. Energy Policy. **128**, 393–401 (2019). https://doi.org/10.1016/j.enpol.2018.12.064
22. Kanyako, F., Lamontagne, J., Baker, E., Turner, S., Wild, T.: Seasonality and trade in hydro-heavy electricity markets: a case study with the West Africa Power Pool (WAPP). Appl. Energy. **329** (2023). https://doi.org/10.1016/j.apenergy.2022.120214
23. O'Connell, Voisin, N., Macknick, Fu: Sensitivity of Western U.S. power system dynamics to droughts compounded with fuel price variability. Appl. Energy. **247**, 745–754 (2019). https://doi.org/10.1016/j.apenergy.2019.01.156
24. Hatamkhani, A., Moridi, A., Haghighi, A.T.: Incorporating ecosystem services value into the optimal development of hydropower projects. Renew. Energy. **203**, 495–505 (2023). https://doi.org/10.1016/j.renene.2022.12.078
25. Nowak, B., Andrzejak, A., Filipiak, G., Ptak, M., Sojka, M.: Assessment of the impact of flow changes and water management rules in the dam reservoir on energy generation at the Jeziorsko Hydropower Plant. Energies. **15**(20) (2022). https://doi.org/10.3390/en15207695

26. Saucedo-Ramírez, O.A., Mahlknecht, J., González-Bravo, R.: Optimization of water allocation networks in highly engineered basins: the case of Guandu River basin, Rio de Janeiro State, Brazil. J. Clean. Prod. **358** (2022). https://doi.org/10.1016/j.jclepro.2022.131787
27. Cuartas, L.A., et al.: Recent hydrological droughts in Brazil and their impact on hydropower generation. Water (Switz.). **14**(4) (2022). https://doi.org/10.3390/w14040601
28. Borowski, P.F.: Water and hydropower—challenges for the economy and enterprises in times of climate change in Africa and Europe. Water (Switz.). **14**(22) (2022). https://doi.org/10.3390/w14223631
29. Brás, T.A., Simoes, S.G., Amorim, F., Fortes, P.: How much extreme weather events have affected European power generation in the past three decades? Renew. Sustain. Energy Rev. **183** (2023). https://doi.org/10.1016/j.rser.2023.113494
30. Liu, Z., Zhang, Z., Zhuo, R., Wang, X.: Optimal operation of independent regional power grid with multiple wind-solar-hydro-battery power. Appl. Energy. **235**, 1541–1550 (2019). https://doi.org/10.1016/j.apenergy.2018.11.072
31. Gonzalez-Salazar, M., Roger Poganietz, W.: Making use of the complementarity of hydropower and variable renewable energy in Latin America: a probabilistic analysis. Energy Strategy Rev. **44** (2022). https://doi.org/10.1016/j.esr.2022.100972
32. Espinel, A., Díaz, I., Vega, A.: Distributed electrical resources with micro hydroelectric power plants in Colombia — study case. Energy Rep. **7**, 169–176 (2021). https://doi.org/10.1016/j.egyr.2021.08.059
33. Ferraz de Campos, É., Pereira, E.B., van Oel, P., Martins, F.R., Gonçalves, A.R., Costa, R.S.: Hybrid power generation for increasing water and energy securities during drought: exploring local and regional effects in a semi-arid basin. J. Environ. Manag. **294** (2021). https://doi.org/10.1016/j.jenvman.2021.112989
34. Zhao, X., Huang, G., Li, Y., Lu, C.: Responses of hydroelectricity generation to streamflow drought under climate change. Renew. Sustain. Energy Rev. **174** (2023). https://doi.org/10.1016/j.rser.2022.113141
35. Wörman, A., et al.: Virtual energy storage gain resulting from the spatio-temporal coordination of hydropower over Europe. Appl. Energy. **272** (2020). https://doi.org/10.1016/j.apenergy.2020.115249
36. Paltán, H.A., Pant, R., Plummer Braeckman, J., Dadson, S.J.: Increased water risks to global hydropower in 1.5 °C and 2.0 °C Warmer Worlds. J. Hydrol. **599** (2021). https://doi.org/10.1016/j.jhydrol.2021.126503
37. Blakers, A., Stocks, M., Lu, B., Cheng, C.: A review of pumped hydro energy storage. Prog. Energy. **3**(2) (2021). https://doi.org/10.1088/2516-1083/abeb5b
38. Kenawi, M.S., Alfredsen, K., Stürzer, L.S., Sandercock, B.K., Bakken, T.H.: High-resolution mapping of land use changes in Norwegian hydropower systems. Renew. Sustain. Energy Rev. **188** (2023). https://doi.org/10.1016/j.rser.2023.113798
39. Okang, Q.A., Bakken, T.H., Bor, A.: Investigation of the hydroelectric development potential of nonpowered dams: a case study of the Buyuk Menderes River Basin. Water (Switz.). **15**(4) (2023). https://doi.org/10.3390/w15040717
40. Ayik, A., Ijumba, N., Kabiri, C., Goffin, P.: Preliminary assessment of small hydropower potential using the Soil and Water Assessment Tool model: a case study of Central Equatoria State, South Sudan. Energy Rep. **9**, 2229–2246 (2023). https://doi.org/10.1016/j.egyr.2023.01.014
41. Nautiyal, H., Goel, V.: Sustainability assessment of hydropower projects. J. Clean. Prod. **265** (2020). https://doi.org/10.1016/j.jclepro.2020.121661
42. Song, C., O'Malley, A., Zydlewski, J., Mo, W.: Balancing fish-energy-cost tradeoffs through strategic basin-wide dam management. Resour. Conserv. Recycl. **161** (2020). https://doi.org/10.1016/j.resconrec.2020.104990
43. Rai, A., Zangpo, D., Feuerbacher, A.: Land cover changes and hydropower in Bhutan: analyzing trade-offs in economic and ecosystem impacts. Energy Sustain. Dev. **78** (2024). https://doi.org/10.1016/j.esd.2023.101371

44. Garrett, K.P., McManamay, R.A., Witt, A.: Harnessing the power of environmental flows: sustaining river ecosystem integrity while increasing energy potential at hydropower dams. Renew. Sustain. Energy Rev. **173** (2023). https://doi.org/10.1016/j.rser.2022.113049
45. Huang, X., Hayashi, K., Fujii, M., Villa, F., Yamazaki, Y., Okazawa, H.: Identification of potential locations for small hydropower plant based on resources time footprint: a case study in Dan River Basin, China. Renew. Energy. **205**, 293–304 (2023). https://doi.org/10.1016/j.renene.2023.01.079
46. Chymyrov, A.U., Chontoev, D.T., Abdrakhmatov, K.E., Zhakeev, B.M.: Modeling and study of reservoir parameters of the designed Kambarata-1 HPP based on space data. Izv. Atmos. Ocean Phys. **58**(12), 1509–1518 (2022). https://doi.org/10.1134/S0001433822120106
47. McManamay, R.A., Larson, K., Tagestad, J., Jager, H.I., DeRolph, C.R., Bevelhimer, M.S.: Mutually beneficial outcomes for hydropower expansion and environmental protection at a basin scale. Sci. Total Environ. **896** (2023). https://doi.org/10.1016/j.scitotenv.2023.165298
48. Zahedi, R., Ahmadi, A., Gitifar, S.: Reduction of the environmental impacts of the hydropower plant by microalgae cultivation and biodiesel production. J. Environ. Manag. **304** (2022). https://doi.org/10.1016/j.jenvman.2021.114247
49. Bahati, H.K., Ogenrwoth, A., Sempewo, J.I.: Quantifying the potential impacts of land-use and climate change on hydropower reliability of muzizi hydropower plant, Uganda. J. Water Clim. Change. **12**(6), 2526–2554 (2021). https://doi.org/10.2166/wcc.2021.273
50. Yue, Q., Li, S., Hu, X., Zhang, Y., Xue, M., Wang, H.: Sustainability analysis of electricity generation technologies based on life-cycle assessment and life-cycle cost—a case study in Liaoning Province. Energy Technol. **7**(7) (2019). https://doi.org/10.1002/ente.201900365
51. Hočevar, M., Novak, L., Drešar, P., Rak, G.: The status quo and future of hydropower in Slovenia. Energies. **15**(19) (2022). https://doi.org/10.3390/en15196977
52. Shrestha, P., Shrestha, A., Shrestha, N.T., Papadakis, A., Maskey, R.K.: Assessment on scaling-up of mini-grid initiative: case study of mini-grid in rural Nepal. Int. J. Precis. Eng. Manuf. Green Technol. **8**(1), 217–231 (2021). https://doi.org/10.1007/s40684-020-00190-x
53. Tellefsen, T., Van Putten, J., Gjerde, O.: Norwegian hydropower: connecting to continental Europe. IEEE Power Energy Mag. **18**(5), 27–35 (2020). https://doi.org/10.1109/MPE.2020.3001417
54. Iweh, C.D., Semassou, G.C., Ahouansou, R.H.: Optimization of a hybrid off-grid solar PV - hydro power systems for rural electrification in Cameroon. J. Electr. Comput. Eng. **2024** (2024). https://doi.org/10.1155/2024/4199455
55. Jovan, D.J., Dolanc, G., Pregelj, B.: Utilization of excess water accumulation for green hydrogen production in a run-of-river hydropower plant. Renew. Energy. **195**, 780–794 (2022). https://doi.org/10.1016/j.renene.2022.06.079
56. Rodríguez-Pérez, Á.M., Pérez-Calañas, C., Pulido-Calvo, I.: Energy recovery in pressurized hydraulic networks. Water Resour. Manag. **35**(6), 1977–1990 (2021). https://doi.org/10.1007/s11269-021-02824-4
57. Castro, P.W., Mantilla, C.A.: Implementation of strategies for the management of dams with sedimented reservoirs. Water Resour. Manag. **35**(13), 4399–4413 (2021). https://doi.org/10.1007/s11269-021-02956-7
58. Fagerström, J., Das, S., Klyve, Ø.S., Olkkonen, V., Marstein, E.S.: Profitability of battery storage in hybrid hydropower–solar photovoltaic plants. J. Energy Storage. **77** (2024). https://doi.org/10.1016/j.est.2023.109827
59. Parida, A., Chatterjee, D.: Remote area power supply through suitable solar PV augmented micro-hydro generation: a case study. Int. Trans. Electr. Energy Syst. **30**(2) (2020). https://doi.org/10.1002/2050-7038.12247
60. Sasges, G., Ziegler, A.D.: We have eaten the rivers: the past, present, and unsustainable future of hydroelectricity in Vietnam. Sustainability. **15**(11) (2023). https://doi.org/10.3390/su15118969

61. Dolter, B., Fellows, G.K., Rivers, N.: The cost effectiveness of new reservoir hydroelectricity: British Columbia's Site C project. Energy Policy. **169** (2022). https://doi.org/10.1016/j.enpol.2022.113161
62. Calder, R.S.D., Robinson, C.S., Borsuk, M.E.: Total social costs and benefits of long-distance hydropower transmission. Environ. Sci. Technol. **56**(24), 17510–17522 (2022). https://doi.org/10.1021/acs.est.2c06221
63. Almeida, R.M., et al.: Climate change may impair electricity generation and economic viability of future Amazon hydropower. Glob. Environ. Change. **71** (2021). https://doi.org/10.1016/j.gloenvcha.2021.102383
64. Ramião, J.P., Carvalho-Santos, C., Pinto, R., Pascoal, C.: Hydropower contribution to the renewable energy transition under climate change. Water Resour. Manag. **37**(1), 175–191 (2023). https://doi.org/10.1007/s11269-022-03361-4
65. Zanoli, S.M., Pepe, C., Astolfi, G., Luzi, F.: Reservoir advanced process control for hydroelectric power production. Processes. **11**(2) (2023). https://doi.org/10.3390/pr11020300
66. Li, X., Ouyang, T., Lin, Y., Zhu, Z.: Interstage difference and deterministic decomposition of internal unsteady flow in a five-stage centrifugal pump as turbine. Phys. Fluids. **35**(4) (2023). https://doi.org/10.1063/5.0150300
67. Balacco, G., et al.: PaT-ID: a tool for the selection of the optimal pump as turbine for a water distribution network. Energy. **282** (2023). https://doi.org/10.1016/j.energy.2023.128366
68. Trinh, V.L., Chung, C.K.: Renewable energy for SDG-7 and sustainable electrical production, integration, industrial application, and globalization: review. Clean. Eng. Technol. **15** (2023). https://doi.org/10.1016/j.clet.2023.100657
69. Sambou, M.H.A., Liersch, S., Koch, H., Vissin, E.W., Albergel, J., Sane, M.L.: Synergies and trade-offs in water resources management in the Bafing Watershed under climate change. Water (Switz.). **15**(11) (2023). https://doi.org/10.3390/w15112067
70. Novitasari, D., Sarjiya, Hadi, S.P., Budiarto, R., Deendarlianto: The climate and land-use changes impact on water availability for hydropower plants in Indonesia. Energy Strategy Rev. **46** (2023). https://doi.org/10.1016/j.esr.2022.101043
71. Yildiz, V., Vrugt, J.A.: A toolbox for the optimal design of run-of-river hydropower plants. Environ. Model. Softw. **111**, 134–152 (2019). https://doi.org/10.1016/j.envsoft.2018.08.018
72. Kumar, K., Saini, R.P.: Application of artificial intelligence for the optimization of hydropower energy generation. EAI Endorsed Trans. Ind. Netw. Intell. Syst. **8**(28), 1–12 (2021). https://doi.org/10.4108/EAI.6-8-2021.170560
73. Rai, A.K., Kumar, A., Staubli, T.: Optimization of hydropower plants regarding hydro-abrasive erosion. Int. J. Fluid Mach. Syst. **12**(2), 119–127 (2019). https://doi.org/10.5293/IJFMS.2019.12.2.119
74. Ilseven, E., Göl, M.: Incorporation of generator maintenance scheduling with long-term power sector forecasting and planning studies. IET Gener. Transm. Distrib. **14**(13), 2581–2591 (2020). https://doi.org/10.1049/iet-gtd.2019.1545
75. Dourado, G.F., Rallings, A.M., Viers, J.H.: Overcoming persistent challenges in putting environmental flow policy into practice: a systematic review and bibliometric analysis. Environ. Res. Lett. **18**(4) (2023). https://doi.org/10.1088/1748-9326/acc196
76. Ebhota, W.S.: Power accessibility, fossil fuel and the exploitation of small hydropower technology in sub-saharan Africa. Int. J. Sustain. Energy Plan. Manag. **19**, 13–28 (2019). https://doi.org/10.5278/ijsepm.2019.19.3
77. Mohale, V., Sonandkar, S., Chelliah, T.R.: Large asynchronous hydro generating systems connected to TCSC compensated transmission lines (Simulation with experimental validation): A review on SSO perspectives. Electr. Power Syst. Res. **219** (2023). https://doi.org/10.1016/j.epsr.2023.109257
78. Gajic, A., Stevanovic, V., Pejovic, S.: Pumped-hydro storages are balancing electric energy production of wind and solar reducing average costs and pollution. Int. J. Fluid Mach. Syst. **12**(1), 47–55 (2019). https://doi.org/10.5293/IJFMS.2019.12.1.047

79. Gasore, G., Ahlborg, H., Ntagwirumugara, E., Zimmerle, D.: Progress for on-grid renewable energy systems: identification of sustainability factors for small-scale hydropower in Rwanda. Energies. **14**(4) (2021). https://doi.org/10.3390/en14040826
80. Mauro, M.A.S., Pimenta, C.S., Garcete M, R.A., Rocha E, J.J.: The evolution of the maintenance processes increases the operational availability and contribute to the operational efficiency of Itaipu Power Plant. CIGRE Sci. Eng. **2021**(23) (2021)

Part IV
Low Carbon Energy Technology

Enabling Long-Term Investments in Low-Carbon Energy Systems in Galapagos Islands

Janeth Carolina Godoy, Angel Recalde, Ricardo Cajo, Mustafa Arman, and José Jara

1 Introduction

The threat of climate change is pushing global efforts toward renewable energy, but transitioning remote islands like the Galapagos remains complex due to their dependence on fossil fuels. The Galapagos Islands, located about 1000 km from mainland Ecuador, consist of 19 islands, four of which are inhabited by approximately 33,000 residents. These islands are not connected to the continental power grid, and their isolated power systems primarily rely on diesel generators, contributing to environmental degradation and impacting the islands' unique ecosystems [1].

Despite the establishment of the Galapagos National Park to protect 97% of the land and the islands' UNESCO World Heritage status due to their unique biodiversity and high endemism, increasing energy demands from tourism and everyday activities are challenging the transition to renewable energy [1]. For example, the 2001 Jessica oil spill released around three million liters of diesel, causing significant ecological damage [2, 3]. In response, various initiatives were launched. The *e8* (now the Global Electricity Partnership) constructed the first wind-diesel system to reduce fossil fuel consumption on San Cristobal Island [4]. Additionally, the

J. C. Godoy · M. Arman
Renewable and Sustainable Energy Systems, Technical University of Munich, Garching bei München, Germany

A. Recalde (✉) · R. Cajo
Escuela Superior Politécnica del Litoral, ESPOL, Facultad de Ingeniería en Electricidad y Computación, Campus Gustavo Galindo, Guayaquil, Ecuador
e-mail: arecalde@espol.edu.ec

J. Jara
Universidad del Azuay, Grupo Investigación Galapagos Research Network/CIENER, Facultad de Ciencia y Tecnología, Cuenca, Ecuador

M. Espinoza-Andaluz et al. (eds.), *Congress on Research, Development and Innovation in Renewable Energies*, Green Energy and Technology,
https://doi.org/10.1007/978-3-031-88995-0_13

Government introduced the "Zero Fossil Fuel Initiative" and deployed wind and solar projects, but as of 2022, only about 11.8% of the islands' electricity generation comes from renewables.

The islands have an electricity mix composed of renewable sources (wind and solar PV) and thermal generation run by diesel generators. Each island has an independent power system and there's no transmission interconnection between the islands, except Santa Cruz and Baltra that have a short underwater cable interconnection. Electricity is mainly consumed in residential, public, and commercial (touristic) sector [5]. For more detailed information about the Galapagos electricity reports please refer to [6, 7].

The scope of this chapter is to develop a comprehensive strategy to eliminate fossil fuel dependence in the Galapagos using open-source tools. This strategy will assess environmental risks, the technical feasibility of new technologies, and economic factors such as investment and operational costs. By examining these aspects, the study aims to answer critical questions about achieving a zero-carbon energy system and balancing economic sectors while maintaining the island's ecological integrity. The goal is to propose a realistic and holistic plan for transitioning to cleaner energy, enhancing both the power system and transportation, ultimately contributing to the sustainable development and conservation of the Galapagos Islands.

1.1 Literature Review

The Galapagos Islands hold significant ecological value globally. Decarbonizing their energy system is crucial for sustainable development [8]. The Ecuadorian government, alongside international organizations, intensified efforts to achieve a zero-emission energy system [1, 9]. This has led to developing model-based project portfolios and assessing their feasibility with available technology.

Foundational work includes the National Energy Efficiency Plan (Plan Nacional de Eficiencia Energética (PLANEER) 2016–2035), published in 2015 [10], and the Electricity Master Plan (Plan Maestro de Electricidad), published in 2018 [11], which focus on enhancing energy efficiency through appliance substitution. Additionally, [12] analyses achieving 100% renewable energy by 2031, proposing a hybrid system with renewable technologies and storage, forecasting demand with ANNs, and optimizing configurations with HOMER Pro. This study demonstrates reduced costs, lower emissions, and increased reliability, offering a model for sustainable energy planning in isolated regions. Furthermore, the "Galapagos 2030: Galapagos Islands" report by the Galapagos Government Council [5] examines social and economic aspects to enhance development, while the Galapagos Islands Energy Transition Plan (Plan de Transición Energética de las Islas Galapagos) [13] discusses new regulations and enhancements for grid digitalization and transportation systems.

Public and private alliances have also explored the future of the Galapagos. The Bariloche Foundation collaborated with the government on civil engineering regulations, resulting in the "Consultancy for Survey and Development of Sustainable Behaviour Standards of Buildings in the Galapagos Islands" [14]. This work aimed to enhance residential energy efficiency. Additionally, the study "Consulting for the Construction of the Action Plan for the Sustainable Energy Transition of the Galapagos Islands Archipelago—Period 2020–2040 (Consultoría para la Construcción del Plan de Acción para la Transición Energética sostenible del Archipiélago de las Islas Galápagos—Periodo 2020–2040)" [7] examined complex energy transition scenarios using LEAP modeling. Numerous studies address energy transitions in island power systems. For example, [15] explores renewable energy potential with multi-criteria decision analysis, while works like #GalápagosGoesGreen [16] and Carbon Footprint of the Galapagos Islands [17] propose strategies for sustainable tourism. D. Morales et al. [18] focus on Demand Side Management (DSM) with a combined Demand Response (DR) and Time of Use (TOU) pricing program, and J.-M. Clairand et al. [19] advocate for renewable microgrids to benefit remote communities. Daniel Icaza-Alvarez et al. present a plan for decarbonizing by 2050 using EnergyPLAN software to assess a fully renewable system.

Despite existing models' benefits, limitations persist in simulating current operations, as they were designed for future scenarios rather than present-day modelling [20]. This work explores feasible energy transition routes, highlighting extensive efforts to address the complex challenges of achieving a sustainable energy paradigm in the Galapagos Islands. Finally, it is important to highlight that this work was developed before the existence of a green hydrogen roadmap in Ecuador [21] (recently published in 2023). Therefore, from an academic perspective, this study offers an alternative vision on how we could advance the decarbonization of the Galapagos Islands.

2 Methodology

The approach in this paper develops a long-term investment plan that is composed of three main steps. It begins with the data acquisition stage, i.e., investigation, compilation, and analysis of the energy system behavior across all the economic sectors. This step entails the renewable resources assessment and collecting valuable data from previous works. The second step corresponds to defining the scenarios. It requires refinement of the inputs for each technological stage that would serve to reach the decarbonization goal considering the unique energy flow aspects of Galapagos. In this work, three different scenarios are proposed; each one increases the level of ambition in energy transition toward zero carbon emissions, which produces a change in the demand requirements. Each scenario encompasses strategy definition, new policies, upgrades, and initial investments for phasing out non-sustainable resources. Finally, the third step involves formulating an optimization problem for solving energetic transition with *urbs* tool. Before problem solving, it is imperative to conduct energy model validation. After simulations are run, results are analyzed, and the economic viability is validated.

2.1 Renewable Energy Assessment

Renewable energy assessment on the Galapagos Islands covers the utilization of solar PV and wind energies, as well as hydrogen, for the Isabela, Santa Cruz and Baltra, San Cristobal and Floreana islands. An info chart map of Galapagos is shown in Fig. 1.

Solar and Wind

The Galapagos Islands possess significant potential for renewable energy, including geothermal, solar, wind, and biomass. However, this project primarily focuses on solar photovoltaic (PV) and onshore wind energy resources. Due to the unique characteristics of the archipelago, multiple factors must be evaluated when selecting appropriate sites for renewable energy power plants. To address this, we referenced the study by Villacreses et al. [15], which employed a multi-criteria analysis (MCA) for resource assessment. This study used geographic data to identify suitable locations for wind and solar power generation, considering factors such as priority influences and restrictions related to political, environmental, or natural risks. The identified sites exhibit average wind speeds of 4–7 m/s and receive 5.70 kWh/m^2/day of solar radiation without limitations. The work highlighted prime locations, identifying a solar potential area of 524.5 km^2 on the main inhabited island, excluding Santiago Island. Given that a 1 MWp PV plant requires 1800 m^2, the Galapagos has a technical potential of 291.38 GW.

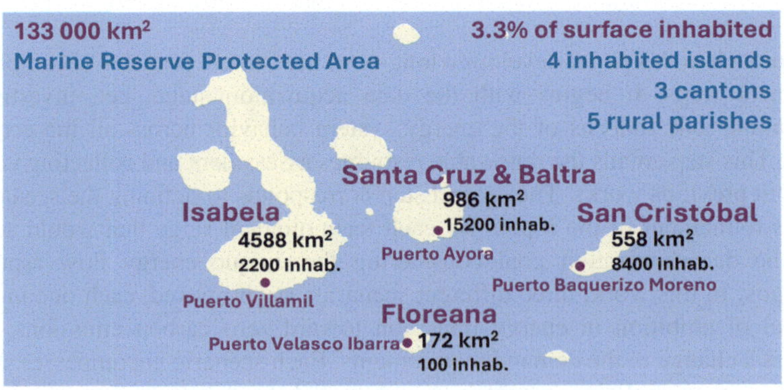

Fig. 1 Galapagos Island marine reserve

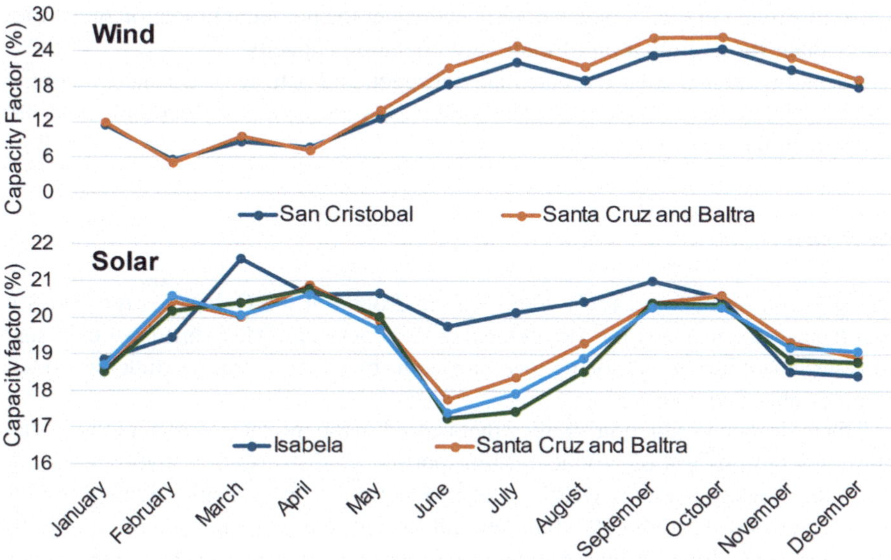

Fig. 2 Monthly wind and solar capacity factor per island in monthly average values

Utilizing the *Renewable.ninja* online web tool developed by Pfenninger and Staffell [22] and factoring in a 15% system performance loss in PV plants due to various factors such as wire losses, soiling, mismatching, and inverter efficiency, an hourly availability factor is calculated for each island. Figure 2 illustrates the solar capacity factors for the islands of Isabela, Santa Cruz and Baltra, San Cristobal, and Floreana throughout a typical year. The capacity factor, which measures the ratio of actual energy produced to the maximum potential energy, shows a consistent seasonal pattern across all islands, with higher values observed from March to May and lower values from June to August. Among the islands, Isabela consistently demonstrates the highest capacity factor, indicating more efficient solar energy utilization, while Santa Cruz and Baltra, San Cristobal, and Floreana follow similar but slightly lower trends.

The Galapagos Islands' proximity to the equator results in a near-zero Coriolis force, leading to very low wind speeds. Consequently, there are only a few suitable locations with mean wind speeds meeting the criteria, primarily on the two major islands, Santa Cruz and Baltra, and San Cristobal. These islands already host existing wind plants at the recommended locations, validating the GIS results using the MCA. The usable area on these two islands is 147.2 km^2, with a shared area for solar energy of 48.1 km^2 [15].

Wind data analysis indicates that only wind turbines with IEC III classification are suitable for these low mean wind speeds. Simulation using similar turbine dimensions in [22] determines the hourly availability factor for these islands with wind potential, considering a system efficiency of 95%. The monthly capacity factor is also illustrated in Fig. 2, which does not consider the potential limitation of a diesel power plant for its limitation of minimum operating capacity.

Beyond wind and solar PV potential, hydrogen and battery-based energy storage system (ESS) can be implemented as well. ESSs are essential to complement the intermittent nature of renewables.

Hydrogen

The substitution of fossil fuels with green hydrogen in the Galapagos represents a significant enhancement for the archipelago's ecosystem. This shift ensures a sustainable and enduring well-being for its residents by enabling transportation powered by renewable fuel sources.

The hydrogen production strategy that is used in the model is shown in Fig. 3 that illustrates the process of producing, transporting, and distributing hydrogen using renewable energy sources (RES). The production process begins with seawater being processed through a desalination unit, which removes salts and other impurities, resulting in deionized water suitable for hydrogen production. This step ensures that the water used does not negatively impact the island's hydrology, maintaining environmental balance. Following desalination, the deionized water is fed into a proton exchange membrane (PEM) electrolyzer. The PEM electrolyzer, powered entirely by electricity from a 100% renewable energy mix, splits the water into hydrogen and oxygen. This renewable energy mix is crucial for stabilizing the grid, especially in cases of overproduction due to the variability inherent in renewable energy sources. As a result, higher capacities are often installed to ensure availability, and hydrogen production serves as a flexible means to absorb excess energy.

The hydrogen produced by the PEM electrolyzer is in compressed gas form at medium pressure. It is then transported through a dedicated pipeline infrastructure. These pipelines are essential for efficiently moving the hydrogen from the production site to areas where it will be stored, further processed, or used. The infrastructure supports both inland and maritime mobility on the larger islands, where hydrogen serves as a primary fuel. Finally, the hydrogen is delivered to fueling stations or retail points, where it may be further compressed to high pressure for end-use applications,

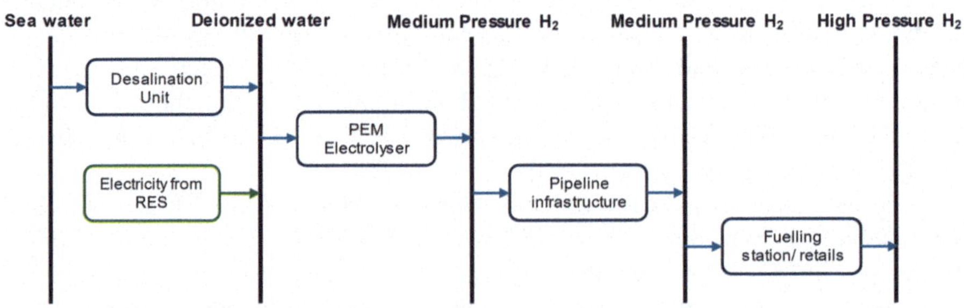

Fig. 3 Hydrogen production strategy for Galapagos Islands model

such as hydrogen fuel cell vehicles. This entire process underscores the sustainable integration of renewable energy with hydrogen production, transportation, and use, facilitating a clean and efficient energy ecosystem on the islands.

Estimating electrolytic hydrogen production involves two main steps. First, we determine the available excess renewable electrical power on each island for electrolysis. This includes power from existing renewable energy plants and additional plants connected specifically to the hydrogen production facility to ensure it runs continuously based on demand. Next, we calculate the amount of H_2 produced, considering its potential to replace diesel and gasoline in the transportation sector. This calculation accounts for the energy equivalence based on the lower heating value (LHV) and the efficiency of the production and consumption systems.

2.2 Scenario's Generation

Considering the model's scope extends until 2050, it is important to account for both the rising demand and population alongside advancements in technology efficiency and prices. To create the scenarios, each sector was individually examined, dividing the total energy demand of the four islands into distinct categories: domestic (that includes residential, commercial, and public lighting), land transportation, and maritime transportation (apart from fishing and islander commuting, touristic use plays the biggest role in the energy consumption in the transport sector). Each sector will be examined across various scenarios, with the reference scenario acting as a point of comparison. In the reference scenario, the future is depicted without deliberate interventions, with rising demand met predominantly by fossil fuels and the current proportion of renewables. This scenario serves as the standard against which other scenarios will be evaluated. Furthermore, each sector exhibits unique consumption patterns, growth forecasts, and strategies for substitution and decarbonization.

Main Drivers for the Scenario's Generation

The main drivers for the scenario's generation are growth forecast and the transport sector trends with respect to inland mobility and maritime fleet. Population growth is based on the National Institute for Statistics and Census (INEC) [23], see Table 1.

The main socioeconomic driver is tourism, as approximately 60% of the residents are engaged in tourism activities and tourism contributes about 80% [24] to the local economy. Regulations are in place to limit the number of tourists, avoid overcrowding, and preserve the ecosystem. The scenario foresees a gradual increase in annual tourists, reaching 400,000 in 2030 and maintaining this level throughout the study period, as shown in Table 1 [7, 17].

Table 1 Forecast for 2050

		2018	2030	2040	2050
Population (k inhabitants)	San Cristobal	8.595	9.705	10.42	10.87
	Floreana	0.135	0.152	0.163	0.17
	Isabela	2.911	3.614	4.199	4.74
	Santa Cruz and Baltra	19.96	27.226	34.217	41.783
Tourists (k people)	On Land	207	300	300	300
	On Board	69	100	100	100
Vehicles (units)	Cars/SUV	577	523	580	699
	Van	38	35	39	47
	Truck	735	665	737	888
	Bus	56	51	57	69
	Motorcycle	1396	1263	1399	1685
	Heavy duty vehicle	155	145	163	199
	Total vehicles	**2957**	**2682**	**2975**	**3587**
Vessels (units)	Passenger vessel	316	374	374	374
	Cargo vessel	102	120	120	120
	Fishing boats	264	311	311	311
	Control vessel	15	17	17	17
	Other	103	121	121	121
	Total vessels	**800**	**943**	**943**	**943**

k: thousands

Transport Forecast for Inland Mobility

In 2022, there were 2295 registered vehicles in Galapagos, with an average use of 8 years. The expected lifespan for vehicles in South America is 18 years, and replacements are accounted for. Following this period, electric vehicles (EVs) or fuel cell electric vehicles (FCEVs) may be considered as alternatives, depending on each scenario, with some exceptions for tankers and construction vehicles. As of 2022, there are already 52 electric vehicles, accounting for 2.25% of the total inland mobility. 4.5% of heavy-duty vehicles are accounted for in the islands, with hydrogen production across the scenarios. A forecast of vehicle types is developed and presented in Table 1.

Transport Forecast for the Maritime Fleet

Future trends in maritime transportation are shifting toward using sustainable vessels. These vessels are opting for electricity on shorter routes and hydrogen for longer journeys. While the precise costs of these technologies remain uncertain, as the production costs for electricity and hydrogen decrease, they become more appealing. Additionally, the reduced operational expenses associated with electricity and hydrogen contribute to their attractiveness as eco-friendly options.

Currently, hydrogen ferries are approximately 30% more expensive than diesel ones of similar capacity [25]. However, it is expected that this price difference will decrease over the next 15 years due to ongoing advancements and increased production [26]. Eventually, prices are projected to equalize as economies of scale lower costs in a high-volume market. Without regulations controlling vessel numbers, there is a projected continuous increase in vessel count, expected to match Galapagos' tourist capacity by 2030, as shown in Table 1. In 2020, there were approximately 800 vessels registered [27], estimated to increase to 943 vessels, with projections following the tourism ratio. Limiting is essential to protect the ecosystem from further harm. Furthermore, fossil fuel prices are notoriously unpredictable due to factors such as market manipulation and political turmoil. To manage costs and avoid escalating subsidies, it is crucial to determine the appropriate price of fossil fuels in Galapagos to make electricity and transportation affordable. In addition, shipping fuel to the islands via tankers increases costs and carbon emissions. By analyzing data from 2016 to 2019 in [17, 28] and using Energy Outlook forecasts [29], future fossil fuel prices were estimated.

Strategies for Substitution and Decarbonization

Within this framework, three different scenarios emerge, each with varying levels of ambition in energy transition regarding achieving zero carbon emissions based on the willingness for investments and efforts that are required to achieve this goal. Each scenario encompasses comprehensive strategies, policy implementations, replacements, and fresh investments; all geared toward mitigating emissions and phasing out non-sustainable resources across energy sectors. The three scenarios are detailed as follows.

1. Scenario 1, or *conservative transition* scenario, involves a modest approach, relying primarily on transportation policies to gradually shift toward sustainable sources. Additionally, this scenario implicates a shift to e-Mobility in road transportation, specifically passenger transport such as cars, motorcycles, and buses.
2. Scenario 2, also known as *mild transition* scenario, adopts a moderate stance, introducing hydrogen to facilitate a gradual transition in energy consumption.
3. Scenario 3, or *advanced transition* scenario, represents a more ambitious approach, emphasizing rapid adoption of clean electricity sources and widespread electrification of inland mobility. In this scenario, hydrogen is utilized early in the maritime sector to aid in decarbonizing vessels.

The strategies for substitution and decarbonizing the three sectors (domestic, inland transport, maritime transport) are shown in Table 2. The strategies for domestic sector include (1) percentage (%) of penetration of the induction stove for replacing liquefied petroleum gas stoves (LPG), (2) percentage of penetration of solar water/electric heater to replace LPG water heater, (3) percentage of household appliances replaced by more energy efficient units. For the domestic sector it is

Table 2 Strategy's projections (in %) for decarbonization (all scenarios)

		Conservative			Mild			Advanced		
		2030	2040	2050	2030	2040	2050	2030	2040	2050
Domestic	**1** LPG to IS	50	75	100	50	100	100	100	100	100
	2 LP to S/E water heater	25 S 10 E	50 S 20 E	75 S 25 E	40 S 15 E	75 S 25 E	75 S 25 E	75 S 25 E	75 S 25 E	75 S 25 E
	3 Efficient app.	50	100	100	100	100	100	100	100	100
Inland	**1** Different means	100	80	75	75	50	50	70	40	40
	2 PKM to EB	100	120	125	125	150	150	130	160	160
	3 ICEV to BEV	100	80	75	80	70	50	80	40	0
	4 HD to FCEV	100	100	100	100	100	50	100	50	0
Maritime	**1** Policies	25	50	100	50	75	100	50	100	100
	2 Electric prop.	0	0	10	0	25	50	10	50	75
	3 H2 prop.	0	0	0	0	0	50	0	50	75

S: solar heater, E: electric heater, IS: induction stove, EB: electric buses, HD: heavy duty

assumed that the 100% of compact fluorescent lamps (CFL) are replaced by LED lamps in all scenarios. The strategies for inland transport sector include (1) change in percentage (Δ%) of motorized passenger-kilometer (PKMs) attended by the different means of transport, (2) change in percentage of motorized PKMs attended by electric buses relative to the reference year, (3) percentage of ICEV for the different means of transport that are replaced with BEV, and (4) percentage of ICEV for heavy duty vehicles replaced by fuel cell electric vehicles (FCEV). Notice that diesel buses with internal combustion engines (ICEV) that are replaced with a battery electric vehicle (BEV) are not considered because of hydrogen-based transportation. Finally, the strategies for maritime transport sector include: (1) implementing policies for the replacement of existing fleet, such that a decrease in fuel consumption is obtained (slow steaming, solar energy auxiliary engine with battery system), (2) switching to electric propulsion motors for the suitable fleet (fishing boats, speed boats, passenger vessel with capacity from 12 to 35, yacht), and (3) switching to hydrogen powered ships for the suitable fleet (passenger vessels with high capacity and cargo ships).

Based on the projections shown in Table 2, the electricity demand for each scenario per island is shown in Table 3. Table 3 illustrates projected electricity and hydrogen demand (in GWh) across four islands—San Cristobal, Floreana, Isabela, and Santa Cruz and Baltra—from 2022 to 2050. The projections are shown under various scenarios: reference, conservative, mild, and advanced, each reflecting different assumptions about future energy needs.

In San Cristobal (see Table 3), electricity demand steadily increases across all scenarios, growing from 15.72 GWh in 2022 to 22.08 GWh by 2050 in the conservative scenario. The mild and advanced scenarios predict even higher growth, with demand reaching 42.3 GWh and 55.2 GWh by 2050, respectively. Notably, these scenarios also introduce hydrogen demand, which reaches 13.32 GWh by 2050 in the mild scenario and 21.44 GWh in the advanced scenario. Numbers in green

Table 3 Electricity demand (in GWh) per island for all scenarios

	Reference			Conservative			Mild			Advanced		
Base year 2022	2030	2040	2050	2030	2040	2050	2030	2040	2050	2030	2040	2050
Santa Cruz and Baltra												
38.1	42.2	49.0	56.9	45.4	61.2	84.9	47.5	79.8	**144.4**	58.0	**130.2**	**187.1**
Isabela												
5.6	6.2	7.1	8.3	6.0	7.5	9.7	6.4	9.8	12.7	7.7	11.7	14.8
Floreana												
0.26	0.28	0.33	0.38	0.23	0.25	0.3	0.22	0.34	0.41	0.25	0.38	0.45
San Cristóbal												
15.7	18.2	21.1	24.5	19.0	18.6	22.1	16.9	24.3	**42.3**	20.4	**42.4**	**55.2**

See text for description of bold and underlined numbers

color means a combination of electricity and hydrogen, i.e., 29 GWh of electricity and 13.3 GWh of hydrogen in 2050 in mild scenario, 29 GWh of electricity and 13.5 GWh of hydrogen in 2040 (advanced), and 33.8 GWh of electricity and 21.4 GWh of hydrogen in 2050 (advanced). Floreana shows relatively low but consistent growth in electricity demand across all scenarios, reflecting its smaller population and lower consumption compared to the other islands. Electricity demand increases modestly from 0.26 GWh in 2022 to 0.45 GWh by 2050 in the advanced scenario, with no hydrogen demand projected. In Isabela, electricity demand also rises steadily, starting from 5.56 GWh in 2022 and reaching 9.7 GWh by 2050 in the conservative scenario.

The advanced scenario projects a more substantial increase, with electricity demand expected to reach 14.82 GWh by 2050, although hydrogen demand remains absent. Santa Cruz and Baltra exhibit the largest growth in energy demand among the islands. Electricity demand surges across all scenarios, increasing from 38.1 GWh in 2022 to 84.87 GWh by 2050 in the conservative scenario. Numbers in green color means a combination of electricity and hydrogen, i.e., 111.4 GWh of electricity and 33.1 GWh of hydrogen in 2050 in mild scenario, 95.1 GWh of electricity and 35.1 GWh of hydrogen in 2040 (advanced), and 129.8 GWh of electricity and 57.3 GWh of hydrogen in 2050 (advanced).

The mild and advanced scenarios anticipate significant increases in both electricity and hydrogen demand, with hydrogen demand projected to reach 33.05 GWh by 2050 in the mild scenario and 57.29 GWh in the advanced scenario by the same year. The emergence of hydrogen as a significant energy source by 2050 highlights a potential strategic shift toward hydrogen in future energy planning.

2.3 Modelling Tool: Linear Optimization Model for Distributed Energy Systems **urbs**

The Galapagos power system was modelled using *urbs*, a tool designed to optimize both capacity expansion and unit commitment. This optimization is achieved by minimizing the total annual costs, which encompass investment, operational, and environmental factors. Moreover, *urbs* facilitates the inclusion of multiple input and output commodities, allowing for a detailed examination of energy conversion processes. With a high temporal resolution of 8760 h/year, *urbs* provides a chronological perspective on supply and demand dynamics. The model ensures continuous coverage of demand by considering input commodities and technological processes at each time step. Additionally, it treats the expansion of energy and power capacities independently while maintaining a linear relationship between them [30, 31]. A scheme showing the model structure is depicted in Fig. 4.

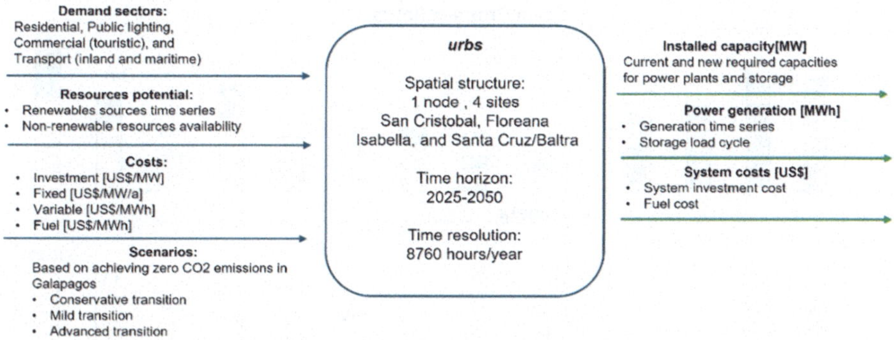

Fig. 4 *urbs* structure for the Galapagos Islands model

3 Results and Discussion

In this section, the results and discussion are presented. First, generation installed capacity mix and electricity generation are discussed. Then, feasibility in hydrogen production is shown. Finally, analysis in emission reduction and cost analysis is presented.

3.1 Installed Capacity and Electricity Generation

In this section, we present the results for installed capacity and electricity generation. Figure 5 represents the installed capacity of different energy sources in the Galápagos Islands—specifically in San Cristobal, Floreana, Isabela, and Santa Cruz and Baltra—from 2022 to 2050 under various scenarios. The energy sources analyzed include diesel plants, PV systems, wind parks, and hydrogen electrolyzers.

Santa Cruz, connected to Baltra Island, is the largest energy consumer in the Galápagos due to its population and tourism. In 2022, their combined capacity is 17.26 MW, mostly from diesel, with 58% on standby. The reference scenario predicts a decrease to 12.03 MW by 2050, still using diesel. The conservative scenario projects an increase to 68 MW, the mild scenario to 113.97 MW with hydrogen electrolyzers, and the advanced scenario to 152.03 MW by 2050 with PV, wind, and hydrogen. Santa Cruz and Baltra will be the main centers for hydrogen production, requiring 133 MW of capacity and a 19 MW electrolyzer. By 2050, the advanced scenario expects 262 GWh of generation and 383 MWh of storage, with hydrogen production using 30% of the total energy and wind helping reduce storage needs. Isabela Island has geothermal potential that is not utilized due to costs and environmental concerns. In 2022, its capacity is 4.70 MW, mainly from diesel. The reference scenario reduces this to 1.49 MW by 2050, still using diesel. The conservative scenario increases capacity to 7.02 MW by 2050 with PV systems, the mild scenario to 8.94 MW, and the advanced scenario to 10.29 MW, all primarily driven

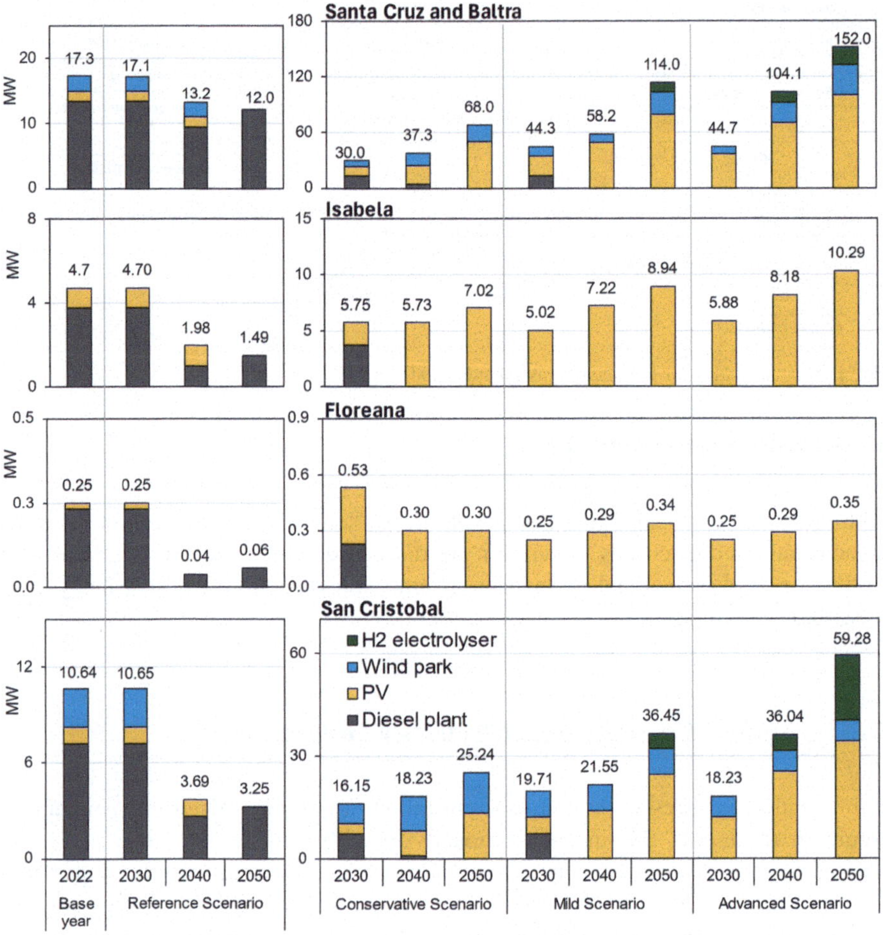

Fig. 5 Forecast of the installed capacity (in MW) per island for all scenarios

by PV. Isabela plans to meet its energy needs with 62.5 MWh of storage and 21.28 GWh of total generation by 2050.

Floreana's base capacity is 0.25 MW, mainly from diesel. The reference scenario shows no change by 2030. The conservative scenario predicts a decline from 0.53 MW in 2030 to 0.30 MW by 2050 as PV replaces diesel. The mild scenario maintains capacity between 0.29 MW and 0.35 MW from PV, while the advanced scenario keeps it between 0.34 MW and 0.35 MW with more PV use. Floreana aims to be fully reliant on renewable energy by 2040, boosting self-sufficiency and benefits like tourism and public services. Wind energy is not feasible due to low demand, and there are no plans for hydrogen production. The advanced scenario includes 2.1 MWh of storage and 0.7 GWh of annual solar generation. In San Cristobal, the installed capacity in 2022 is 10.64 MW, mostly from diesel. The

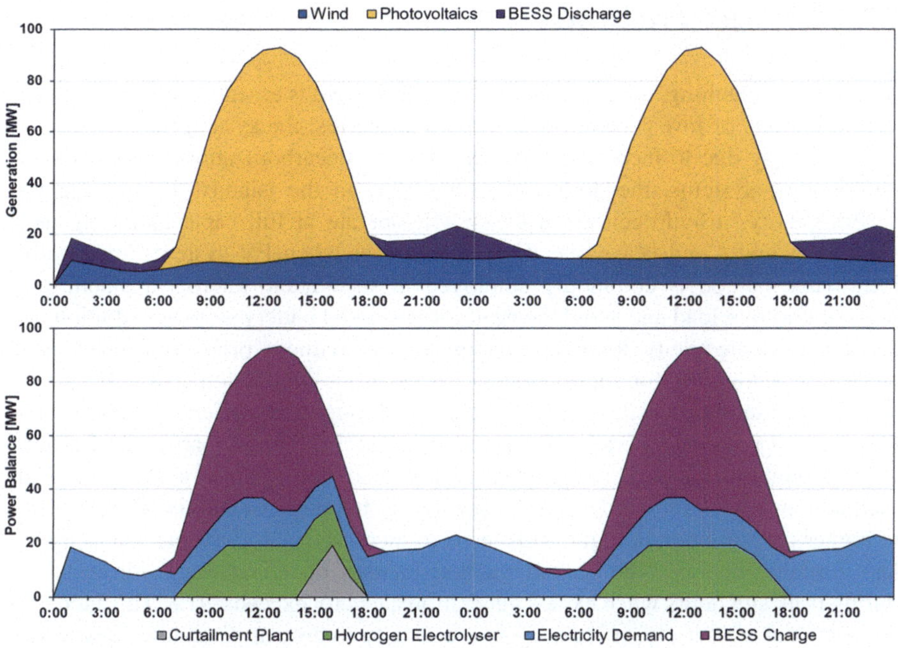

Fig. 6 Power generation profiles per source (upper figure) and power balance (lower figure) for Santa Cruz and Baltra during 2 days in September 2050, advanced scenario

reference scenario shows a slight increase to 10.65 MW by 2030 and then a drop to 3.25 MW by 2050, mainly from diesel. The conservative scenario sees capacity rise from 16.15 MW in 2030 to 25.24 MW in 2050, mainly from PV and wind. The mild scenario projects growth from 19.71 MW in 2030 to 36.45 MW by 2050 with more PV and wind farms. The advanced scenario forecasts an increase from 18.23 MW in 2030 to 59.28 MW by 2050, supported by PV, wind, and hydrogen. Energy storage is expected to reach 103 MWh in the mild scenario and 130 MWh in the advanced scenario.

Figure 6 shows the power generation and energy consumption over a 48-h period. The top plot displays power generation from wind, PV, and BESS discharge, while the bottom plot shows power balance, including energy use, curtailment, hydrogen electrolyzer, and BESS charging. PV generation peaks around noon, decreases in the evening, and stops at night, while wind provides a steady output. BESS discharges in the evening and early morning when PV output is low and charges during the day when PV output exceeds demand. The hydrogen electrolyzer runs mostly during the day using excess solar power. The advanced scenario for Santa Cruz and Baltra demonstrates effective use of renewable energy, minimizing curtailment and optimizing storage to manage fluctuations in solar and wind power.

3.2 Hydrogen Production

In microgrid planning, a reliable energy storage system is essential to supply energy during periods of low generation. In certain scenarios, the system may experience excess energy due to increased capacity aimed at decarbonization, primarily from photovoltaic systems, the dominant technology on the islands. To manage this excess energy, a hydrogen electrolyzer will operate at full capacity during peak generation hours and cease operation after sunset when PV systems are inactive [32]. This method is commonly employed in sustainable microgrid planning to prevent grid overload and avoid the need for oversized battery storage. Additionally, using excess electricity for hydrogen production reduces production costs, with electricity accounting for approximately 27.5% of the production cost [33].

The production of hydrogen was evaluated through two different methods. The first method involves operating the electrolyzer at a minimum part-load of around 20% throughout the day, ramping up to full load during solar peak hours. This method reduces the required electrolyzer capacity but requires additional storage due to higher consumption at night. This leads to increased energy storage capacity but less curtailed energy. The second method focuses on generating hydrogen only during peak sun hours when excess generation capacity exceeds 20%. This increases the size of the electrolyzer capacity to meet daily hydrogen demands. However, it eliminates the need for battery storage, resulting in lower energy costs.

Upon comparison in the advanced scenario for the year 2050 for hydrogen production, the optimization model yielded the results shown in Table 4. Consequently, the second method was selected for this model.

For a hydrogen production facility producing between 1 and 10 MW, the electrolyzer investment cost reaches approximately to $400.000 per MW in 2022 to 2025, $300.000 per MW in the decade of 2030, $200.000 in 2040, and $100.000 in 2050 [33, 34]. The fixed operation and management costs are $20.000 per MW per year. The production is forecasted to be between 18.75 and 25 kg H_2 per hour per MW installed. Correspondingly, the energy consumption is expected to be between 53 and 40 kWh per kg H_2 produced per MW installed. Finally, the investment of the desalination plant is around $7.0 per kg H_2 produced per day [33, 34]. Now we will continue with the emissions analysis.

Table 4 Hydrogen production for the advanced scenario for 2050 in Santa Cruz and Baltra

Description	Part load operation without solar generation	Solar generation exceeds 20% capacity
Electrolyzer capacity (MW)	14	19
Energy storage capacity (MWh)	526	383

3.3 Emission Analysis

The emissions are solely derived from operational "tank-to-wheel" emissions, excluding the entire life cycle "well-to-wheel" emissions. This approach specifically applies to emissions from diesel power plants and the transport sector, calculated based on fuel consumption rather than considering emissions over the entire life cycle. Consequently, in this model, renewable energy power plants are assumed to produce zero emissions during operation. Figure 7 shows the emissions for inland and maritime transportation.

Emissions in Inland Transportation

In the reference scenario, emissions rise steadily, reaching about 30 $kTCO_2eq$ by 2050. In the conservative scenario, the reduction is relatively small, and emissions tend to increase again. This is primarily due to the assumption that heavy-duty vehicles, a significant emission source, will naturally increase in number to meet service demands, economic development, and the construction of new energy projects and infrastructure. The mild scenario features a gradual decline in emissions, dropping to approximately 10 $kTCO_2eq$ by 2050. The advanced scenario exhibits the most significant reduction, with emissions nearly reaching 0 $kTCO_2eq$ by 2050.

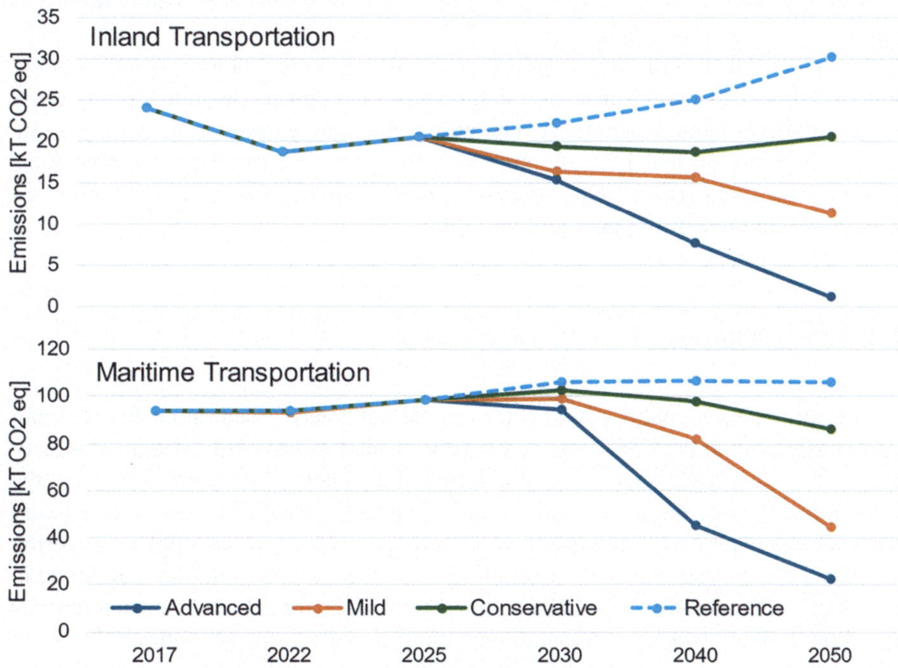

Fig. 7 Emissions for inland and maritime transportation

The inland fleet's transformation begins in 2030 with the full electrification of buses, improving public transportation and increasing passenger-kilometers travelled. Introducing hydrogen as a fuel reduces diesel demand and aids in transitioning heavy-duty vehicles to cleaner energy, significantly cutting emissions. Electrification brings efficiency benefits, with electric motors outperforming internal combustion engines. Aggressive scenarios yield greater emission reductions, while the reference and conservative scenarios result in higher and growing emissions over time.

Emissions in Maritime Transportation

Despite the increasing number of vessels required to meet tourism and local demands, achieving zero emissions in this sector remains challenging across all scenarios. Nevertheless, a notable decrease in total energy consumption is observed, attributed to the adoption of electric motors, with hydrogen and electricity playing important roles in transitioning the fleet to renewable-based fuels. This transition results in a significant reduction in fossil fuel reliance. In the reference scenario, emissions initially increase slightly and then remain around 105 $kTCO_2eq$. The conservative scenario shows a stable trend with emissions hovering near 100 $kTCO_2eq$ after a small rise in 2025. The mild scenario presents a gradual decline in emissions, dropping to about 40 $kTCO_2eq$ by 2050. The advanced scenario demonstrates the most significant reduction, with emissions decreasing sharply to nearly 20 $kTCO_2eq$ by 2050.

Complete decarbonization requires practical solutions that necessitate gradual adoption time, especially for ferries, cargo ships, and tankers, as transitioning these vessels involves considerable cost. The proposed strategy suggests switching at the end of their operational life, aligning with the goal of ensuring renewable fuels become cost competitive. This approach aims to bridge the price gap between sustainable and fossil fuel-powered vessels.

3.4 *Cost Analysis*

A cost optimization analysis has been assessed not solely based on profitability but also considering energy security, sustainability, and affordability. Hence, the economic feasibility has been discussed in terms of the energy investment costs and the transition infrastructure costs, both contributing to the total cost. The total annuity cost includes capital expenditure (CapEx) and operation expenses (OpEx), covering maintenance and variable costs (excluding fuel). CapEx and OpEx are combined to calculate annual annuity costs. The levelized cost of energy (LCOE) is then derived based on annual electricity demand, excluding curtailed energy, with a WACC of 7%, typical for such projects. Transition infrastructure costs, which do not directly impact LCOE, are offset by savings from reduced energy expenses. In 2022, a

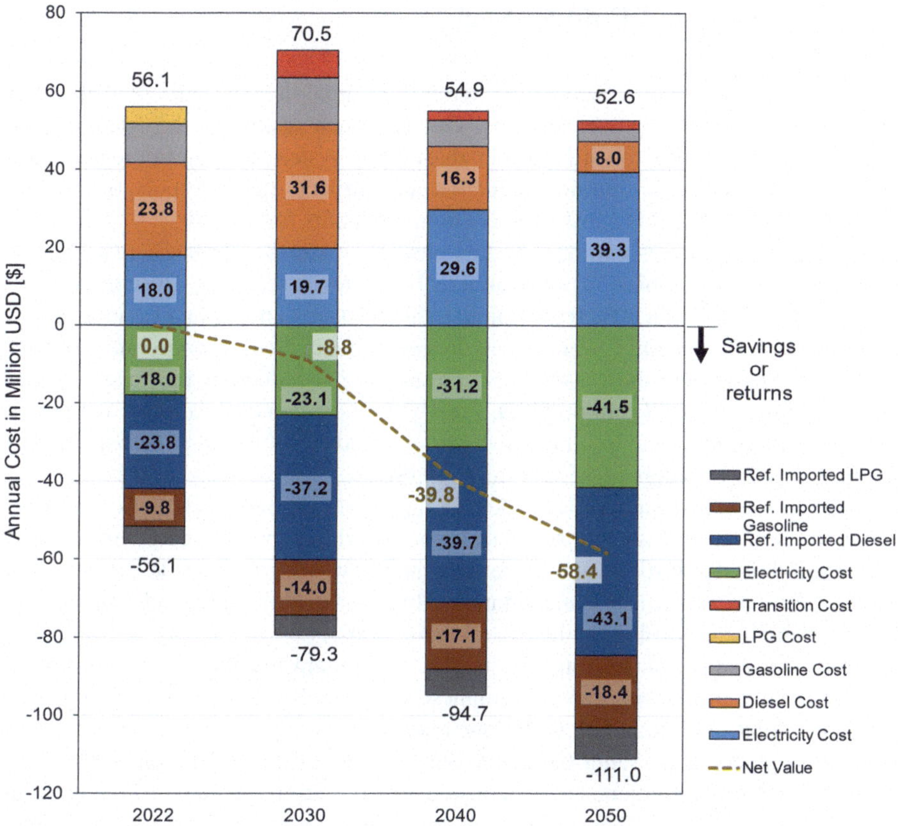

Fig. 8 Annual cash flow for the advanced transition scenario compared to reference scenario

government initiative aims to upgrade 11,500 households at around $2850 per home. Additional costs cover e-Mobility, including home chargers, motorcycle adapters, and charging stations for buses and ferries.

Regarding the cash flow for the advanced scenario shown in Fig. 8, it compares the reference and advanced scenario costs. Below the x-axis (negative cash flow values or savings), there are costs for electricity, diesel, gasoline, and LPG for cooking. Above the x-axis (positive cash flow values), there are costs for electricity, fossil fuels, and transition costs, including electrification and charging stations.

The net value becomes negative in the lapse 2022–2030, demonstrating the feasibility from the first year of implementation, thus resulting in annual savings of 58,38 USD million by 2050. Figure 8 suggests a scenario where the transition to new energy sources incurs significant costs initially, particularly due to the expenses related to transitioning away from traditional fuels like diesel, gasoline, and LPG. These transition costs peak around 2030 and then gradually decrease as the transition process advances. However, even as transition costs decline, electricity costs increase substantially by 2050, becoming the dominant cost factor.

4 Conclusion and Future Work

Achieving the ambitious zero-carbon emissions goal is not only plausible but imperative in islanded power systems. Our approach results in the feasibility of implementing transition toward a sustainable energy system. A clear implementation path not only achieves emission reduction and protects the archipelago's unique ecosystem but also yields substantial economic benefits for society.

In all three scenarios explored in this chapter, it can be observed that there is tangible reduction in emissions compared to the reference scenario. The conservative scenario demonstrates a 50% reduction in total emissions and annual savings of $39 million USD. In the mild scenario, hydrogen production is incorporated at later stages, and hydrogen showcases an impressive 72% reduction in emissions and annual savings of $50.7 million USD. Finally, in the advanced scenario, with early incorporation of hydrogen production, hydrogen leads results with an 89% reduction in emissions and savings of $58.38 million USD.

The energy transition approach taken in this work not only revitalizes the socio-economic landscape, fostering sustainable tourism development and environmental protection, but also fortifies energy security and paves the way for decarbonization across inland and maritime transportation. However, technological adoptions must be complemented by robust policies and regulations.

From the results obtained in this chapter, three main insights have been found. First, Floreana and Isabela have the highest sustainability figures, leveraging their abundant solar PV potential with low energy consumption. Second, the wind generation potential in Santa Cruz and Baltra offers grid-stabilization without the need for ESS deployment, even considering nighttime demand. Finally, San Cristobal's role in hydrogen production could be proportionally scaled to supply increasing retail demand.

The Galapagos data model developed in this research encompass all sectors. It also serves as a valuable foundation for future energy transition analyses and studies. It not only articulates the needs for each sector but also proposes solutions with feasible results. The proposed timeline derived in this research represents a foundational scheme toward implementing a successful transition and enabling long-term investments in low-carbon energy systems.

Future research must overcome data restrictions and energy management limitations presented in this study. First, addressing data limitations needs to be explored. Access to accurate and updated data is crucial for avoiding underestimation or overestimation. Second, focusing on energy dynamics in land and maritime transportation could provide important insights for designing tailored solutions, e.g., energy market for transportation. Third, evaluating project locations and simulating microgrid energy flow would help in identifying opportunities to circumvent flow congestion. Fourth, exploration of biofuels is recommended for further investigation because it could lead to ESS capacity reduction and a potential enhancement of grid stability. Finally, considering maximization of renewable integration, it is important to study demand shifting and demand behavior control, such that consumers' elastic load could be shifted toward renewable peak generation motivated by lower electricity price, especially for consumers relying on e-Mobility.

References

1. Orellana, D., Smith, F.: Assessing geographic isolation of the Galapagos Islands. Int. Arch. Photogramm. Remote. Sens. Spat. Inf. Sci. **XLI-B8**, 733–737 (2016). https://doi.org/10.5194/isprs-archives-XLI-B8-733-2016
2. reliefweb: Galapagos: chronological account of the Tanker Jessica. Galápagos National Park Press Releases. https://reliefweb.int/report/galapagos-islands-ecuador/galapagos-chronological-account-tanker-jessica (2001). Accessed 21 June 2024
3. Sanderson, W., Tierceline, C., Villanueva, J.: Accident of the oil tanker Jessica off the Galápagos Islands (Ecuador). (2001)
4. Global Sustainable Electricity Partnership: Galápagos San Cristóbal Island wind project 2003–2016. https://www.globalelectricity.org/projects/wind-galapagos/ (2007). Accessed 21 June 2024
5. Elecgalapagos, S. A.: Galapagos Electrical Utility: Business plan 2023. Santa Cruz. https://www.elecgalapagos.com.ec/public/lotaip/2023/ANEXOS/Plan_Negocios_2023.pdf (2023)
6. Elecgalapagos, S. A.: Portal web Elecgalapagos S.A. https://www.elecgalapagos.com.ec/ (2022). Accessed 22 May 2024
7. Bariloche, F.: Consultoría para la construcción del Plan de Acción para la Transición Energética Sostenible del Archipiélago de las Islas Galápagos, período 2020–2040. https://fundacionbariloche.org.ar/wp-content/uploads/2022/02/Ficha-Prospectiva-Energetica-Galapagos.pdf (2020)
8. Herremans, I.M., Echegaray, A.M.: Biofuels in the energy mix of the Galapagos Islands. In: Sustainable Energy Mix in Fragile Environments, pp. 49–56. Springer (2018)
9. Recalde, Á., Acosta, V., Ortiz, G., Cajo, R., Godoy, C.: Assessment of green hydrogen production from hydropower in Ecuador. In: Congress on Research, Development, and Innovation in Renewable Energies, pp. 99–119 (2024)
10. MEER: Plan Nacional de Eficiencia Energética 2016–2035. Quito. https://cdn.flipsnack.com/widget/v2/flipsnackwidget.html?hash=fdz94jcze&bgcolor=EEEEEE&t=1496354811 (2016)
11. MERNNR: Plan Maestro de Electricidad. Quito. https://www.recursosyenergia.gob.ec/plan-maestro-de-electricidad/ (2020)
12. Arévalo, P., Eras-Almeida, A.A., Cano, A., Jurado, F., Egido-Aguilera, M.A.: Planning of electrical energy for the Galapagos Islands using different renewable energy technologies. Electr. Power Syst. Res. **203**, 107660 (2022). https://doi.org/10.1016/j.epsr.2021.107660
13. MEM: Plan de Transición Energética de las Islas Galápagos. Quito. https://www.recursosyenergia.gob.ec/wp-content/uploads/2023/11/BID-GALAPAGOS-PUBLICACION-v7-fotos-comprimido.pdf (2023)
14. Fundación Bariloche: Consultancy for survey and development of Sustainable Behaviour Standards of buildings in the Galapagos Islands. https://fundacionbariloche.org.ar/wp-content/uploads/2019/04/5.-Executive-Summary-Galápagos.pdf (2018)
15. Villacreses, G., Jijón, D., Nicolalde, J.F., Martínez-Gómez, J., Betancourt, F.: Multicriteria decision analysis of suitable location for wind and photovoltaic power plants on the Galápagos Islands. Energies. **16**(1), 29 (2022). https://doi.org/10.3390/en16010029
16. Penn Institute for Urban Research: #GalápagosGoesGreen: Galápagos green mobility strategy resiliency and CO2 emission reduction pathways via green transportation for recovering the tourist industry in the aftermath of COVID-19. https://penniur.upenn.edu/uploads/media/Galapagos_Goes_Green.pdf (2021)

17. Cordoba-Vallejo, X., Yang, X., Ponce-Cueto, E.: Carbon footprint of the Galapagos Islands – quantifying the environmental impact of tourist activities. J. Clean. Prod. (2012). https://www.semanticscholar.org/paper/Carbon-Footprint-of-the-Galapagos-Islands-—the-of-Cordova-Vallejo-Blanco/f64a65302a1e7608c7881e47010b2a7e342efc8e
18. Morales, D.X., Besanger, Y., Sami, S., Alvarez Bel, C.: Assessment of the impact of intelligent DSM methods in the Galapagos Islands toward a smart grid. Electr. Power Syst. Res. **146**, 308–320 (2017). https://doi.org/10.1016/j.epsr.2017.02.003
19. Clairand, J.-M., Arriaga, M., Canizares, C.A., Alvarez-Bel, C.: Power generation planning of Galapagos' microgrid considering electric vehicles and induction stoves. In: 2019 IEEE Power & Energy Society General Meeting (PESGM). IEEE (2019). https://doi.org/10.1109/PESGM40551.2019.8973897
20. Icaza-Alvarez, D., Jurado, F., Tostado-Véliz, M., Arevalo, P.: Decarbonization of the Galapagos Islands. Proposal to transform the energy system into 100% renewable by 2050. Renew. Energy. **189**, 199–220 (2022). https://doi.org/10.1016/j.renene.2022.03.008
21. Ministerio de Energia y Minas: Hoja de Ruta del Hidrogeno Verde en el Ecuador. Quito. https://www.recursosyenergia.gob.ec/wp-content/uploads/2023/08/Hoja-de-Ruta-del-Hidrogeno-Verde-en-Ecuador.pdf (2023)
22. Pfenninger and Staffell: Welcome to Renewables.ninja. https://www.renewables.ninja/. Accessed 22 May 2024
23. Secretaria Nacional de Planificacion y Desarrollo: Proyecciones Poblacionales Totales Provinciales 2010 – 2050. Quito. https://www.ecuadorencifras.gob.ec/proyecciones-poblacionales/ (2010)
24. Consejo de Gobierno del Regimen Especial de Galapagos: PLAN GALÁPAGOS 2030 Plan de Desarrollo Sustentable y Ordenamiento Territorial del Régimen Especial de Galápagos. Santa Cruz. https://unidosporgalapagos.wordpress.com/wp-content/uploads/2021/05/plan-galacc81pagos-2030-2.pdf (2023)
25. Becker, B.: Calculating boat fuel consumption. https://www.boatingmag.com/calculating-fuel-consumption/ (2019). Accessed 24 May 2023
26. Dokso, A.: Hydrogen ferry sea change arrives in San Francisco. Energy News. https://energynews.biz/hydrogen-ferry-sea-change-arrives-in-san-francisco/ (2023). Accessed 24 May 2024
27. Ministerio de Transporte y Obras Publicas: INFORME DE RENDICIÓN DE CUENTAS SUBSECRETARÍA DE PUERTOS Y TRANSPORTE MARÍTIMO Y FLUVIAL ENERO-DICIEMBRE 2019. Quito. https://www.obraspublicas.gob.ec/rendicion-de-cuentas/ (2019)
28. UDLA: Petróleo al día. Quito (2018)
29. U.S. Energy Information Administration (EIA): Short-term energy outlook. November 2022 and EIA, AEO2023 National Energy Modeling System. https://www.eia.gov/outlooks/aeo/ (2022)
30. Dorfner, J.: urbs: a linear optimisation model for distributed energy system. https://urbs.readthedocs.io/en/latest/ (2023). Accessed 24 May 2024
31. Candas, S., et al.: Code exposed: review of five open-source frameworks for modeling renewable energy systems. Renew. Sust. Energ. Rev. **161**, 112272 (2022). https://doi.org/10.1016/j.rser.2022.112272
32. Enapter: Ditching the diesel: hydrogen microgrids. https://www.enapter.com/blog/hydrogen-microgrids/ (2020). Accessed 22 May 2024
33. IRENA: Green hydrogen cost reduction. Scaling up electrolysers to meet the 1.5C climate goal. https://www.irena.org/publications/2020/Dec/Green-hydrogen-cost-reduction (2020)
34. IRENA: Making the breakthrough: green hydrogen policies and technology costs. https://www.irena.org/-/media/Files/IRENA/Agency/Publication/2020/Nov/IRENA_Green_Hydrogen_breakthrough_2021.pdf (2021)

Design and Evaluation of a Low-Cost Methane Concentration Measurement Prototype Using Sensor Integration and Gas Chromatograph Calibration

Gabriel Bobadilla-Carreño, Santiago Lomelí-García, Erick Eduardo Guzmán-Quezada, Cecilia del Carmen Puertas-Donoso, Óscar Aguilar-Juárez, and José Luis Zavala-Aguirre

1 Introduction

Previous studies have demonstrated the development of efficient and cost-effective systems and machinery for methane concentration detection as alternatives to gas chromatography equipment [1, 2]. Nagane et al. [1] show a low-cost methane analyzer evaluated under field conditions; also, Yang et al. [2] show the feasibility of constructing a low-cost and portable device for the analysis of methane concentration. These approaches are good alternatives to the use of chromatographic equipment. These are versatile tools capable of analyzing the composition of a variety of materials, but their accessibility and maintenance requirements remain prohibitive for widespread use among the broader population who could benefit from such technology.

There are few studies focused on evaluating methane monitoring systems constructed from low-cost sensors [1]. Commercially available methane sensors are efficient at detecting low concentrations of gas, which makes them suitable for use in the alarm industry. However, these sensors are not recommended for use in environments with high methane concentrations, as prolonged exposure can lead to sensor damage [3]. For instance, the MQ-4 methane sensor from Henan Hanwei

G. Bobadilla-Carreño · S. Lomelí-García · E. E. Guzmán-Quezada · J. L. Zavala-Aguirre (✉)
Universidad Autónoma de Guadalajara, Zapopan, Mexico
e-mail: jzavala@edu.uag.mx

C. del Carmen Puertas-Donoso
Universidad Internacional del Ecuador, Quito, Ecuador

Ó. Aguilar-Juárez
Centro de Investigación y Asistencia en Tecnología y Diseño del Estado de Jalisco, Guadalajara, Mexico

M. Espinoza-Andaluz et al. (eds.), *Congress on Research, Development and Innovation in Renewable Energies*, Green Energy and Technology,
https://doi.org/10.1007/978-3-031-88995-0_14

Electronics datasheet [4] shows a detection range of 0.02–1%. The problem to solve in this project was to design a system using a commercial sensor, as its main component, that can evaluate a wide range of methane concentrations efficiently and at an accessible price. The objective was to design a methane measurement prototype using economical sensors that allow reliable readings to be standardized under a variety of pressure, temperature, and humidity conditions.

The primary attribute of this equipment is the low cost, which makes it more accessible to a wider range of users. For instance, in the field of renewable energy, it could be employed in methods of biogas extraction from wastewater or other kinds of waste. Furthermore, the compact size of the prototype would make it highly portable and adaptable to different geographical and climatic conditions. This project therefore aimed to design a prototype for measuring methane using low-cost sensors that allow standardized and reliable readings under different conditions of pressure, temperature, humidity, and laboratory or field conditions.

2 Methods

2.1 Preliminary Tests

The mechanism developed by Nagahage et al. [1] was used as a reference, leading to the construction of a unit for evaluating methane concentration. This unit was developed by coupling a methane gas sensor (Flying Fish®, MQ-4) within a 150 mL total volume chamber (DELTALAB^R Cat. 409702). The chamber was equipped with two check valves, positioned inversely to regulate inflow and outflow, and a septum for sample injection (Sigma®, Thermogreen® LB-2 Septa). The unit was connected to an Arduino® controller, which provided values in [mV] according to gas concentration. Successive volumes of methane (CH_4) were injected, followed by successive volumes of air to evaluate the linearity of the MQ-4 response and to have a base line for the design of a module with complementary sensors.

2.2 Module Design

A printed circuit board (PCB) was designed, incorporating a set of sensors to detect operating conditions during the measurements: barometric pressure (Bosch®, BMP180), methane gas concentration (Flying Fish®, MQ-4), temperature, and humidity (Asair®, DHT11). These sensors were connected to an Arduino Mini Wi-Fi board (Arduino®, Arduino® Mini Wi-Fi), a mercury switch (GEN®, Mercury Switch), and a disposable 9 V battery. The integration of all sensors and accessories enabled completely wireless readings of gas concentration, pressure, and temperature, facilitating easier maintenance of hermetic conditions inside a chamber.

2.3 Prototype Calibration

Methane gas samples were taken using a chromatographic syringe to sequentially measure methane concentration, first with the module and then with the gas chromatograph (Perkin Elmer®, Clarus-580). The procedure started with sequential injections of methane, from the container (Tedlar® 1 L bag) to the module chamber for reading [mV] raw values with the Arduino-Computer interface, according to Table 1. The procedure continued until no more scaling up values, meaning that the system reached the original methane concentration (≈85%). At this point a sample of 1.0 mL taken from the module was injected into the Clarus-580 to read the actual concentration. This was the first pair of data shown in Table 2. After that, the gas in the prototype chamber started to be diluted with successive air injections of 20, 20, 40, 60, 100, 100, and 140 mL, with corresponding lectures from the module and the Clarus-580, to complete a total of eight data pairs. Lectures were enough to generate a range of concentrations and to evaluate the correlation index between the data from both devices. The Clarus-580 was operated with Nitrogen as carrier gas because its presence is ignored by the chromatographer and by the MQ-4 sensor.

2.4 Reliability Evaluation

Given the potential for sensor degradation over time, which could lead to erroneous readings, we constructed a total of five identical modules. To ensure the reliability of the modules, they were tested simultaneously with the same methane injections. The testing aimed to determine the lowest and highest values that each module could detect with the gas and to compare the data across them, allowing for precise calibration and validation of their responses.

3 Results

3.1 Preliminary Test

To ensure the correct sealing of the chamber, a series of submersion tests were conducted. These tests involved pumping air into the container and observing for the presence of bubbles, which would indicate potential leaks. Following the successful identification and elimination of leaks, we proceeded with methane injection tests within the chamber. The results showed a proportional response of the MQ-4 according to the methane injected. The same proportional response was observed with dilution of methane inside the chamber due to injection of air (see Table 1).

Table 1 Results of the preliminary test evaluating the linearity of the MQ-4 sensor

Trial	1	2	3	4	5	6	7	8	9	10	11	12	13	14	15	16
Methane (mL)	5	5	10	15	15	15	20	20	20	–	–	–	–	–	–	–
Air (mL)	–	–	–	–	–	–	–	–	–	20	20	40	60	100	100	140
MQ-4 ([mV])	280	310	340	370	395	410	420	435	405	345	322	285	242	196	160	140

Table 2 Gas concentration given by the PCB in [mV] and the values after using the equation

Analog data	Conversion with equation	Real gas concentration	MAE	MSE
[mV]	ppm	ppm		
395	844,853.65	862,400		
345	842,491.65	833,200		
322	816,129.46	788,200		
285	740,300.85	734,700		
242	600,389.94	622,400		
196	389,086.74	408,800		
160	179,279.60	177,500		
140	45,859.60	31,100		
Totals			14,818.38	287,464,469.96

MAE mean absolute error, *MSE* mean squared error

3.2 *Module Design*

The module was built to be compatible with a variety of container volumes (chambers), thereby integrating the different sensor types in a more adaptable and mobile manner.

As illustrated in Fig. 1, two electronic boards were necessary. The first one features an ESP32 processor and supports different communication protocols, including Wi-Fi and Bluetooth (D). This board was used as the main one, with its primary task being to gather information from the second electronic circuit, which included several types of sensors. Notable among these sensors are the MQ-4 gas sensor (A), the DHT-11 module that measures temperature and humidity (B), and the BMP180 barometric pressure sensor.

As illustrated in Fig. 1, there are two printed circuit boards (PCBs), each equipped with an MQ-4 gas sensor (A), a DHT-11 temperature and humidity sensor (B), an Arduino Mini Wi-Fi board (D), and a BMP180 barometric pressure sensor (C).

To ensure the chamber's impermeability, where the sensors conduct their measurements, the system was designed with comprehensive wireless capabilities. Consequently, this configuration eliminates the need for wired connections, thus maintaining the integrity and impermeability of the sensor chamber.

3.3 *Prototype Calibration*

The different lectures at high concentrations were compared using a Clarus-580 chromatograph (Perkin Elmer®) located at the CIATEJ facilities. The specifications include a HayeSep® D column (3 × 3.2 mm) with a 100/120 mesh size. The oven, injector, and detector were maintained at temperatures of 40 °C, 75 °C, and 110 °C, respectively, using nitrogen as the carrier gas at a flow rate of 30 mL/min.

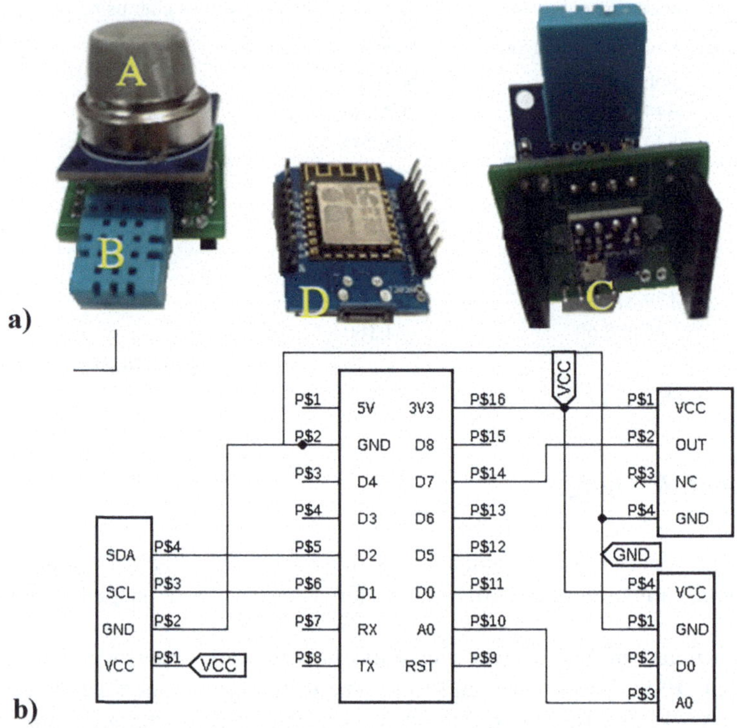

Fig. 1 (**a**) Image of the module already built. Left: frontal view; right: inferior view; center: Arduino card. (**b**) Connection scheme

The relation of the chromatographic methane concentration vs. raw [mV] lectures of the MQ-4 is shown as "original data" series in Fig. 2. The projected values of methane concentrations based on raw [mV] lectures are shown as "estimated data" series in the same figure. The estimated values are calculated with Eq. (1) (below).

The correlation model (Eq. 1) was obtained by comparing the analog data from the sensor with the actual gas concentrations using statistical functions integrated in MS Excel®. The determination coefficient (R^2) is presented in Eq. (2).

$$\text{ppm} = -15.054([\text{mV}])^2 + 11{,}187([\text{mV}]) - 1{,}225{,}290 \tag{1}$$

$$R^2 = 0.9967 \tag{2}$$

In Table 2, we observe the estimated values of methane concentrations using the obtained correlation equation.

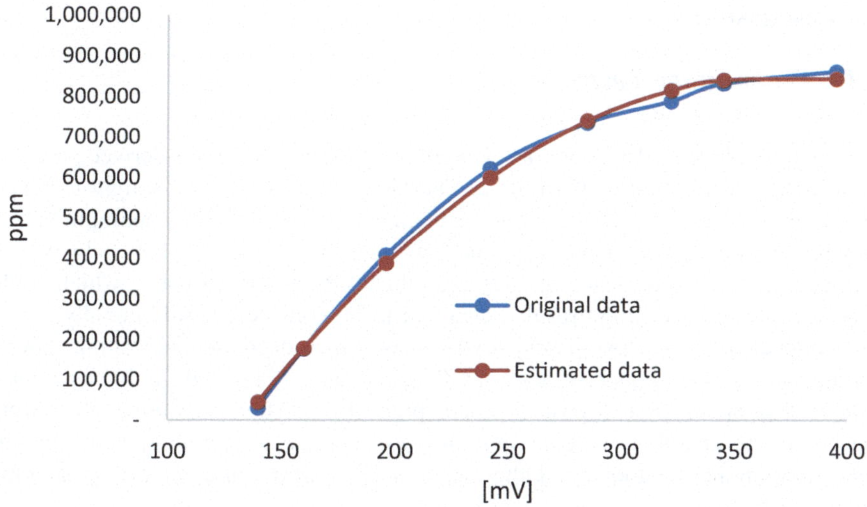

Fig. 2 Graph showing the relation between the raw [mV] values of the MQ-4 sensor against methane concentrations: original data by the GC and estimated data

Table 3 Raw [mV] values indicated by each one of the MQ-4 sensors at minimum and maximum methane concentrations

PCB	Environmental values	Saturation values	Environmental average	Saturation average
A	124 111 112	401 450 459	115.66	436.66
B	176 145 151	693 706 705	157.33	701.33
C	169 146 128	601 605 607	147.66	604.33
D	243 146 137	542 604 738	175.33	628.00
E	46 46 50	606 580 605	47.33	597.00

3.4 Reliability Evaluation

These tests produced experimental data to show possible differences between the five MQ-4 sensors used. Table 3 shows that sensors B, C, and D presented relatively parallel responses in the different averages (see Table 3).

4 Discussion

4.1 Preliminary Tests

It is not possible to find a laboratory without the presence of carbon-derived gases, as even human respiration adds to these. Therefore, in all cases, we sought to find a minimum average for each sensor at the beginning, as well as the maximum values they recorded during methane injection. Since the MQ-4 sensor is not exclusively a methane sensor, the presence of CO_2 and other carbon-derived gases, which could generate interference in the results, also had to be considered. Additionally, when first connected, the sensors display high values, which decrease and stabilize after a certain period. Consequently, each PCB tested was connected 24 h in advance (MQ-4 data sheet, [4]) to ensure proper preheating. After preheating, they were placed in the chamber with air pumping to generate a minimum value of gas concentration in the module. Other authors [7] report quite inferior preheating times, <2 min.

Raw values after preheating and air flushing resulted in notable differences (Table 3) between each PCB tested (52.67 standard deviation; Kruskal-Wallis $P = 0.02$), being PCB_D the more different. Differences increased when methane gas was injected (98.28 standard deviation; Kruskal-Wallis $P = 0.05$), being PCB_A the more different. We attribute the differences in the lectures to the accumulated variability of the components. To avoid damage to the sensor, air was also pumped to return the values as close as possible to the initial value after each test.

4.2 Module Design

Incorporating a set of sensors to detect various conditions during the measurements is because these sensors are sensitive to relative humidity and temperature [1, 5]. Our design also included the incorporation of a pressure sensor, so the series of data captured by the Arduino system could be used to calculate the real concentration of methane, based on a thermodynamic model.

Each evaluated chamber had three perforations: an injection septum and two check valves, one for input and the other for output. The proper sealing of each component had to be verified individually. Depending on the material, these components could be damaged due to pressure changes, being necessary to have replacements. The damage sustained by each module could lead to different responses to each gas injection, potentially affecting the readings.

The raw [mV] values generated during gas dilutions within the module were compared by injecting concentrated gas of known composition. The gas was allowed to vent to stabilize the internal pressure, and once the MQ-4 reading stabilized, 1 mL of gas was sampled and directly injected into the gas chromatograph to determine the direct relationship between the values. The correlation between the sensor readings and the chromatograph began to diverge as the sensor approached its maximum values, splitting farther from the real concentration. To evaluate higher methane concentrations above the upper limit of the MQ-4 sensor, it will be necessary to use bigger chambers in which the module is installed and use the relation between the containers volumes as a conversion factor to calculate the real methane concentration.

4.3 Prototype Calibration

Nagane et al. [1] reported a high average normalized root mean squared error (NRMSE = 0.5517) for the MQ-4 methane sensor. Contrary to this, in the present study it was found that the MQ-4 sensor generated adequate methane concentration values, validated by the chromatographic readings (NRMSE = 0.0204). Fakra et al. [6] report that the fidelity of the MQ series sensors depends on the method adopted to achieve gas dilution; the authors present three different methods with different efficiency, so the present proposal generates a fourth option to be considered by potential users. The observed differences (see Table 3) may be attributed to the age of some of the sensors, with the MQ-4 sensor in Module A being the oldest, used for tests prior to those described in this article. This suggests the possibility of wear-related damage, mainly due to overheating of the external film of the sensor [8]. Conversely, the sensor in Module E exhibited markedly different behavior from the other modules from the outset, consistently showing the lowest values during the minimum value tests. These variations highlight the importance of considering sensor age and prior usage when interpreting data, as well as the necessity for regular calibration and maintenance to ensure accuracy and reliability in measurements.

According to the MQ-4 sensor datasheet [4], the sensor can take measurements of methane within a range of 200–10,000 ppm. However, as shown in Fig. 2 and Table 2, an extended range can be observed, roughly from 100,000 to 800,000 ppm. Falkra et al. [6] present alternatives to extend the range of the sensor up to a value of 200,000 ppm with a value of $R^2 = 0.9973$ linear correlation. The extension of the measurement range up to 800,000 ppm could be explained by the located nonlinear but quadratic model ($R^2 = 0.9967$, Eq. 1).

5 Conclusion

There is a very good correlation (Eq. 1) between the raw values shown by the MQ-4 sensor under normal pressure and temperature conditions compared to the gas chromatograph (GC), producing reliable and replicable results: $R^2 = 0.9967$.

Five modules were created, each with its own set of sensors, costing approximately \$100 USD in total. Each component is easily found in the market and is interchangeable and reprogrammable, making it an economical and replicable prototype for use in areas where access to chromatographic equipment is limited. This opens up possibilities for work on biogas production in more environments outside of specialized research laboratories.

It will be necessary to explore different scenarios of operating the MQ-4 methane sensor (and others with similar characteristics) to try to extend their use to a complete range of methane concentrations with high confidence.

Although the module can be calibrated with a deterministic thermodynamic approach, we present here a simple method based on a quadratic regression between MQ-4's raw values and GC lectures. Furthermore, the flexibility of the prototype allows for adaptation to different biogas production scenarios, from small-scale experiments to larger industrial applications. The ability to easily replace sensors and reprogram the module ensures long-term usability and cost-efficiency. By making advanced biogas monitoring technology more accessible, this prototype can facilitate broader research and practical applications, promoting sustainable energy solutions in diverse settings.

Acknowledgments This manuscript is part of the project "Compost production based on biomass from exotic plants at Galapagos Islands" with code UIDE-DGIP-MAT-PROY-EXTR-22-01 financed by the Competitive Funds Program, Extraordinary Call 2022 of the International University of Ecuador—UIDE and by the BBI fund of the United Nations. The intercalibration of the module was achieved through the joint work between the "Universidad Autónoma de Guadalajara (UAG)" and "Centro de Investigación y Asistencia en Tecnología y Diseño del Estado de Jalisco (CIATEJ)". The publication costs were covered by the UAG.

References

1. Nagahage, I.S.P., Nagahage, E.A.A.D., Fujino, T.: Assessment of the applicability of a low-cost sensor-based methane monitoring system for continuous multi-channel sampling. Environ. Monit. Assess. **193**, 509 (2021). https://doi.org/10.1007/s10661-021-09290-w
2. Yang, S., Liu, Y., Wu, N., Zhang, Y., Svoronos, S., Pullammanappallil, P.: Low-cost, Arduino-based, portable device for measurement of methane composition in biogas. Renew. Energy. **138**, 224–229 (2019). https://doi.org/10.1016/j.renene.2019.01.083
3. Daugela, I., Visockiene, J.S., Kumpiene, J., Suzdalev, I.: Measurements of flammable gas concentration in landfill areas with a low-cost sensor. Energies. **14**(13), 3967 (2021). https://doi.org/10.3390/en14133967
4. MQ-4 Sensor technical data. https://www.sparkfun.com/datasheets/Sensors/Biometric/MQ-4.pdf. Accessed 18 Aug 2024
5. Eugster, W., Kling, G.W.: Performance of a low-cost methane sensor for ambient concentration measurements in preliminary studies. Atmos. Meas. Tech. **5**(8), 1925–1934 (2012). https://doi.org/10.5194/amt-5-1925-2012
6. Fakra, D.A.H., Andriatoavina, D.A.S., Razafindralambo, N.A.M.N., Amarillis, K.A., Andriamampianina, J.M.M.: A simple and low-cost integrative sensor system for methane and hydrogen measurement. Sens. Int. **1**, 100032 (2020). https://doi.org/10.1016/j.sintl.2020.100032

7. Adhim, F., Priananda, C., Rahayu, L., Pramudijanto, J., Tsauri, D., Susila, J., Kindhi, B.: Carbon monoxide and methane gas identification system. In: International Conference on Advanced Mechatronics, Intelligent Manufacture and Industrial Automation (ICAMIMIA), pp. 263–267. IEEE (2019). https://doi.org/10.1109/ICAMIMIA47173.2019.9223367
8. Filipovic, L., Selberherr, S.: Performance and stress analysis of metal oxide films for CMOS-integrated gas sensors. Sensors. **15**, 7206–7227 (2015). https://doi.org/10.3390/s150407206

Construction of a Low-Cost Thermal Comfort Monitoring Station: A Case Study of Ecuador

Maria Gabriela Cando, Daniel Moreira, José León, Gabriela Salcan-Reyes, and Guillermo Soriano

1 Introduction

According to the Intergovernmental Panel on Climate Change (IPCC), the world is undergoing a progressive temperature increase. Climate change is a reality faced by all countries [1]. The rise in ambient temperatures in cities presents challenges ranging from increased electricity demand to modifying existing buildings [2, 3]. Several studies have focused on the relationship between thermal comfort and climate change in the last decade [4–8]. Hot temperatures and relative humidity characterize Ecuador throughout the year, mainly in its coastal and Amazonian regions. Guayaquil is in Ecuador at 2°10'S latitude, with a humid climate and an average maximum temperature of 33 °C. During the austral winter, the ambient temperature drops slightly due to the Humboldt cold current near the coast. For this reason, air conditioning systems are necessary to maintain thermal comfort.

Thermal comfort models are tools to predict and evaluate individuals' thermal comfort in each environment. These models consider various environmental and personal variables to determine whether the thermal conditions are comfortable.

Two globally recognized standards, ANSI/ASHRAE Standard 55 [9], addressing thermal comfort in North America, and the European Standard EN 15251 [10], encompassing thermal comfort and additional indoor environmental parameters, have adopted evaluative methods rooted in an adaptive approach, presuming it adequately accounts for the impact of all other variables such as air temperature, relative humidity, airspeed, mean radiant temperature, clothing insulation, and metabolic activity.

M. G. Cando · D. Moreira · J. León · G. Salcan-Reyes · G. Soriano (✉)
ESPOL Polytechnic University, ESPOL, Guayaquil, Ecuador
e-mail: gsorian@espol.edu.ec

M. Espinoza-Andaluz et al. (eds.), *Congress on Research, Development and Innovation in Renewable Energies*, Green Energy and Technology,
https://doi.org/10.1007/978-3-031-88995-0_15

Thermal comfort monitoring stations are systems of sensors and equipment designed to measure the environmental conditions affecting thermal comfort in a specific setting. Gabriel et al. [11] conducted an experimental study about thermal comfort in classrooms of three different universities in Ecuador. A heat stress monitor was used to measure indoor air temperature (Ta), relative humidity (RH), and air velocity (Va). The mean radiant temperature (MRT) was calculated using the ASHRAE handbook based on the globe temperature (Tg). The authors concluded that the students evaluated preferred to receive classes in air-conditioned environments compared to classrooms at room temperature. They also compared the neutral temperature of each case study with the literature.

In Ecuador, accessing technology is costly due to the reliance on imported equipment, as local development is lacking. This situation increases prices, especially for monitoring equipment that must be certified and calibrated according to international standards. The high costs associated with these devices hinder their acquisition.

However, a low-cost thermal monitoring station can be easily constructed with pre-manufactured components, as in the ClimateBOX Project [12], which uses open-source documentation and does not require experience in microelectronic hardware—allowing the development of low-cost equipment to evaluate indoor environmental factors.

In this study, we built an adapted prototype from the design of the Comfort Monitoring Sensor Station (COMOS) done by Kimmling, Mathias et al. [13]. We made measurements with the prototype and compared it with commercial equipment built under international standards. The proposed simple design allows easy construction from standard parts and common tools. Additionally, it offers the possibility of being 3D printed, facilitating its reproduction and adaptation. This design does not require an internet connection, as the data is stored locally on a microSD card. Moreover, this monitoring approach can serve as a model for other institutions and residential developments, encouraging other researchers to create low-cost, locally manufactured designs.

2 Methodology

To design the thermal comfort monitoring station, it was necessary to highlight the guidelines to carry it out. The project was based on the requirements listed in Table 1, which we carefully considered in the design process.

Table 1 Criteria for design implementation

Requirements
The necessary equipment and parts for assembly were purchased locally
Being able to measure temperature and relative humidity in the conditions of the city of Guayaquil
The sensors must be configurable on an Arduino board

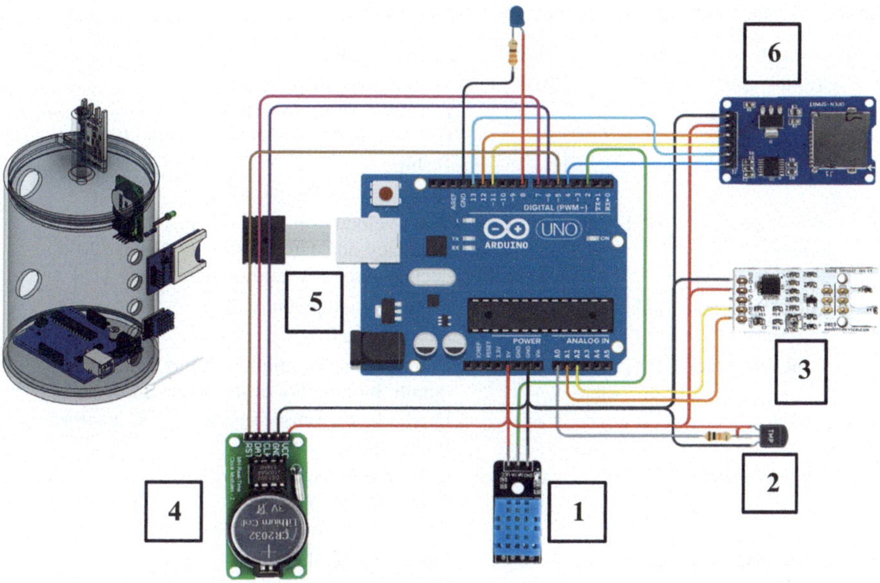

Fig. 1 Schematic circuit of the station and 3D model

The connections, sensor configuration, as well as the basic calibration method are based on the open-source design of Kimmling, Mathias et al. [13]. The structure was printed using an ANYCUBE I3 Mega S printer, and PLA was chosen as the material. The CAD design was created using Autodesk Studio software under an educational license. Sensor programming was performed using the Arduino IDE. The selected board was the Arduino Uno R3 (Fig. 1, Table 2).

We used the educational test bench from Armfield R3 for prototype testing, which allows controlled air circulation, setting velocity, temperature, and relative humidity [21] (Fig. 2).

2.1 Calibration Procedure

We iterated the calibration procedure until we achieved a stable adjustment within the planned measurement range. We used the reference sensors from Table 3 for simultaneous measurements throughout. Initially, we took measurements with the prototype without performing any mathematical operations on the analog and digital data beyond those recommended by the manufacturers. We tested different constant measurement points and established mathematical models to adjust the measurements between the prototype and the reference sensors.

Table 2 Technical details of sensors used in the prototype

Item	Variable	Reference	Model	Specification
1	Temperature and humidity sensor	[14]	DHT11	Temperature range: 0–50 °C Humidity range: 20–90% Resolution: Temperature and humidity both are 16-bit. Accuracy: ±1 °C and ±1%
2	Wind velocity sensor	[15]	MD0550	Wind velocity range: 0–27 m/s Accuracy not stated but Dunn, Davis (2017) declare assess by calibration as ±4% [16]
3	Globe temperature sensor	[17]	DS18B20	Operating temperature range: −55 °C to +125 °C Accuracy: ±0.5 °C
4	Real time clock	[18]	DS3231	Counts seconds, minutes, hours, date of the month, month, day of the week, and year Digital temp. Sensor output: ±3 °C accuracy Input voltage: 3.3 V
5	Microcontroller board	[19]	Arduino UNO	ATmega328P microcontroller Input voltage: 5 V 14 digital input output pins
6	Micro SD card module	[20]	–	Input voltage: 4.5~5.5 V SPI interface

Fig. 2 Sensor used for verification process

Next, we measured under varying temperature, air velocity, and humidity conditions. We calculated the error between the prototype measurements and the reference sensors using this data. The measured data were subjected to BIAS and RMSE calculations using the following equations:

Table 3 Description of sensor used for verification process

Model	Description	Reference
Testo 635	Device capable of connecting external sensors to measure air temperature, humidity, and pressure	[22]
ONSET HOBO U14-001 with S-THB-M008 sensor	Measures temperature and humidity, with data logging capability and LCD display	[23, 24]
Testo black globe thermometer	Measures radiant heat from surfaces	[25]
Testo 405i anemometer	Measures air velocity and volume flow	[26]

$$\mathrm{BIAS} = \frac{1}{n}\sum_{i=1}^{n}(y_i - \hat{y}_i) \tag{1}$$

$$\mathrm{RMSE} = \sqrt{\frac{1}{n}\sum_{i=1}^{n}(y_i - \hat{y}_i)^2} \tag{2}$$

Nomenclature

y_i represents the measured value.
$\hat{y}_i$ represents the reference value.
n is the number of data points.

3 Results

The construction of the equipment was completed. The CAD model was successfully printed on the 3D printer. The printed prototype exhibits adequate mechanical strength to properly support the sensors. It took around 5 h to print the whole design. It showed no noticeable deterioration under laboratory conditions. Figure 3 illustrates the assembled prototype and the sensor circuit during an operational test.

We conducted the local acquisition of the sensors to arrange components into the prototype. The only sensor not available locally is the wind speed sensor, which must be imported and took 2 weeks from the order. The total invoice amount for locally sourced materials was $150 USD.

We successfully configured all acquired sensors with the Arduino Uno board. We found the necessary libraries by searching for the respective sensors which are freely available. The Arduino programming code, developed for this study based on the guidelines outlined in the work of Kimmling, Mathias et al. [13], can be found in GitHub repository [27].

The monitoring station was allowed to record 1 h of data in an uncontrolled environment. Simultaneously and in the exact location, selected sensors were placed for comparison. The recorded data is shown from Figs. 4, 5, 6, and 7.

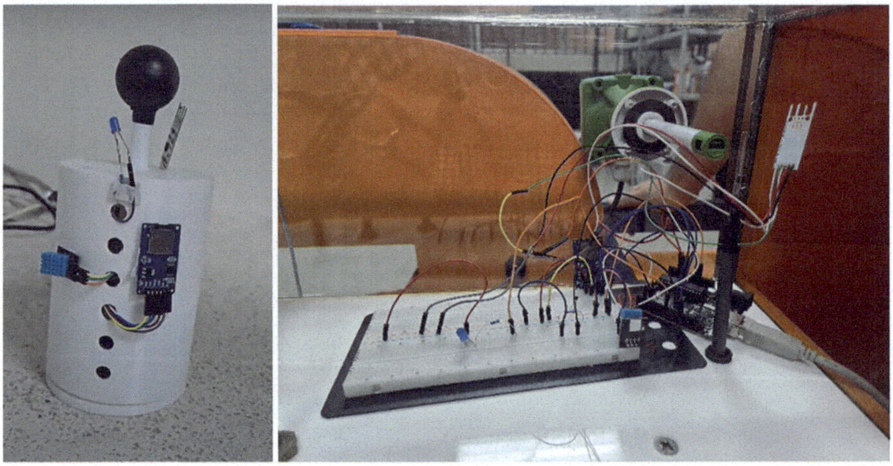

Fig. 3 Prototype of the thermal station assembled. Sensor calibration process according to the recommendation of [1]

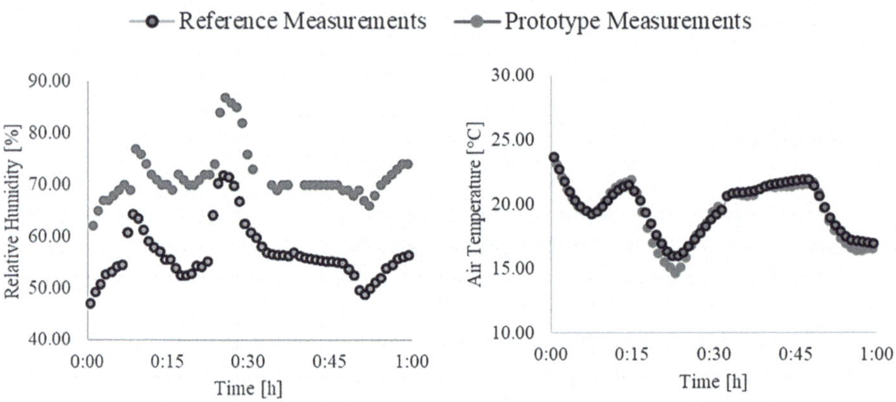

Fig. 4 The scatterplot shows different response behavior between these measurements

3.1 Raw Measurements

Measurements were carried out on the air conditioning test bench and the prototype was able to measure variable temperatures and relative humidity for 1 h. The equipment was used in cooling mode without air recirculation operation and finally, simultaneous measurements were performed with the reference sensors. Figure 4 shows the trend of temperature and relative humidity during this test. The principal finding in this performance is that these measurements showed the time-dependent behavior of the prototype's temperature and humidity sensors under unstable conditions.

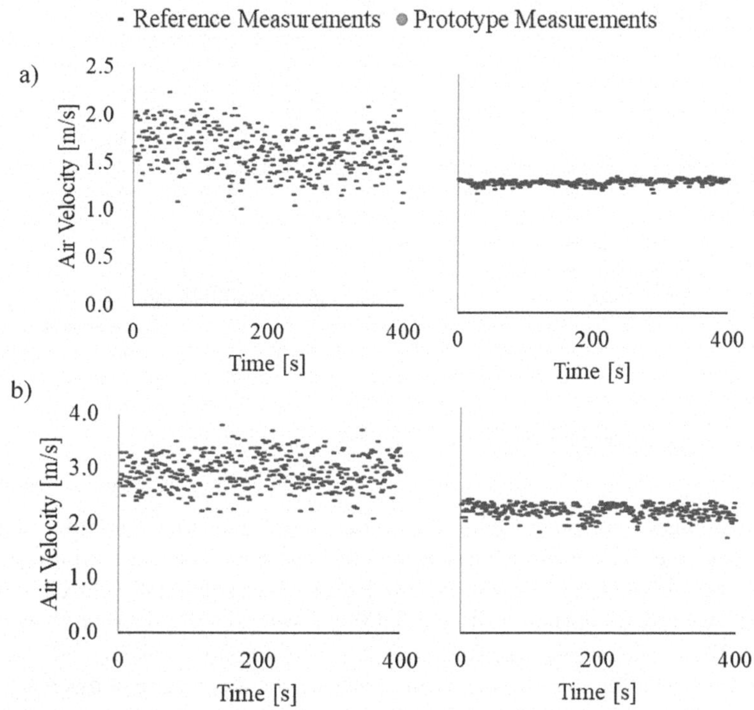

Fig. 5 Depict measurements from MD0550 and Testo 405i under constant airflow conditions within the test bench duct

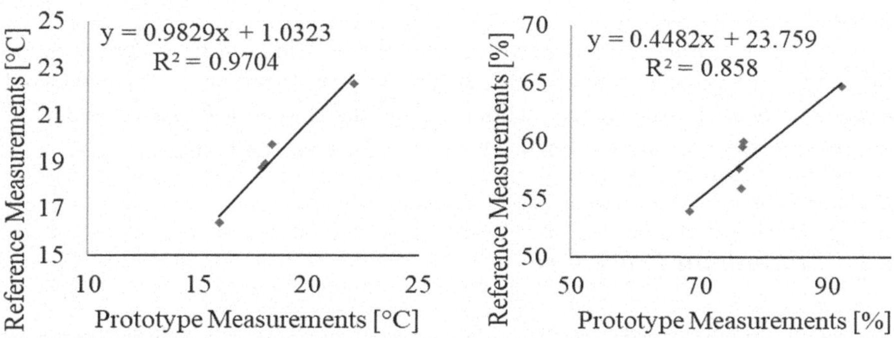

Fig. 6 Correlation between the data measured by the DHT11 and S-THB-M008 sensor

The obtained curves show that the variations in temperature and humidity are proportional between the prototype sensors and the reference sensors, validating that the sensor was accurately recording the air conditions. Notably, there is a difference between the prototype relative humidity sensor and the reference sensor, indicating that the default configuration needed to be improved, and calibration was necessary.

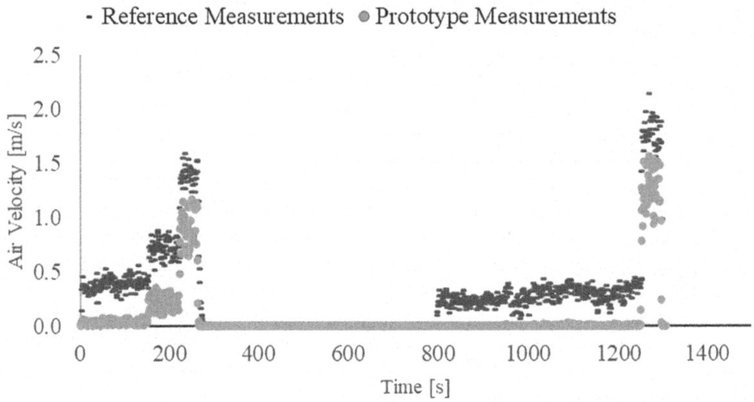

Fig. 7 Data measured by MD0550 and the Testo 405i

A preliminary zero-point calibration process was performed for the air sensor according to the manufacturer's recommended method. Then, the prototype wind speed sensor and the reference wind speed sensor were sequentially placed in the test bed duct. Also, two constant airflow conditions were evaluated in the tunnel: 50% (a) and 100% (b) fan speed, resulting in wind speeds of approximately 1.4 and 2 m/s, measured by the reference sensor. Figure 5 shows the airspeed readings at the exact location, taken at 1-s intervals, with the prototype's wind sensor and the reference sensors. Since mounting and dismounting the sensors took time, we recorded measurements for 10 min for each condition at a consistent air temperature of 22 ° C. Given the constant fan operating conditions, we used the central points of the records.

It was observed that the air speed recorded by the wind sensor of the prototype differs from the value recorded by the reference sensor. Therefore, it was considered appropriate to perform a custom calibration and develop a mathematical model to adjust the readings of the prototype through the Arduino board code.

3.2 Calibration Procedure

Several measurements were carried out under constant set point conditions on the air conditioning test rig. Although the test bench does not allow direct control of the cooling set point or the relative humidity of the air, it does contain resistors before and after the cooling heat exchanger, as well as a manually adjustable humidification system. The procedure involved adjusting these parameters until the test bench sensors recorded stable conditions, with the temperature set to a relative humidity condition of 60%.

The mean of each experiment was estimated using the data obtained to perform a linear regression between the means recorded by the humidity and temperature

Table 4 Averages of the air temperature and relative humidity measurements

Sample	Air temperature [°C]	Relative humidity [%]	Air temperature [°C]		Relative humidity [%]	
			PM[a]	RM[b]	PM[a]	RM[b]
1	16	60	15.97	16.43	68.5	54.02
2	16	60	15.93	16.42	76.14	57.64
3	19	60	17.9	18.77	76.58	55.92
4	19	60	18.09	18.93	76.8	59.55
5	20	60	18.37	19.74	76.87	59.56
6	22	60	22.07	22.38	92.17	64.78

[a]Prototype measurement
[b]Reference measurement

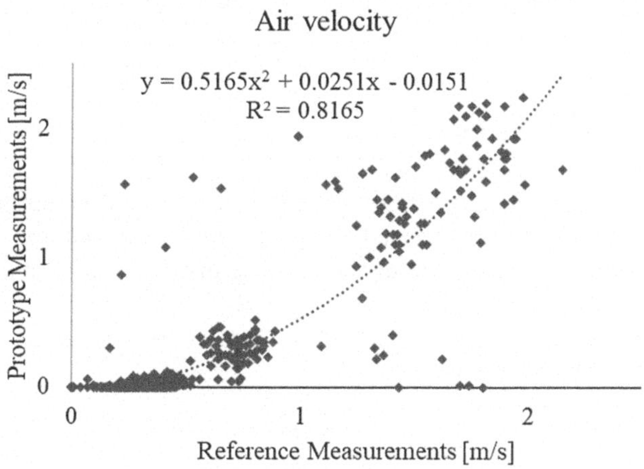

Fig. 8 Correlation between the data measured by the MD0550 and the Testo 405i

sensors of the prototype and those of the humidity and temperature sensors of the reference instrument (Table 4).

Figure 6 shows the result of the correlation. The temperature sensor selected for the prototype was found to correlate closely with the temperature recorded by the reference sensor. However, the relative humidity sensor exhibits a lower correlation with the data obtained from the reference sensor.

Simultaneous measurements were performed to calibrate the air velocity sensor in an open environment by generating airflow with the air duct fan. The sensors were placed side by side and airflow configurations were recorded ranging from no airflow (0 m/s), an intermediate value of about 0.5 m/s, and a high value at maximum fan speed. Figure 7 shows the response of the sensors to these varying conditions. Readings were recorded at 1-s intervals at a constant temperature of 22 °C.

Figure 8 shows a scatterplot of the correlation between the wind speed measured with the prototype and the wind speed measured with the reference sensor. A quadratic fitting model was found and used to calibrate the data read by the wind sensor.

3.3 Error Estimation

Once the regression models were established, a new measurement was performed under unsteady conditions to estimate the BIAS and RMSE.

Figures 9 and 10 show the measurements taken with the prototype after calibration in an uncontrolled but air-conditioned environment. The BIAS of the humidity and temperature sensors was −4.5% and 0.66 °C, respectively. The RMSE was also calculated, which was 8.05% and 0.78 °C (Fig. 11).

Once the temperature sensor was calibrated, a black painted sphere was mounted to capture the mean radiant temperature. The COMOS station study inspired this setup. The temperature of the sphere was measured in an operational boiler room. Figure 10 shows the data log, with measurements taken every 2 s for 30 min. In this measurement, a BIAS of −1.40 °C and an RMSE of 1.78 °C were achieved.

4 Discussion

The equipment was built using the guidelines of Kimmling's free model [13]. Difficulties were encountered when trying to reproduce the exact model, as it suggested using a transparent material for the cylindrical tops of the structure. It was decided to

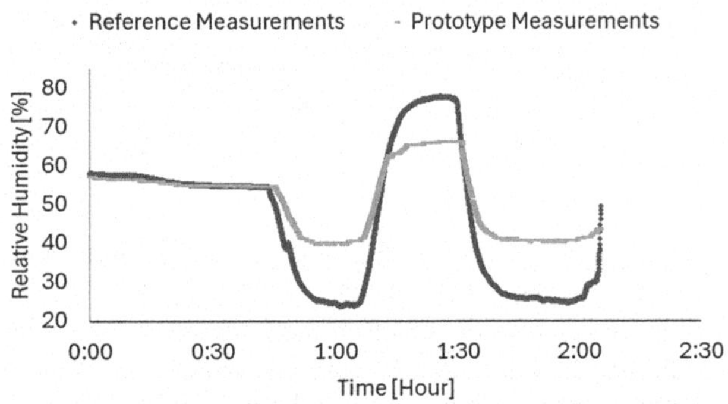

Fig. 9 DHT11 and S-THB-M008 sensor measurement after calibration

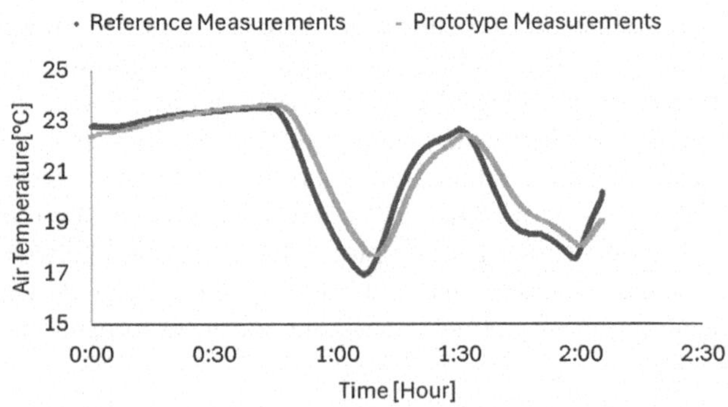

Fig. 10 Measurements from the DHT11 and S-THB-M008 sensors after calibration

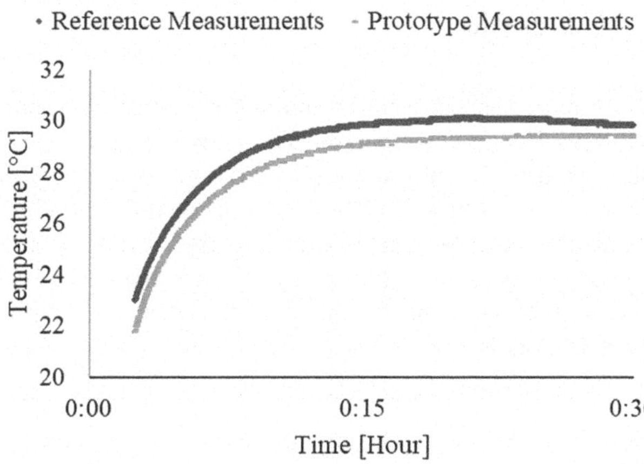

Fig. 11 Globe temperature measurement after calibration of temperature sensor

use a single material, which simplified the construction process. There were also problems with the option of wireless data transmission to a local network. To speed up the design, the information was recorded on an SD card.

The design of the measuring station was selected based on what was proposed in the Kimmling study. This open-source design was chosen for local construction, and we ensured replicability by following the provided instructions and guidelines, which are available in a GitHub repository. This choice allowed us to provide a framework for future work implementing this design. Minor modifications were made during the validation process, such as using a single opaque material instead of transparent during 3D printing. Instead of using the same sensors to measure temperature, average radiant temperature and relative humidity as recommended by Kimmling, we opted for a basic Arduino kit and calibrated the sensors for greater accuracy. The Arduino kit was chosen because of its easy availability in local electronics shops, rather than importing the suggested sensors from Kimmling. The only sensor that remained unchanged is the MD0550 for measuring wind speed, as there are no local alternatives available. This sensor was used in two previous studies [12, 13].

Calibration of the temperature, global temperature, relative humidity, and air velocity sensors has been completed. The readings obtained during field measurements are adequate to characterize the environmental conditions of an indoor environment. Direct measurements with the sensors, without a calibration procedure, were found to show discrepancies compared to readings from laboratory-calibrated reference sensors. Therefore, a calibration methodology was proposed, and linear regression models were identified to improve the accuracy of the measured data.

The calibration of the wind sensor proved to be particularly challenging. As expected, the sensor is sensitive to changes in air speed and, according to the manufacturer, requires real-time air temperature correction. It is recommended that future work include a specialized calibration of the air sensor using a different test rig than the one used in this study.

5 Conclusion

A prototype thermal comfort monitoring station was built, which costs $150 USD on the local Ecuadorian market. The design was based on open-source guidelines, allowing it to be built and replicated on a local scale. The station records physical parameters over time, and the trends match those of reference sensors. However, it is not recommended to use the data directly for processing in thermal comfort models. Instead, a process of calibration and error characterization of the selected sensors is proposed to ensure that the thermal parameter readings are suitable for use in thermal comfort models.

References

1. Pachauri, R.K., et al.: Climate Change 2014: Synthesis Report. Contribution of Working Groups I, II and III to the Fifth Assessment Report of the Intergovernmental Panel on Climate Change. IPCC (2014)
2. Holmes, M.J., Hacker, J.N.: Climate change, thermal comfort and energy: meeting the design challenges of the 21st century. Energ. Buildings. **39**(7), 802–814 (2007). https://doi.org/10.1016/j.enbuild.2007.02.009
3. Filippín, C., Ricard, F., Flores Larsen, S., Santamouris, M.: Retrospective analysis of the energy consumption of single-family dwellings in Central Argentina. Retrofitting and adaptation to the climate change. Renew. Energy. **101**, 1226–1241 (2017). https://doi.org/10.1016/j.renene.2016.09.064
4. Bienvenido-Huertas, D., Sánchez-García, D., Rubio-Bellido, C.: Comparison of energy conservation measures considering adaptive thermal comfort and climate change in existing Mediterranean dwellings. Energy. **190**, 116448 (2020). https://doi.org/10.1016/j.energy.2019.116448
5. Barbosa, R., Vicente, R., Santos, R.: Climate change and thermal comfort in Southern Europe housing: a case study from Lisbon. Build. Environ. **92**, 440–451 (2015). https://doi.org/10.1016/j.buildenv.2015.05.019
6. Jafarpur, P., Berardi, U.: Effects of climate changes on building energy demand and thermal comfort in Canadian office buildings adopting different temperature setpoints. J. Build. Eng. **42**, 102725 (2021). https://doi.org/10.1016/j.jobe.2021.102725
7. Saeed Khan, H., Paolini, R., Caccetta, P., Santamouris, M.: On the combined impact of local, regional, and global climatic changes on the urban energy performance and indoor thermal comfort—the energy potential of adaptation measures. Energ. Buildings. **267**, 112152 (2022). https://doi.org/10.1016/j.enbuild.2022.112152
8. Kong, Q., Zheng, J., Fowler, H.J., Ge, Q., Xi, J.: Climate change and summer thermal comfort in China. Theor. Appl. Climatol. **137**(1), 1077–1088 (2019). https://doi.org/10.1007/s00704-018-2648-5
9. American Society of Heating, Refrigerating and Air-Conditioning Engineers: ASHRAE Standard 55: thermal environmental conditions for human occupancy
10. European Committee for Standardization, EN 15251: Indoor environmental input parameters for design and assessment of energy performance of buildings addressing indoor air quality, thermal environment, lighting and acoustics
11. Guevara, G., Soriano, G., Mino-Rodriguez, I.: Thermal comfort in university classrooms: an experimental study in the tropics. Build. Environ. **187**, 107430 (2021). https://doi.org/10.1016/j.buildenv.2020.107430

12. Kramer, T., Garcia-Hansen, V., Omrani, S.: climateBOX: a low-cost and open-source monitoring device for personal thermal comfort evaluation. Energ. Buildings. **283**, 112830 (2023). https://doi.org/10.1016/J.ENBUILD.2023.112830
13. Kimmling, M., Lauenroth, K., Hoffmann, S., LivingLabKL: How to build a comfort monitoring sensor station. Instructables (2018). https://www.instructables.com/How-to-Build-an-Indoor-Comfort-Monitoring-Sensor-S/. Accessed 24 Jun 2024
14. Arduino Project Hub: Using DHT11. https://projecthub.arduino.cc/arcaegecengiz/using-dht11-12f621. Accessed 24 Aug 2024
15. Modern Device: Wind Sensor Rev. C. https://moderndevice.com/products/wind-sensor. Accessed 24 Aug 2024
16. Dunn, P.F., Davis, M.P.: Measurement and Data Analysis for Engineering and Science. CRC Press (2017)
17. Adafruit Industries: DS18B20 Digital temperature sensor + extras. https://www.adafruit.com/product/374. Accessed 24 Aug 2024
18. Digital Temp Sensor. Register for Aging Trim. RST. Simple Serial Interface Connects to Most Microcontrollers. www.maximintegrated.com. Accessed 24 Aug 2024
19. Arduino Official Store: Arduino Uno Rev3. https://store.arduino.cc/products/arduino-uno-rev3. Accessed 24 Aug 2024
20. Last Minute Engineers: In-depth tutorial to interface micro SD card module with Arduino. https://lastminuteengineers.com/arduino-micro-sd-card-module-tutorial/. Accessed 24 Aug 2024
21. Armfield: RA3 - computer controlled recirculating air conditioning unit. https://armfield.co.uk/product/ra3-recirculating-air-conditioning-unit/. Accessed 24 Jun 2024
22. testo 635 Humidity/temperature/pressure dew point measuring instrument Instruction manual en
23. CETM Singapore: ONSET HOBO U14-001 LCD - temperature/relative humidity (RH) data logger. https://www.cetm.com.sg/home/831-onset-hobo-u14-001-lcd-temperaturerelative-humidity-rh-data-logger.html. Accessed 24 Aug 2024
24. ONSET: Datasheet S-THB-M008 Sensor ONSET.
25. Testo: Globe thermometer (TC type K) for radiant heat. https://www.testo.com/en-UK/globe-probe-o-150mm-tc-type-k-for-measuring-radiant-heat/p/0602-0743. Accessed 24 Aug 2024
26. Testo: Anemómetro térmico testo 405i para smartphone. https://www.testo.com/es-ES/anemometro-testo-405i/p/0560-1405. Accessed 24 Aug 2024
27. GitHub: ComfortStation danielmorzam/ComfortStation@894a463. https://github.com/danielmorzam/ComfortStation/commit/894a463cc7ae55a0734879acbaae7b48ccd15d59. Accessed 24 Jun 2024

Part V
Smart Grid and Wind Energy

Voltage Control in Distribution Network via Vehicle-to-Grid Concept: Implementation and Experimental Validation in a Microgrid Laboratory

Andrew Quezada, Alexander Nivicela, Danny Ochoa-Correa, Edisson Villa-Ávila, and Vinicio Iñiguez-Morán

1 Introduction

The traditional reliance on fossil fuels for transportation, a key driver of human development, has significantly contributed to environmental degradation, particularly in urban areas. This reliance results in air pollution and raises concerns about fossil fuel depletion, prompting a shift toward sustainable transportation models like electromobility [1]. In light of these challenges, international efforts are promoting the integration of electric vehicles (EVs) into urban infrastructures to mitigate air pollution and reduce greenhouse gas emissions, thus supporting the development of healthier, sustainable cities [2, 3].

Despite the environmental benefits associated with the widespread adoption of EVs, concerns have arisen among distribution utilities regarding the potential impacts on the electrical grid. The mass uptake of EVs, particularly in urban areas, poses challenges to the stability and reliability of the grid, particularly during peak charging periods. Issues such as overload on distribution lines and transformers and power quality degradation regarding voltage stability have become pressing concerns for grid operators [4]. Introducing fast-charging stations exacerbates these challenges, placing additional strain on the grid and increasing the risk of infrastructure overload. In response, researchers and industry stakeholders have begun exploring innovative solutions to mitigate the impacts of EV integration on the grid

A. Quezada · A. Nivicela
Faculty of Engineering, Universidad de Cuenca, Cuenca, Ecuador

D. Ochoa-Correa (✉) · E. Villa-Ávila · V. Iñiguez-Morán
Department of Electrical Engineering, Electronics and Telecommunications, Universidad de Cuenca, Cuenca, Ecuador
e-mail: danny.ochoac@ucuenca.edu.ec

M. Espinoza-Andaluz et al. (eds.), *Congress on Research, Development and Innovation in Renewable Energies*, Green Energy and Technology,
https://doi.org/10.1007/978-3-031-88995-0_16

[5]. One such solution gaining traction is vehicle-to-grid (V2G) integration. By enabling bidirectional energy flow between EVs and the grid, V2G systems have the potential to alleviate grid congestion and enhance system flexibility. Moreover, V2G holds promise to leverage EV batteries as distributed energy storage assets, further enhancing grid resilience and stability [6, 7].

In this regard, the work reported in [8] proposed a robust V2G voltage control scheme for distribution networks (DNs) that addresses cyber-attacks and customer interruptions by utilizing time-varying uncertain matrices and signum variables. Their approach, verified through case studies on modified IEEE test feeders, demonstrated effectiveness in maintaining voltage stability even under significant measurement anomalies. Reference [9] introduced a kernel-based predictive model that integrates distributed optimization with ensemble predictions to manage V2G capacity and control reactive power. This framework was designed to be robust against stochastic variations and aimed to minimize the need for grid reinforcement. Authors in [10] focused on the design and experimental validation of control strategies for V2G inverters, emphasizing the provision of ancillary services like virtual inertia and frequency control. Their experimental results confirmed the potential of V2G inverters to stabilize the grid. Research presented in [11] reviewed the rapid expansion of the EV market driven by V2G technology, discussing its impact on grid load balancing, frequency and voltage control, and integrating renewable energy sources. The study provided a comprehensive overview of the current challenges and opportunities in V2G systems and suggested future research directions. Finally, reference [12] explored using a boost inverter-based EV charger for V2G reactive power compensation, addressing issues such as second-order harmonic ripple currents and the impact on EV battery life. Their experimental prototype demonstrated the viability of providing reactive power compensation without compromising battery integrity. Table 1 summarizes the findings of the state of the art presented above.

Unlike previous studies that primarily rely on simulations or controlled environments, this chapter contributes to the field by validating the proposed voltage control method in an actual distribution network using commercially available equipment. This chapter explores the potential of EV chargers for voltage control within real distribution networks, emphasizing the practical challenges and solutions associated with integrating V2G. Starting with an overview of EV and V2G technologies, it details our experimental study's components and control methods. The structure includes a methods section, experimental results, detailed discussions, and concludes with the study findings.

2 Materials and Methods

2.1 Electric Vehicles

EVs represent a transformative shift in urban mobility, offering a sustainable option that addresses air quality concerns. However, integrating EVs into existing power

Table 1 State of the art on V2G technology for voltage control and grid stabilization

Study	Method	Scope	Experimental validation
An et al. (2024), [8]	Robust V2G voltage control using linear matrix inequality techniques	Cyberattack resistance and customer interruptions	No (case studies simulated on IEEE test feeder)
Haghi and Qu (2018), [9]	Distributed optimization integrating stochastic processes and ensemble predictions	Real-time voltage control and reactive power management	No
Luh and Blank (2023), [10]	Controller structures for ancillary services (virtual inertia, frequency control) with bidirectional AC/DC inverters	Grid stabilization and ancillary services	Yes (prototype testing)
Biswas et al. (2024), [11]	Comprehensive review of V2G technology, focusing on grid load balancing and control	Grid balancing, renewable integration, and reactive power adjustment	No
Abeywardana et al. (2018), [12]	Boost inverter-based EV charger for reactive power compensation	Reactive power management without battery degradation	Yes (experimental prototype)

networks presents challenges, particularly regarding the impact of EV charging on grid infrastructure. This study focuses on battery electric vehicles, which connect to the grid and have energy storage systems enabling vehicle-to-grid technology [13]. Charging of EV batteries is typically done through two methods: "off-board" charging at external stations that convert grid AC to DC and "on-board" charging using the vehicle's internal electronics for energy conversion [14].

2.2 *Vehicle-to-Grid (V2G) Technology*

Vehicle-to-grid (V2G) technology transforms EVs into interactive mobile energy storage units that engage in bidirectional energy transfer with the grid, enhancing grid stability, resilience, and efficiency and offering revenue opportunities for EV owners [3]. Key V2G functionalities include reactive power support, active power regulation, load balancing, imbalance compensation, and harmonic current filtering [15]. These features enable V2G-equipped vehicles to dynamically stabilize energy supply and demand fluctuations, improving grid reliability. The design of EV charging stations, crucial for V2G, involves advanced control systems for managing bidirectional power flows essential for voltage control at grid connection points (see Fig. 1). Although fast-charging stations offer rapid battery charging, they can demand high currents from the grid, potentially leading to voltage drops.

The interface in an EV charging station connects the components, organizes the required equipment for effective vehicle linkage, and regulates the power supply,

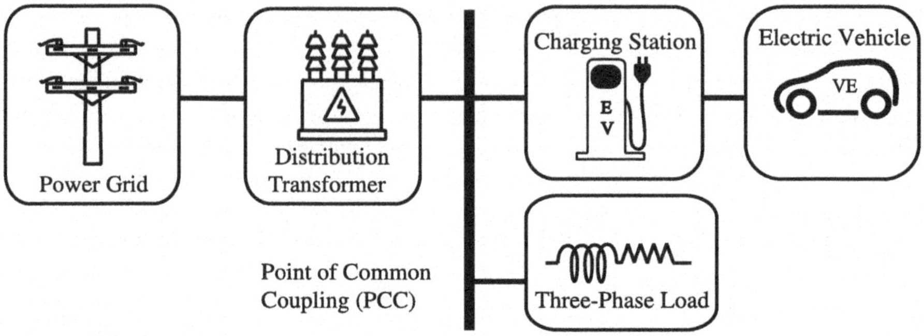

Fig. 1 Typical connection scheme of low-voltage electrical network elements with the incorporation of an EV charging station

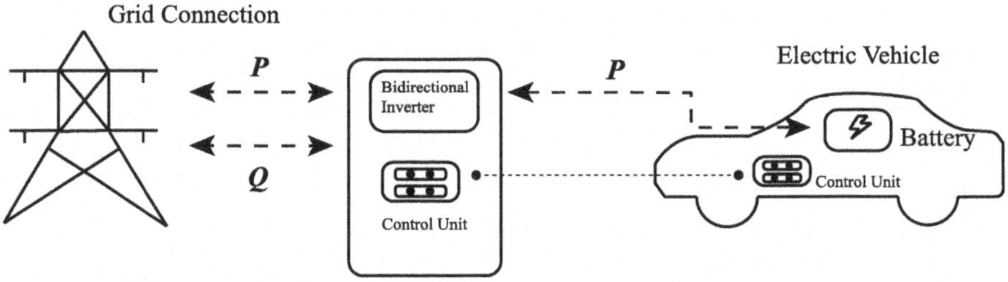

Fig. 2 Simplified schematic of a V2G-implemented EV charging station

both active and reactive. Essentially, this station functions as a bidirectional inverter or electronic power converter. Figure 2 shows a typical connection setup between the electrical grid and the charging station, where P and Q indicate the bidirectional active and reactive powers managed by the station electronic interface [3].

2.3 Power Control of Bidirectional Three-Phase Inverter

Bidirectional three-phase inverters are crucial in power electronics for managing active and reactive power exchange between EV charging stations and the grid, particularly in V2G systems. Understanding power control theories is critical for improving inverter performance and ensuring grid stability. The functionality of a three-phase electronic power converter to dispatch or absorb power relies on the currents along the direct and quadrature axes, i_d and i_q, within the $dq0$ rotating reference frame [16]. Aligning the direct axis of the $dq0$ system with the voltage vector of one phase in a three-phase system defines the operation of these inverters. Eqs. (1) and (2) describe these dynamics, considering P and Q for active and reactive power, v_d for the direct axis voltage, and i_d and i_q for the current $dq0$ components.

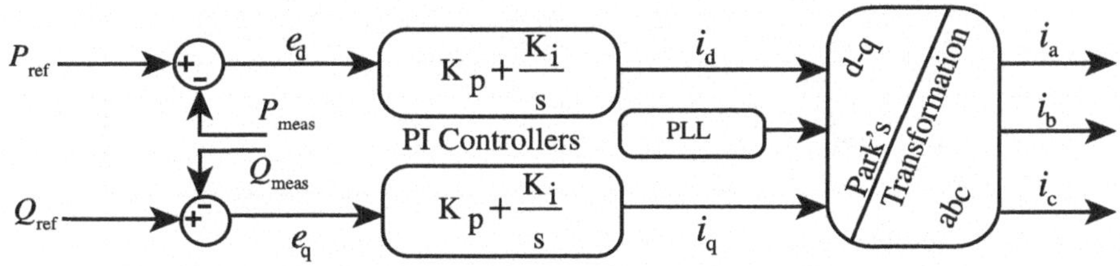

Fig. 3 Simplified PI control scheme of active and reactive power for the electronic power converter incorporating an EV charging station

$$P = \frac{3}{2}(v_d i_d) \tag{1}$$

$$Q = \frac{3}{2}(-v_d i_q) \tag{2}$$

Therefore, the control of active and reactive power of a three-phase power electronic converter (PEC) is performed in a decoupled manner, as shown in Fig. 3.

The active and reactive power setpoints, P_{ref} and Q_{ref}, are set and compared with the measured power P_{meas} and Q_{meas} at the point of common coupling (PCC). This comparison yields errors e_d and e_q which are used as inputs to a proportional-integral (PI) controller that generates the references, I_d and I_q. These currents are transformed into the *abc* reference frame using the Park transform, necessitating a system frequency reference from a phase-locked loop (PLL)-based control system [17]. The resulting currents, I_a, I_b, and I_c, are then injected at the PCC. In Eq. (3), I_d, I_q, and I_0 represent currents in the $dq0$ frame, ω denotes electrical frequency, and t denotes the time.

$$\begin{bmatrix} I_d \\ I_q \\ I_0 \end{bmatrix} = \frac{2}{3} \begin{bmatrix} \cos \omega t & \cos\left(\omega t - \frac{2\pi}{3}\right) & \cos\left(\omega t + \frac{2\pi}{3}\right) \\ -\sin \omega t & -\sin\left(\omega t - \frac{2\pi}{3}\right) & -\sin\left(\omega t + \frac{2\pi}{3}\right) \\ \frac{1}{2} & \frac{1}{2} & \frac{1}{2} \end{bmatrix} \begin{bmatrix} I_a \\ I_b \\ I_c \end{bmatrix} \tag{3}$$

2.4 Voltage Control Scheme Proposed for V2G

Ensuring stable voltage at the grid connection point is critical in EV charging stations with V2G capability. A common approach involves injecting controlled reactive power to adjust the quadrature current in the station's bidirectional inverter.

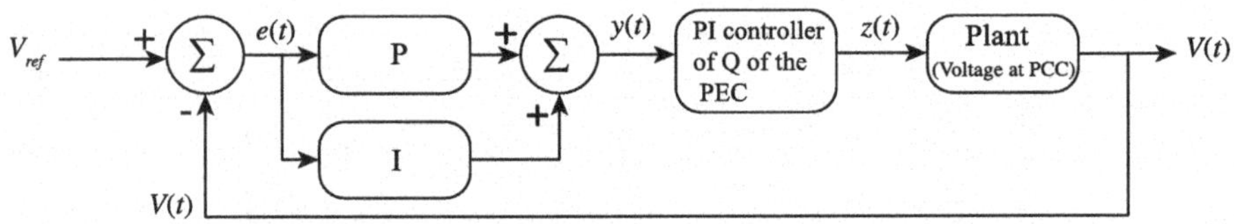

Fig. 4 Proposed voltage control scheme for V2G

This study introduces a feedback-based proportional-integral (PI) system to determine the necessary reactive power injection, as shown in Fig. 4.

The system dynamically sets reactive power setpoints Q_{ref} that the PEC aims to meet, responding to voltage control needs at the connection point, while the reference active power P_{ref} remains constant to align with the vehicle's load demands.

The control system uses two main inputs: the reference voltage V_{ref} and the real-time grid voltage $V(t)$. The difference between these, $e(t)$, is processed by a PI controller. The controller's output, $y(t)$, determines the reactive power setpoint for the PEC, represented as Q_{ref} in Fig. 3. The output control signal $z(t)$ from the PEC specifies the quadrature current to inject at the PCC. In this scheme, the block "Plant" represents the effect of the reactive power injection at the PCC by the PEC, which directly influences the voltage $V(t)$ at that point.

Voltage measurement involves capturing the positive sequence component of the three-phase voltages in a balanced system, typically using symmetrical component transformation in steady states. For transient or real-time operations, instantaneous symmetrical components are often more practical, as described in Eq. (4). In this equation, a is a complex number equal to $1 \angle 120°$.

$$\begin{vmatrix} v_0(t) \\ v_+(t) \\ v_-(t) \end{vmatrix} = \frac{1}{\sqrt{3}} \begin{vmatrix} 1 & 1 & 1 \\ 1 & a & a^2 \\ 1 & a^2 & a \end{vmatrix} \cdot \begin{vmatrix} v_a(t) \\ v_b(t) \\ v_c(t) \end{vmatrix} \tag{4}$$

In the practical implementation of the method, it is necessary to introduce physical limits on reactive power based on the capabilities of the energy storage system and the bidirectional inverter. Thus, limits for $y(t)$, or equivalently Q_{ref}, are determined by calculating the maximum reactive power the PEC can manage using Eq. (5). Here, P_{max} represents the maximum active power that the inverter can handle, and S_{max} is the maximum apparent power capacity of the inverter. This calculation ensures that the reactive power settings do not exceed the inverter's operational capabilities.

$$Q_{\text{max}} = \sqrt{S_{\text{max}}^2 - P_{\text{max}}^2} \tag{5}$$

The experimental implementation of the scheme in Fig. 4 requires a discrete expression of the control loop and the development of a digital controller, which is mathematically represented as a discrete equation. The resulting equation is as follows:

$$y(t_k) = k_p e(t_k) + k_i \sum_{k=1}^{n} e(t_k)\Delta t \tag{6}$$

where:

$y(t_k)$ is the corrective signal of the controller, corresponding to the reference reactive power that the PEC, Q_{ref} should follow.

$e(t_k)$ represents the error between the reference and measured values at a time t_k.

Δt denotes the sampling time for control tasks and data acquisition.

k_p, k_i are the proportional and integral constants.

The V2G voltage control routine is implemented using MATLAB on the SCADA system in the laboratory, as outlined in the flowchart provided in Fig. 5.

The procedure involves calculating the positive sequence component of the voltage at the PCC using the SCADA system. This value is then used to compute the real-time error signal for the digital PI controller. Equation (6) calculates and transmits the reactive power setpoints to the inverter connected to the SCADA. A "deadband" feature is included to prevent unnecessary controller actions for minor voltage deviations. The reactive power limits of the inverter are set at $\pm Q_{max}$, ensuring that it does not exceed 30 kVAr, aligning with the maximum capacity of the lithium battery and component design. Figure 5 provides a flowchart of the V2G voltage control algorithm.

2.5 *Experimental Test Bench for Proposal Validation*

An experimental test bench was set up in the Microgrid Laboratory at the University of Cuenca, Ecuador, to validate the proposed V2G voltage control method. Figure 6 illustrates the setup, which includes a bidirectional inverter (50 kW grid-following converter) connected to a 44 kWh lithium-ion battery storage system, used to emulate the reactive power response of the EV charging station for the implementation of V2G. The PCC comprises a 220 V three-phase (60 Hz) network connected to the 22 kV public grid via a 75 kVA transformer. Due to the current handling limitations of the bidirectional inverter, it is not feasible to emulate both the fast-charging power profile (50 kW) and the reactive power regulation simultaneously with the same equipment in our laboratory. Therefore, the active power load profile of the EV is emulated separately using a programmable AC load (0–150 kW), which allows for controlled manipulation of the charging process. The setup connects to a three-phase busbar for bidirectional power flow with the public grid and features monitoring and control equipment for data acquisition, performance evaluation, and regulatory compliance. Figure 6 also illustrates the communication architecture for efficient control and data handling during experiments. Details of these and all the equipment available in the laboratory for these research purposes can be consulted in [18].

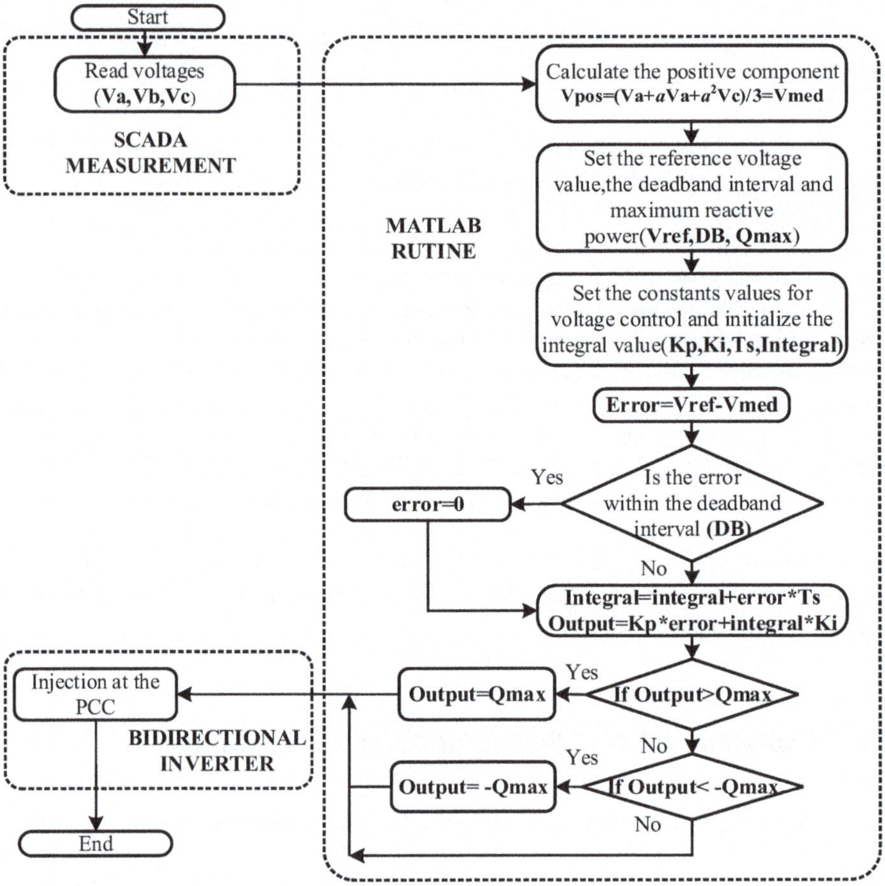

Fig. 5 Flowchart of the experimental implementation of the V2G voltage control method

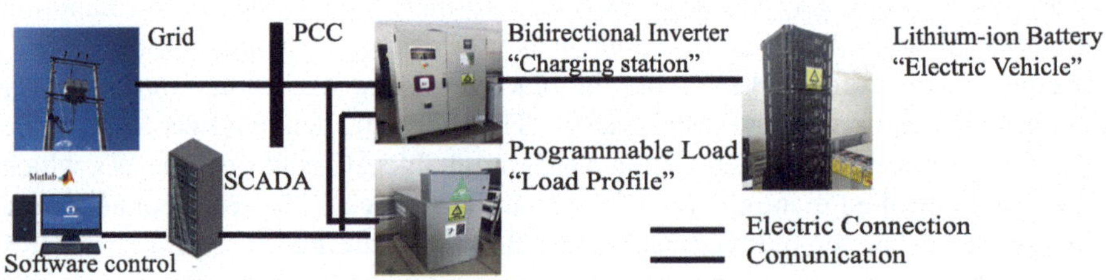

Fig. 6 Schematic representation of the experimental test bench implemented for validation

3 Results

3.1 Electric Vehicle Charging Profile

The validation of the control method was based on the actual demand profile of a van-type SUV manufactured by BYD during fast charging [19]. The charging profile, shown in Fig. 7, progresses from a 40% to 100% state of charge (SOC), adhering to manufacturer guidelines to not discharge the EV below 40% to prevent battery degradation.

The EV charging process starts with a rapid power increase, maintains a constant charging phase, and then gradually declines, demonstrating fast charging capacity. This profile serves as the active power reference for laboratory evaluations. Despite minor discrepancies between the real and emulated profiles due to 5 kW step adjustments, the lab-generated profile effectively supports research. The resistive, software-controlled programmable load causes a significant voltage drop at the connection point, highlighting the need for robust control strategies to ensure stability during fast charging.

Using the nominal line voltage (220 V) and the distribution transformer capacity (75 kVA) as base values for the per-unit calculation of the plotted variables, this criterion was applied to generate Figs. 8, 9, and 10. Figure 8 shows the impact of the charging process on network voltage at the PCC, revealing a 3% voltage drop from the pre-charging level. This finding highlights the importance of developing strategies to mitigate such impacts, which this research aims to address.

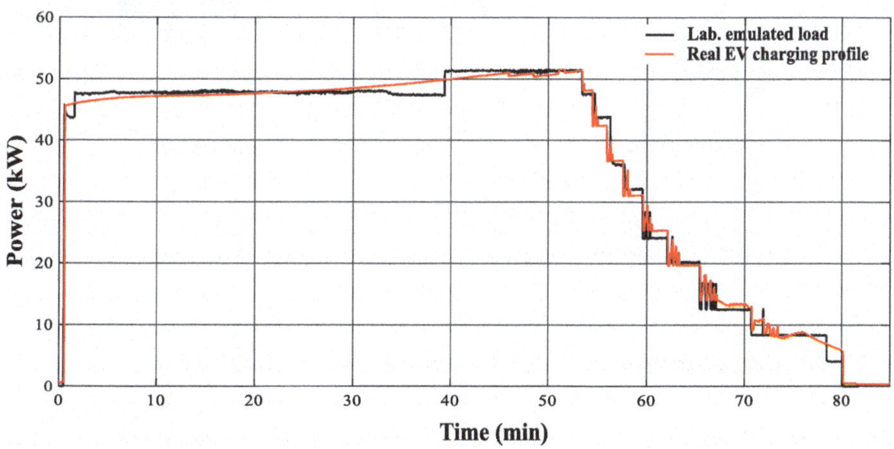

Fig. 7 Fast EV charging profile implemented in the laboratory (real vs. emulated)

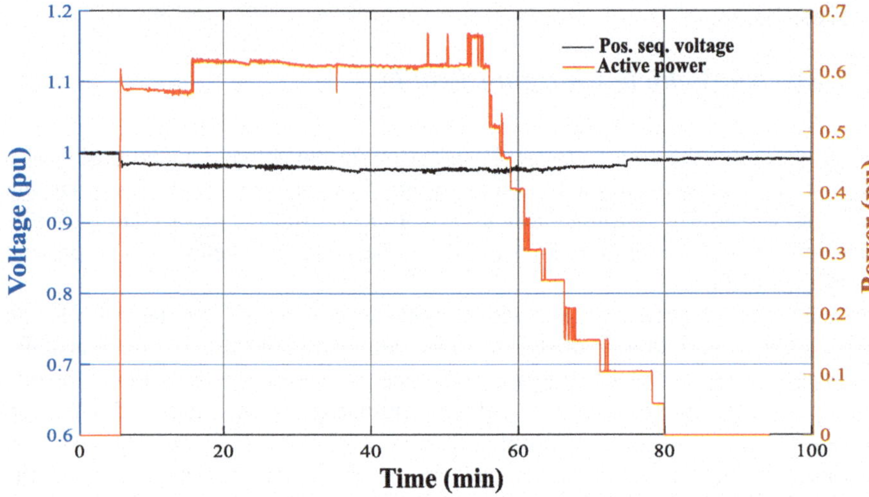

Fig. 8 Impact of fast EV charging process on voltage at the PCC

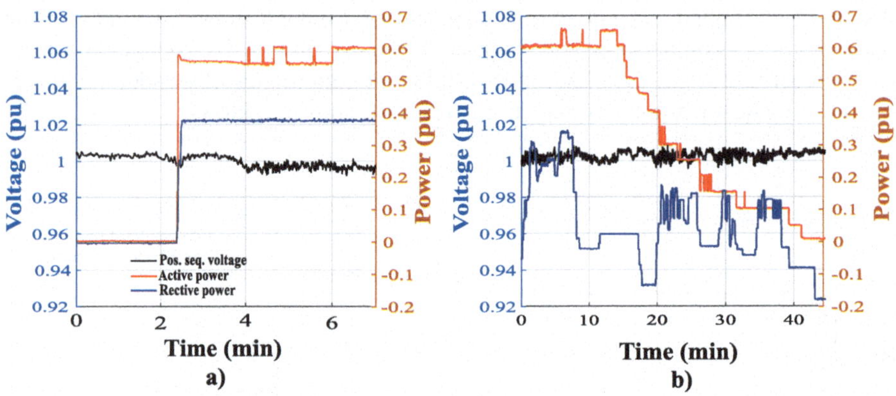

Fig. 9 Experimental validation of the proposal: (**a**) voltage, active, and reactive power at the beginning of the EV charging; (**b**) voltage, active, and reactive power at the final stage of the EV charging

3.2 *Implementation of V2G Voltage Control Method*

This section details the experimental results of the V2G voltage control method described in Sect. 2.4, with algorithm parameters shown in Table 2. The charging process was analyzed in two main stages—initial and final—since power remains constant during the middle stage. This analysis is crucial as it focuses on these variable stages.

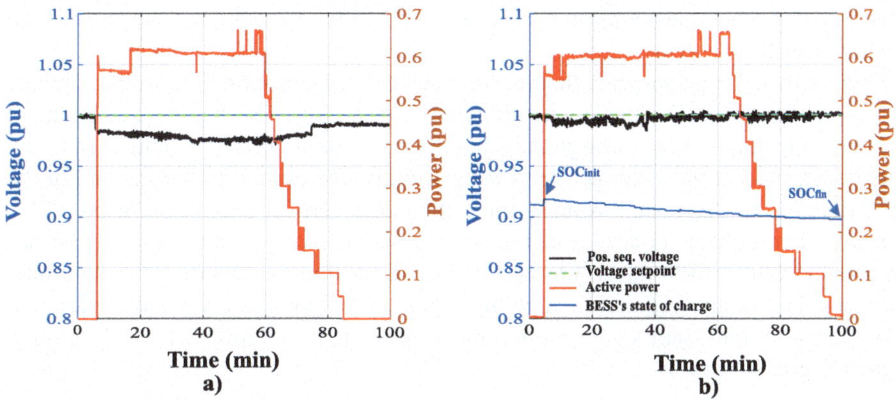

Fig. 10 Complete profile of fast EV charging: (**a**) PCC voltage without compensation; (**b**) PCC voltage by implementing the proposal

Table 2 Parameters assigned to the proposed algorithm

Technical limits	
Deadband	0.5% of rated voltage
Maximum reactive power, voltage reference	30 kVAr, 1 p.u.
Parameters of the digital PI controller	
Proportional and integral gains (K_p, K_i)	5, 80
The time step for calculation (ΔT)	0.1 s

Figure 9 shows the programmable load and voltage profiles. In subplot (a), the voltage remains near its nominal value when the grid conditions allow, though network variability limits complete control. The inverter's output reaches its maximum limit of 30 kVAr of reactive power (equivalent to 0.4 pu on a 75 kVA power base), indicating that the inverter has reached saturation. Subplot (b) depicts more significant variability in reactive power, with the controller effectively injecting both capacitive and inductive types in response to voltage fluctuations from external factors.

This initial test aims to validate the response of the voltage compensation mechanisms through the injection and absorption of reactive power during the most critical stages of the EV charging process. The results demonstrate the controller's ability to dynamically manage reactive power, effectively stabilizing the voltage despite the inherent fluctuations and demands of the charging cycle.

For a comprehensive analysis, Fig. 10 compares the complete voltage profiles before (a) and after (b) implementing the method. Effective voltage compensation at the PCC is observed in the two compared full charging profiles. It is important to note that the compensation efforts are designed to effectively eliminate the voltage dip during the peak power demand of EV charging at the PCC, rather than to restore the voltage to its nominal value. This is because the PCC voltage may already be disturbed by other consumption agents dispersed within the complex structure of the public distribution grid.

Experimental data indicate that reactive power compensation has minimal impact on the battery's SOC, which was obtained directly from the battery management system of the equipment through the laboratory's communication structure. Initially, the battery was at 92% charge, decreasing slightly to around 90% by the end of the test (in Fig. 10, SOC and voltage share the same vertical axis scale in pu). This reveals that voltage control via reactive power delivery from the inverter has negligible effects on battery life and SOC. These findings validate the method's efficacy in voltage regulation and highlight the minor battery charge variation compared to the significant improvement in system stability achieved through reactive power control.

4 Discussion

The experimental results, shown in Figs. 9 and 10, confirm the effectiveness of the proposed algorithm in maintaining voltage levels close to nominal despite significant power demand fluctuations. Figure 10b demonstrates the voltage compensation achieved, highlighting the strategy's ability to mitigate voltage drops from EV charging and other grid stability factors. Additionally, implementing the V2G concept successfully injects reactive power into the grid without harming vehicle battery integrity. This validation suggests potential partnerships between EV owners and power distributors to ease grid loads during peak times.

Furthermore, the experimental setup's ability to replicate real-world conditions strengthens the validity of these findings, as the challenges faced in a live grid environment are more complex and dynamic compared to controlled simulations. This study underscores the need for comprehensive pilot projects integrating V2G systems on a larger scale to evaluate long-term impacts on grid and EV batteries. Such projects could offer valuable insights into optimizing control strategies and developing standards for V2G technology implementation.

These outcomes emphasize that V2G technology's widespread implementation involves more than technical solutions. Political, regulatory, economic, and environmental considerations are crucial for integrating V2G systems into power infrastructures. Future research should explore comparative studies of control strategies and expand V2G applications to more challenging grid scenarios, advancing sustainable energy management practices.

5 Conclusion

The growing adoption of EVs is significantly shaping future energy landscapes, with potential voltage anomalies in electric grids being a primary concern. This study explored leveraging electric vehicle electronic charging devices, such as inverters, to stabilize the grid, particularly addressing voltage variations during sudden EV charging events. While experimental conditions exposed challenges like voltage fluctuations due to external factors and limitations in simulating rapid charging scenarios, the findings were largely positive. Moreover, using commercially available equipment, the successful experimental implementation of the V2G voltage control method in a real distribution network environment highlights this technology's practical applicability and scalability. The experimental setup in the laboratory included a bidirectional inverter connected to a lithium-ion battery storage system for V2G implementation, and a separate programmable AC load was used to emulate the EV's active power profile. This separation was necessary due to the current handling limitations of the inverter, which could not simultaneously manage both the reactive power regulation and the high power demands of fast charging.

The experimental implementation of a voltage control method using V2G technology has yielded promising results. Specifically, the system effectively mitigated a 3% voltage drop at the PCC during peak EV charging periods. The reactive power compensation successfully maintained the technical operating limits of the compensation devices while minimally impacting the EV battery's state of charge, which was obtained directly from the battery management system of the laboratory equipment. This approach effectively enhanced voltage regulation during high power demand periods without negatively affecting the lifespan of EV batteries. These outcomes highlight the potential of V2G technology to contribute significantly to grid stability and resilience, positioning it as a crucial element in adapting to evolving energy demands.

References

1. Tintelecan, A., Dobra, A.C., Martis, C.: LCA indicators in electric vehicles environmental impact assessment. In: 2019 Electric Vehicles International Conference (EV), pp. 1–5. IEEE (2019). https://doi.org/10.1109/EV.2019.8892893
2. Zhang, X., Hu, H., Dong, Z., Huangfu, F., Liu, Y., Strbac, G.: A coordinative strategy for numerous electric vehicles in a resilient urban energy system. In: 2022 IEEE 6th Conference on Energy Internet and Energy System Integration (EI2), pp. 3222–3227. IEEE (2022). https://doi.org/10.1109/EI256261.2022.10116296
3. Chen, G., Zhang, Z.: Control strategies, economic benefits, and challenges of vehicle-to-grid applications: recent trends research. World Electr. Veh. J. **15**(5), 190 (2024). https://doi.org/10.3390/wevj15050190
4. Liu, Y., Li, X., Liang, Y., Zeng, S., Li, M.: Assessment of impacts on integration of disorderly EV charging load to flexible distribution network. In: 2021 11th International Conference on Power, Energy and Electrical Engineering (CPEEE), pp. 139–142. IEEE (2021). https://doi.org/10.1109/CPEEE51686.2021.9383357
5. Manmadha Rao, S., Satputaley, R.J., Keshri, R.K., Patne, N.R., Buja, G.: Voltage Sag Mitigation using DVR in Grid fed EV Fast Charging Station. In: AEIT International Conference on Electrical and Electronic Technologies for Automotive (AEIT AUTOMOTIVE), Nov. 2021, pp. 1–6. IEEE (2021). https://doi.org/10.23919/AEITAUTOMOTIVE52815.2021.9662775

6. Ganne, A., Sahu, L.K.: Performance of single-stage and dual-stage EV battery chargers for G2V and V2G operation. In: 2024 Third International Conference on Power, Control and Computing Technologies (ICPC2T), pp. 486–491. IEEE (2024). https://doi.org/10.1109/ICPC2T60072.2024.10474651
7. Villa-Ávila, E., Arévalo, P., Ochoa-Correa, D., Iñiguez-Morán, V., Jurado, F.: Innovative power smoothing technique for enhancing renewable integration in insular power systems using electric vehicle charging stations. Appl. Sci. **14**(1), 375 (2023). https://doi.org/10.3390/app14010375
8. An, H., et al.: A robust V2G voltage control scheme for distribution networks against cyber attacks and customer interruptions. IEEE Trans. Smart Grid. **15**(4), 3966–3978 (2024). https://doi.org/10.1109/TSG.2024.3365069
9. Haghi, H.V., Qu, Z.: A kernel-based predictive model of EV capacity for distributed voltage control and demand response. IEEE Trans. Smart Grid. **9**(4), 3180–3190 (2018). https://doi.org/10.1109/TSG.2016.2628367
10. Luh, M., Blank, T.: Design and experimental verification of advanced control strategies to provide ancillary services with a bidirectional Vehicle-to-Grid (V2G) inverter. In: 2023 25th European Conference on Power Electronics and Applications. EPE 2023 ECCE Europe (2023). https://doi.org/10.23919/EPE23ECCEEurope58414.2023.10264565
11. Biswas, S., Behera, S., Mekapati, S.R., Choudhury, N.B.D.: Rapid EV market expansion due to V2G technology: a review on V2G grid load balancing and control. Lect. Notes Electr. Eng. **1117**, 381–405 (2024). https://doi.org/10.1007/978-981-99-9235-5_28
12. Abeywardana, D.B.W., Acuna, P., Hredzak, B., Aguilera, R.P., Agelidis, V.G.: Single-phase boost inverter-based electric vehicle charger with integrated vehicle to grid reactive power compensation. IEEE Trans. Power Electron. **33**(4), 3462–3471 (2018). https://doi.org/10.1109/TPEL.2017.2700944
13. Sanguesa, J.A., Torres-Sanz, V., Garrido, P., Martinez, F.J., Marquez-Barja, J.M.: A review on electric vehicles: technologies and challenges. MDPI. **4**(1), 372–404 (2021). https://doi.org/10.3390/smartcities4010022
14. Acharige, S.S.G., Haque, M.E., Arif, M.T., Hosseinzadeh, N., Hasan, K.N., Oo, A.M.T.: Review of electric vehicle charging technologies, standards, architectures, and converter configurations. IEEE Access. **11**, 41218–41255 (2023). https://doi.org/10.1109/ACCESS.2023.3267164
15. Lauinger, D., Vuille, F., Kuhn, D.: A review of the state of research on vehicle-to-grid (V2G): progress and barriers to deployment (2017). https://www.researchgate.net/publication/315144641
16. Ochoa, D., Martinez, S., Arévalo, P.: Extended simplified electro-mechanical model of a variable-speed wind turbine for grid integration studies: emulation and validation on a Microgrid Lab. Electronics (Basel). **11**(23), 3945 (2022). https://doi.org/10.3390/electronics11233945
17. Nicastri, A., Nagliero, A.: Comparison and evaluation of the PLL techniques for the design of the grid-connected inverter systems. In: 2010 IEEE International Symposium on Industrial Electronics. IEEE (2010)
18. Espinoza, J.L., Gonzalez, L.G., Sempertegui, R.: Micro grid laboratory as a tool for research on non-conventional energy sources in Ecuador. In: 2017 IEEE International Autumn Meeting on Power, Electronics and Computing (ROPEC), pp. 1–7. IEEE (2017). https://doi.org/10.1109/ROPEC.2017.8261615
19. Jian-hua, Z.: BYD EV charging challenges solutions. In: 4th International Conference on Power Electronics Systems and Applications, vol. 2011, pp. 1–4. IEEE (2011). https://doi.org/10.1109/PESA.2011.5982911

Mitigating Battery Degradation in Hybrid Energy Storage Systems for Microgrid Applications: A Comparative Analysis of Topologies and Energy Management Strategies

María Alejandra Duque Morales, Camilo Ernesto Montoya Andrade, Andrés Felipe Guerrero Guerrero, and Gustavo Adolfo Lanza Bayona

1 Introduction

Energy sector in Colombia faces significant challenges in electricity coverage and quality, adversely impacting daily life and regional economic development [1]. Microgrids have emerged as a viable solution to mitigate intermittency, interruptions, and the lack of interconnection in remote areas. However, energy storage systems (ESS), typically based on batteries (BESS), experience degradation known as "cycle aging," which shortens their lifespan [2]. The degradation primarily stems from the deterioration of the positive electrode and the growth of the solid electrolyte interface (SEI), these processes are significantly accelerated by elevated charge/discharge rates and fluctuations in the state of charge (SoC), both of which lead to increased thermal stress [3]. Consequently, exclusive reliance on BESS can lead to reduced lifespan, increased costs, and limitations in energy and power density, ultimately compromising supply reliability and increasing long-term maintenance expenses [4].

M. A. Duque Morales · C. E. Montoya Andrade
Programa de Ingeniería Electrónica, Universidad de Cundinamarca, Fusagasugá, Colombia

A. F. Guerrero Guerrero
Programa de Ingeniería Electrónica, Universidad de Cundinamarca, Fusagasugá, Colombia

Grupo de Investigación GITEINCO, Universidad de Cundinamarca, Fusagasugá, Colombia

G. A. Lanza Bayona (✉)
Grupo de Investigación GITEINCO, Universidad de Cundinamarca, Fusagasugá, Colombia

Área de Ciencias Básicas, Universidad de Cundinamarca, Fusagasugá, Colombia
e-mail: galanza@ucundinamarca.edu.co

M. Espinoza-Andaluz et al. (eds.), *Congress on Research, Development and Innovation in Renewable Energies*, Green Energy and Technology,
https://doi.org/10.1007/978-3-031-88995-0_17

Hybrid energy storage systems (HESS), combining lithium-ion batteries (LiB) and ultracapacitors (UC), have recently been deployed to improve efficiency and extend battery lifespan [5]. UCs are capable of efficiently delivering high energy density in very short time frames [6], making them ideal for applications with frequent charge and discharge cycles and reducing stress on battery.

By integrating LiB and UC, HESS provide high-performance energy storage solutions with high energy and power density. These systems aim to minimize fluctuations that cause charge and discharge cycles in batteries, thus preventing degradation and extending their lifespan [7]. Recent studies have shown notable benefits: reductions of up to 33% in battery stress through effective energy management and self-algorithms in wind energy applications [8], a 60.3% improvement in battery lifespan compared to traditional BESS in photovoltaic microgrids [9], and potential lifespan extensions of up to 25 years with optimized hybridization factors for UC [10]. Furthermore, research in [4] laid foundational insights into component selection, system sizing, and control strategies for integrating HESS into microgrids and renewable energy systems. These studies highlight the importance of analyzing HESS configurations based on application needs to optimize system longevity and reliability.

This study evaluates the effectiveness of various HESS configurations in microgrids, focusing on preventing battery degradation. It compares passive, semi-active, and active topologies and assesses the performance of a classic energy management system (EMS) with proportional-integral (PI) control, referenced by a low-pass filter (LPF), against a rule-based EMS. The analysis includes current waveforms and ampere-hour (Ah) degradation coefficients, providing insights into their impact on the longevity and efficiency of battery systems.

2 Methods

The methodology involved simulating scenarios in a HESS, composed of lithium-ion batteries and ultracapacitors, to specifically monitor the battery current's response to sudden changes in load conditions. The performance of HESS varies based on topologies and energy management strategies, leading to the generation of multiple simulation models, each with a different configuration.

The HESS is designed to deliver 700 W to meet system load demands. UCs are selected to handle rapid power fluctuations, with a capacity to provide 14.5 kJ of energy and 1400 W, meeting the requirements for stability during isolated operation. LiBs, with a voltage of 375 V and a capacity of 10 Ah, offer 13.5 MJ of energy and 3000 W, ensuring adequate long-term storage. This sizing ensures effective immediate response and reliable energy supply throughout the microgrid's independent operation. The detailed simulation parameters for each ESS are presented in Table 1.

Table 1 Simulation parameters

	Value (UC)	Value (LiB)
Nominal voltage	300 V	375 V
Nominal capacity	2.9 F	10 Ah
Initial charge	80%	80%
Sampling time	1×10^{-6} s	1×10^{-6} s

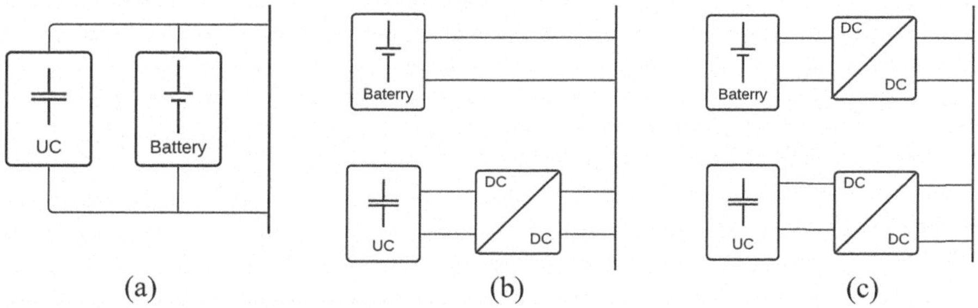

Fig. 1 Topologies of HESS; (**a**) passive topology, (**b**) semi-active topology, (**c**) active topology

2.1 Battery-UC HESS Topologies

Multiple system topologies are proposed for interconnecting the elements when implementing more than one ESS. These topologies can be categorized based on the number of ESS and the number of cells equipped with active control.

We examine three energy storage topologies: passive, semi-active, and active. The passive topology (Fig. 1a), the simplest and most cost-effective, connects storage elements in parallel without system control. The semi-active topology (Fig. 1b) introduces control in one of the elements, typically the ultracapacitor (UC), enabling power management during charge and discharge cycles to reduce battery stress. In contrast, the active topology (Fig. 1c), while being the most complex and expensive, implements control in both elements, providing enhanced system performance by optimizing power distribution.

Topologies incorporating active components require control using a bidirectional DC/DC converter. We utilized the non-isolated buck-boost half bridge converter, following the equations outlined in [11, 12]. Detailed parameters resulting from this design approach are provided in Table 2.

2.2 Energy Management Strategies

To regulate the active components, a proportional-integral (PI) controller is employed for tracking the reference signal, while passive control supplies the remaining current demand. The calculation of system constants is based on the buck-boost converter model, considering bandwidth requirements and phase margin,

Table 2 Buck-boost half bridge converter parameters

	Converter UC	Converter LiB
Input voltage ($V1_{Min} - V1_{Max}$)	300 V	350 V
Output voltage ($V2$)	400 V	400 V
Output current ($I2_{Max}$)	1.74 A	1.74 A
Rated power (P)	700 W	700 W
Efficiency (η)	90%	90%
Inductance (L)	16.16 mH	11.02 mH
Input capacitance (C_1)	1.45 μF	0.35 μF
Output capacitance (C_2)	43.5 μF	43.5 μF
Switching frequency (f_s)	10 kHz	10 kHz

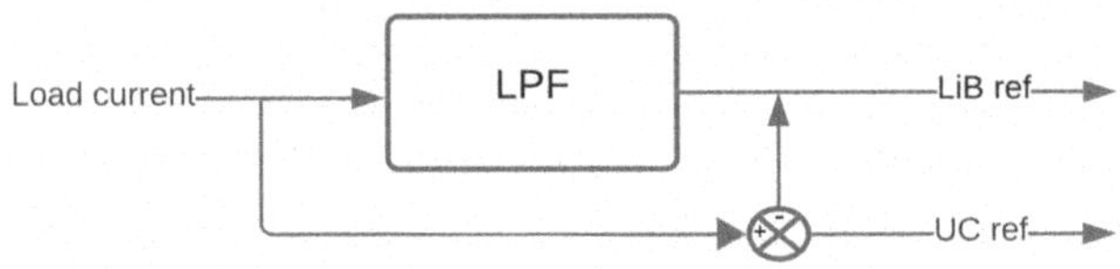

Fig. 2 Low-pass filtering EMS

with Kp = 3.54 and Ki = 155.229. EMS determines the energy contribution of each ESS, involving setting current references for the converters of each component. We explore two approaches: using a low-pass filter (LPF) and a rule-based method (RBM).

In the LPF method, the load current demand is filtered, separating it into high and low-frequency components. These components serve as reference signals for the controllers, one for UC and another for LiB respectively, as shown in Fig. 2. By managing low-frequency reference signals, the LiB experiences reduced stress. This reduction in stress minimizes the physical and chemical transformations within the battery's components, leading to less wear and slower aging. As a result, the battery's cycle life is extended [13].

For implementation, the filter parameters determine the frequency separation between power handling and energy storage in each ESS. The time constant (τ) is crucial: a larger τ narrows the filter's bandwidth, enabling UCs to handle more power. Conversely, a smaller τ increases the bandwidth, leading to greater reliance on LiB [14]. Additionally, the EMS must maintain optimal SoC in both UC and LiB to prevent premature degradation. In this case, a discrete low-pass filter with a τ of 2.5 is employed to achieve this balance.

The RBM strategy focuses on an algorithm based on mathematical and logical models. Building on insights from [10], this method, as depicted in Fig. 3, governs the UC module's power contributions according to its state of charge and the operational mode of charging or discharging. The minimum and maximum voltages are defined to prevent the degradation of the supercapacitor module's capacity.

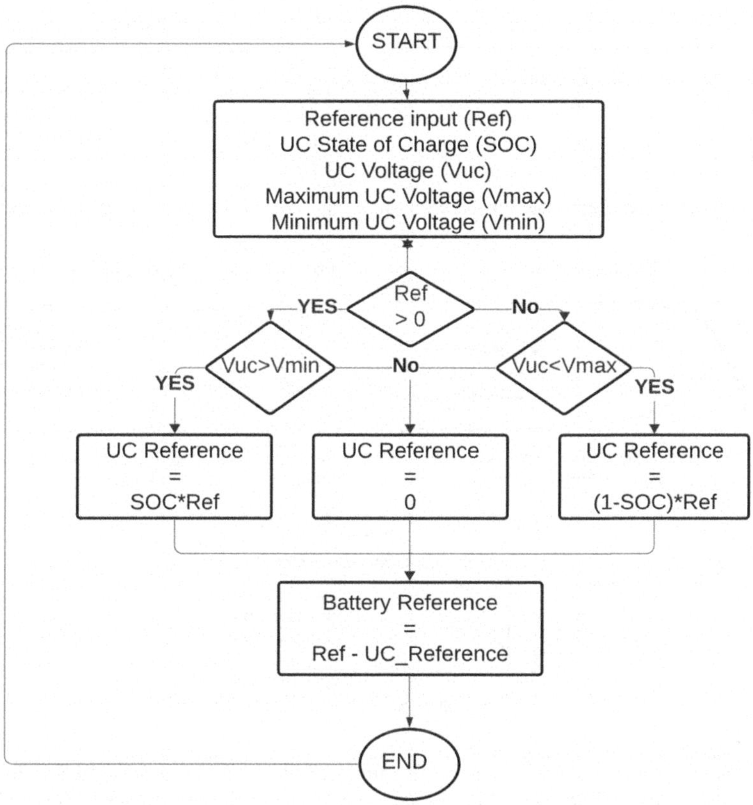

Fig. 3 Rule-based EMS

3 Results

To validate the control and functionality of the battery–UC HESS, specific efficiency test was conducted in simulated environments. To create a situation with a high impact on battery life, step will be used in the simulated load. This is one of the worst-case scenarios for the storage system's lifespan due to the abrupt and instantaneous change in the current demand from it, enabling the measurement and comparison of efficiency across different topologies and EMS configurations considered. The primary objective of the HESS is to alleviate the stress on the battery and enhance overall system performance. Therefore, the simulation primarily focused on the operation of the ultracapacitor to provide support to the battery and ensure optimal charging process. The corresponding current waveforms of HESS are shown in Fig. 4.

Initially, a battery-only energy storage system was evaluated. The battery current exhibited abrupt changes while fully supplying the load demand, indicating the absence of a complementary system to dampen fluctuations (Fig. 4a). Subsequently,

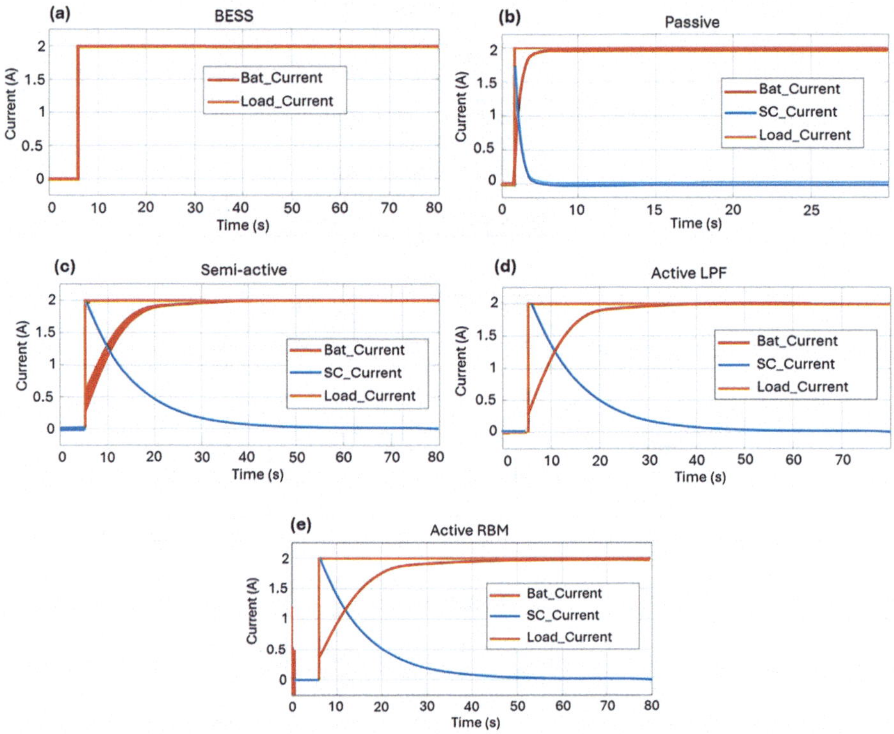

Fig. 4 Dynamic performance of current system during sudden load changes; (**a**) BESS, (**b**) passive HESS, (**c**) semi-active HESS with LPF, (**d**) active HESS with LPF, (**e**) active HESS with RBM

a hybrid system without active control was analyzed, where current distribution among different elements occurred passively without intervention of an EMS. In this scenario, a slight improvement over BESS was observed with energy contribution distribution during disturbance, however, without significant damping (Fig. 4b).

Incorporating control in one component of the HESS, using a low-pass filter as EMS, significantly smooths current fluctuations, leading to a notable reduction in abrupt current changes in the battery. The low-pass filter efficiently distributes energy contributions of each component, thereby reducing the impact on the battery (Fig. 4c). Active control was introduced for both storage components, achieving a response similar to that of a semi-active topology simulation. In a semi-active control, only one component is actively referenced, while the other responds passively to remaining demands. When both components are actively referenced, the now active component supplies energy similarly to the semi-active scenario, resulting in comparable energy distribution behavior (Fig. 4d).

Table 3 Battery lifespan degradation coefficients from simulation results

Topology	EMS	Degradation coefficient	Improvement over BESS
BESS	–	0.55	–
HESS—Passive	–	0.5379	2.25%
HESS—Semi-active	LPF	0.4869	12.96%
HESS—Active	LPF	0.4876	12.8%
HESS—Active	RBM	0.4525	21.55%

Finally, an active hybrid system was evaluated where the EMS managed current based on predefined rules. Here, the best current damping was observed compared to other scenarios considered. The incorporated RBM considers specific parameters of the ultracapacitor, such as SoC and charge/discharge mode, ensuring operation within optimal limits. This approach reduces stress on its components and prolongs its lifespan by intelligently distributing charge and discharge between the battery and ultracapacitor, thereby preventing abrupt current fluctuations on the battery (Fig. 4e).

The resulting current behavior and waveforms obtained from these simulations were further analyzed to determine their impact on battery degradation. By examining the changes in current and the corresponding stress on the battery, the simulations provided critical insights into how different HESS configurations and control strategies influence battery degradation. For this purpose, the battery degradation model described in [15] is proposed, this model emphasizes crack propagation as the primary degradation mechanism and employs relationship between ion loss and stress factors like depth of discharge (DoD), average state of charge (SoC), and temperature, aligning with crack propagation theory. To achieve the objective of this work, we will focus on the part of the model that evaluates the depth of discharge.

In this model, N_{Eff} represents the degradation coefficient over a time interval, according to Eq. (1):

$$N_{\mathrm{Eff}} = \int |i(t)| \mathrm{d}t * \frac{1}{2Q}, \tag{1}$$

where Q denotes the nominal charge capacity of the battery, and $i(t)$ represents the instantaneous current value. A higher value of N_{Eff} indicates a faster rate of SoC change in the battery, leading to accelerated capacity degradation. Results are presented in Table 3.

The implemented degradation coefficient quantifies the Ah-degradation of the battery considering the current values within the analyzed interval. A value of 1.0 represents the worst-case scenario: the battery is completely discharged (100% DoD) in a single cycle, significantly reducing its useful life. In this context, when studying a full cycle with a sudden current change without significant battery discharge, the HESS model with active topology and RBM achieves a 21.55% improvement compared to a storage system with only batteries. Furthermore, this 21.55% improvement compared to the 2.25% of a passive topology highlights the importance of implementing active components and how EMS influences the impact on storage components.

4 Discussion

The simulations involved four different topology configurations: BESS, passive HESS, semi-active HESS, and active HESS. Additionally, two energy management strategies were implemented: one based on a low-pass filter (LPF) and another using a rule-based model (RBM). To establish a baseline for the HESS simulations, the scenario of a battery-only storage system was considered, where the battery current must fully match the load demand, leading to abrupt changes in current. This behavior induces degradation due to repeated charge and discharge cycles over the battery's lifespan. High charge and discharge rates can stress battery materials, hastening their degradation. Therefore, an estimation method was used to assess the impact of current profiles on battery life degradation.

The simulation results evidence that hybridizing energy storage systems, particularly with ultracapacitors, significantly enhances energy delivery stability and efficiency. Comparisons across different topologies and control strategies reveal that while the semi-active topology with LPF provides substantial improvements, incorporating a more advanced EMS like RBM yields the best results in terms of mitigating current fluctuations and extending battery lifespan, achieving a degradation coefficient of 0.4525 compared to the worst-case normalization of 1.0.

Passive hybrid system implementations, while marginally enhancing performance compared to BESS, prove inadequate without active control. Conversely, the active topology with LPF-based EMS shows no significant improvement over the semi-active implementation.

Overall, deploying a HESS for a microgrid involves studying numerous variables. In this study, various topology options and their feasibility were explored based on EMS literature. The best performance was observed with an active topology using RBM; however, a similar level of performance with reduced complexity was achieved by implementing an LPF in a semi-active topology. These findings are considered in view of potential battery degradation associated with hybridization, emphasizing the need for detailed analysis of current profiles in system components.

5 Conclusion

In this work, we explore various topologies and energy management strategies aimed at optimizing the performance of HESS. A key focus is on mitigating abrupt changes in battery current under step input load conditions. Findings highlight two particularly promising scenarios. Firstly, the implementation of an active topology with an EMS based on rules results in a 21.56% improvement in the lifespan cycles of the LiB compared to the natural response of a BESS. In this scenario, preventing degradation of both the LiB and UC is considered due to the establishment of limits and the maximum and minimum values that the UC can operate within the rule-based method. Secondly, the implementation of a semi-active topology with an EMS based on a low-pass filter, resulting in a 12.8% improvement. This scenario, representing lower complexity and eliminating the need for active control in both components, has high viability potential that could indicate a preference under financial feasibility concerns.

The remaining scenarios do not show significant improvement; however, a finding is the utilization of the active topology, which showed comparable performance to the semi-active topology when implementing classical control. In the current study, which focused on mitigating current fluctuations, no significant alteration in the degradation coefficient was observed. Nevertheless, there is potential benefit in employing active control for other microgrid requirements, such as voltage regulation on the DC bus.

The findings of this study contribute to a deeper understanding of HESS, focusing on their ability to mitigate current fluctuations and optimize battery lifespan. These results provide significant insights for selecting, developing, and deploying components in HESS, such as topologies and energy management strategies, which have the potential to enhance operational stability and efficiency within microgrid applications.

Acknowledgments This research was developed as part of the project "Desarrollo de Microredes y almacenamiento (ESS) prestadores de servicios complementarios para incrementar la cobertura, eficiencia y confiabilidad del servicio en el departamento de Cundinamarca" approved in MINCIENCIAS call Number 18 of 2021 and financed with funds from the Department of Cundinamarca, Colombia.

References

1. Fernández Barón, A.: Evaluation of economic and social impact of electrification in non-interconnected zones (NIZ) of rural areas in Colombia. Universidad Autónoma de Bucaramanga. Repositorio Universidad UNAB (2019)
2. Zhang, X., Han, Y., Zhang, W.: A review of factors affecting the lifespan of lithium-ion battery and its health estimation methods. Trans. Electr. Electron. Mater. **22**(5), 567–574 (2021)
3. Roy, P., Shahjalal, M., Shams, T., Fly, A., Stoyanov, S., Ahsan, M., Haider, J.: A critical review on battery aging and state estimation technologies of lithium-ion batteries: prospects and issues. Electronics. **12**, 4105 (2023)
4. Hajiaghasi, S., Salemnia, A., Hamzeh, M.: Hybrid energy storage system for microgrids applications: a review. J. Energy Storage. **21**, 543–570 (2019)
5. Tshiani, C., Umenne, P.: The impact of the electric double-layer capacitor (EDLC) in reducing stress and improving battery lifespan in a hybrid energy storage system (HESS). Energies. **15**(22), 8680 (2022)
6. Akhil, A.A., et al.: DOE/EPRI 2013 electricity storage handbook in collaboration with NRECA (2013). https://www.academia.edu/download/49323464/SAND2013-5131.pdf. Accessed 9 Jan 2024

7. Gee, A., Robinson, F., Dunn, R.: Analysis of battery lifetime extension in a small-scale wind-energy system using supercapacitors. IEEE Trans. Energy Convers. **28**, 24–33 (2013)
8. Jiang, Q., Hong, H.: Wavelet-based capacity configuration and coordinated control of hybrid energy storage system for smoothing out wind power fluctuations. IEEE Trans. Power Syst. **28**(2), 1363–1372 (2013)
9. Jing, W., Lai, C., Wong, S., Wong, M.: Battery-supercapacitor hybrid energy storage system in standalone DC microgrids: a review. IET Renew. Power Gener. **11**(4), 461–469 (2017)
10. Bahloul, M., Khadem, S.: Impact of power sharing method on battery life extension in HESS for grid ancillary services. IEEE Trans. Energy Convers. **34**(3), 1317–1327 (2019)
11. Hauke, B.: Basic calculation of a buck converter's power stage. Texas Instruments Application Report (2011)
12. Hauke, B.: Basic calculation of a boost converter's power stage. Texas Instruments Application Report (2009)
13. Günther, S., Weber, L., Bensmann, A.L., Hanke-Rauschenbach, R.: Structured analysis and review of filter-based control strategies for hybrid energy storage systems. IEEE Access. **10**, 126269–126284 (2022)
14. Millner, A.: Modeling lithium-ion battery degradation in electric vehicles. In: 2010 IEEE Conference on Innovative Technologies for an Efficient and Reliable Electricity Supply, CITRES 2010, pp. 349–356. IEEE (2010)
15. Wikner, E.: Lithium-ion battery aging: battery lifetime testing and physics-based modeling for electric vehicle applications. Licentiate thesis, Chalmers University of Technology, Gothenburg, Sweden (2017)

Prioritizing WAMS Applications for the Modernization of the San Andrés Island Power Grid

Aide J. Urrego Silva, Javier Rosero-Garcia, and Alvaro Zambrano-Pinto

1 Introduction

Electrical distribution systems are undergoing major changes because of the increasing use of distributed energy resources (DERs). These changes are shifting the system from a one-way power flow to a two-way flow, which is especially challenging with renewable energy sources that can be unpredictable and vary in output [1, 2].

To address these challenges, Wide-Area Monitoring Systems (WAMS) have been developed. These systems are crucial for improving the reliability and stability of the power grid. They help by monitoring large areas and providing real-time data, which improves decision-making in control centers and helps with planning and analyzing disturbances [3–5].

State estimation is also vital for managing modern power systems effectively. It provides an accurate, real-time view of how the grid is operating, which helps operators maintain stability and handle the increased use of renewable energy sources. It also helps detect problems early, allowing for quick responses and corrective actions to prevent issues like voltage instability [6].

WAMS also offer tools for managing power use and demand. For example, they use techniques like mean field theory to model and manage energy consumption across whole neighborhoods, benefiting both power grid operators and consumers by improving demand forecasts and energy efficiency [7].

The Adaptive Scheme for Integrity Protection (ASIPS) further enhances real-time control by using data from phasor measurement units (PMUs) and other resources to ensure the grid operates smoothly and securely [8].

A. J. Urrego Silva (✉) · J. Rosero-Garcia · A. Zambrano-Pinto
Universidad Nacional de Colombia, Bogotá, Colombia
e-mail: aurregos@unal.edu.co

M. Espinoza-Andaluz et al. (eds.), *Congress on Research, Development and Innovation in Renewable Energies*, Green Energy and Technology,
https://doi.org/10.1007/978-3-031-88995-0_18

Successfully using these systems requires a clear strategy. This involves evaluating current and future technologies, replacing outdated equipment, and conducting detailed cost-benefit analyses. Experts suggest a strategic approach with 19 categories to prioritize WAMS applications. This chapter also discusses how state estimation can improve monitoring as grid topologies change.

2 Methodology

The following sections describe the approach taken to optimize the electrical system on San Andrés Island. We start by prioritizing applications based on their importance, cost, and complexity, as detailed in the application prioritization section. We then classify these applications according to their criticality and evaluate their cost-benefit ratio to determine the order of implementation.

Additionally, we use phasor measurement units (PMUs) for state estimation. Note that in this study, PMU measurements are simulated. This simulation helps us assess system performance under controlled conditions. The case study of the San Andrés Island network demonstrates how these methods are applied, including voltage estimation and data analysis using simulated PMU data.

2.1 Application Prioritization

To expedite the achievement of set objectives, prioritizing applications offers a structured approach based on their relative importance, considering factors like implementation necessity, infrastructure cost, and complexity. This process tailors solutions to the unique landscape of San Andrés Island, aligning with existing infrastructure and lessons learned. Following the roadmap detailed in [9], the prioritization outlined in Table 1 guides the optimization of the local distribution system. This vital process unfolds in two steps: initial categorization based on general importance and subsequent analysis of cost-benefit dynamics to determine final priorities.

2.2 Application Classification

This step determines the incorporation of phasor measurements to enhance system operation, with the prioritization of applications based on their criticality within the organization. Factors considered include application priority, the value of synchrophasor measurements, and the difficulty of implementation, where application priority ranks applications as urgent, moderate, or unknown depending on their characteristics and potential benefits. Figure 1 illustrates the application value

Table 1 Prioritization results of WAMS applications

Application	Benefit-cost	Benefit-cost numerical value	Priority
Advanced monitoring system	High	1.26	1
Improved load shedding scheme	High	1.25	2
Wide-area visualization	High	1.24	3
High precision fault detection and localization	High	1.22	4
Advanced protection and control system	Medium	1.07	5
Enhanced reliability and resilience analysis	Medium	0.93	6

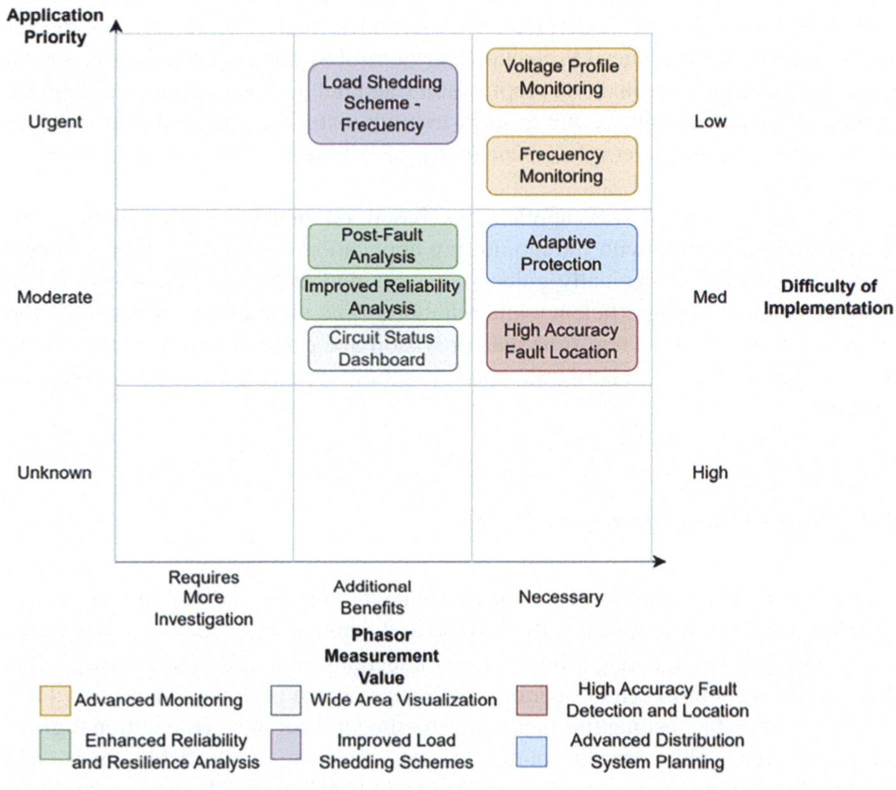

Fig. 1 Value diagram of WAMS applications

diagram, highlighting that monitoring-focused applications generally incur lower costs compared to protection-focused ones. Furthermore, offline applications such as post-fault analysis and enhanced reliability analysis, although not requiring synchrophasor measurements, still benefit from their use.

To proceed with application prioritization, it is crucial to evaluate the cost-benefit ratio to ascertain feasibility and determine whether the benefits outweigh the costs.

2.3 Cost-Benefit Ratio

The ratio indicates the level of value attained for each category, with weights assigned based on the expected organizational benefits from phasor measurement implementation, rated on a scale from 0 to 10. Benefit assessment encompasses factors such as resilience, sustainability, real-time operation, planning, safety, efficiency, innovation, and customer engagement, while cost evaluation considers aspects like complexity, investment, maturity, risk, and readiness, with values assigned according to specific application contexts. Weighting is subsequently applied to both benefits and costs, yielding a weighted average for each category. High-precision fault detection and wide-area visualization emerge as top priorities, underscoring the need to modernize protection infrastructure to enhance efficiency and reliability. The Advanced Protection and Control System and the High-Precision Fault Detection and Localization Application stand out in cost analysis, highlighting significant investment needs due to infrastructure maturity. The final step involves dividing the weighted average of benefits by costs to determine the final prioritization, as shown in Table 1.

The final step prioritizes applications based on benefits outweighing costs, recommending starting with the monitoring application due to its high benefit-cost ratio. This process strategically guides network modernization, minimizing initial costs while maximizing efficiency and reliability. The monitoring application's top priority is justified by its real-time data provision and essential role in state estimation for the distribution network, analyzing key scenarios such as topological changes.

2.4 State Estimation with PMUs

State estimation is crucial for real-time assessment of electrical distribution systems, utilizing phasor measurement units (PMUs) and mathematical models to determine parameters like voltage magnitudes, power injections, and grid configuration. This data empowers control center operators to ensure system stability and security [10].

PMU-based state estimation facilitates the modernization of conventional grids, integrating renewable energy sources, storage systems, controllable loads, and energy transactions. Accurate data obtained from PMUs forms the basis for various functions within the energy management system (EMS), highlighting its importance.

PMUs offer higher accuracy and reliability compared to traditional methods, contingent on factors such as installation numbers, data quality, and algorithm selection.

This section develops a linear state estimator using PMU measurements, focusing on San Andrés Island's distribution network. It analyzes scenarios where synchronized phasor information is vital for maintaining energy quality and efficient services, crucial for societal well-being and economic development.

2.5 Case Study: San Andrés Island Network

This research evaluates state estimation performance with phasor measurement units (PMUs) across critical electrical system scenarios including short circuits, topology changes, load variations, and real data from the San Andrés electrical network. Using a model of the San Andrés Island electrical network in PowerFactory software (Fig. 2), it aims to develop and test a robust state estimator under challenging conditions, starting with load flow analysis and presenting results.

The San Andrés Island electrical distribution system operates radially with three main nodes: Punta Evans Generation Plant (Node 1), Bight Substation (Node 2), and School House Substation (Node 3). Punta Evans Plant, fueled by diesel combustion, has a 51.2 MW nominal capacity at 13.8 kV generation voltage. This voltage is transmitted via 34.5 kV double-circuit lines to the Bight Substation and then redistributed in a single-circuit to the School House Substation. The distribution network comprises 14 circuits operating at 13.2 kV nominal voltage, with seven feds from Bight Substation and five from School House Substation [11].

A state estimation code tailored to San Andrés Island's specific characteristics was developed based on equations from [12], refined to meet island requirements. Figure 3 illustrates the flowchart for the state estimation process applied to the electrical system of San Andrés Island. The process begins with initialization, where the model parameters are set. This is followed by RMS simulation in PowerFactory, using voltage phasors, current phasors from the three main buses, and frequency as input variables. The simulation results are exported to Excel, where they are combined with the values of reactance, resistance, and susceptance of the network in a Python script. This script performs the data processing, generating the

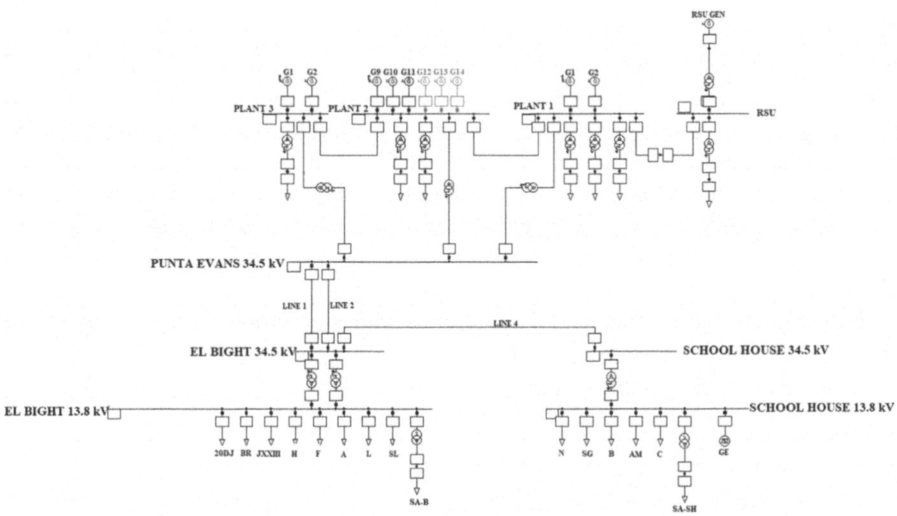

Fig. 2 The San Andrés Island Network model in power factory

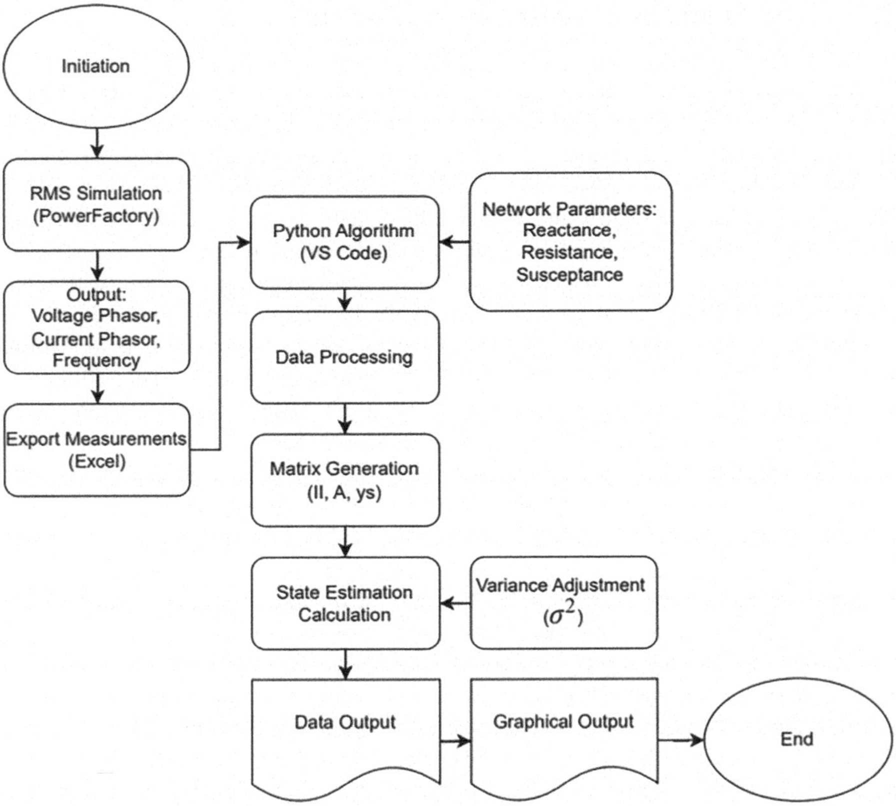

Fig. 3 Process flow for state estimation in the San Andrés Island Network

necessary matrices for state estimation, adjusting the variance to enhance accuracy. Finally, the estimated voltage phasors for the three main buses are obtained and plotted, providing a comprehensive view of the procedure from initial simulation to result visualization.

Based on the three-bus system, shown in Fig. 2, and the detailed data provided in Table 2, the estimation of voltage phasors is then performed at the three main buses on San Andrés Island. Specifically identified as Punta Evans, El Bight, and School House.

In the values for reactance and susceptance presented in Table 2, "$\boldsymbol{j}$ "represents the imaginary unit where $\boldsymbol{j} = \sqrt{(-\mathbf{1})}$.

In summary, the system features:

- Three nodes
- State variables $\boldsymbol{x} = [\boldsymbol{V}_1, \boldsymbol{V}_2, \boldsymbol{V}_3, \boldsymbol{\theta}_{v1}, \boldsymbol{\theta}_{v2}, \boldsymbol{\theta}_{v3}]$
- Six measurements provided by

Table 2 Detailed data for the lines in the three-bus system

Line	Resistance [p.u]	Reactance [p.u]	Susceptance [p.u]
1	0.2132	0.5093j	3.6800e-05j
2	0.2132	0.5093j	3.6800e-05j
4	0.0771	0.1803j	1.3361e-05j

$$z = [V_1, V_2, V_3, \theta_{v1}, \theta_{v2}, \theta_{v3}, I_1, I_2, I_3, \theta_{I1}, \theta_{I2}, \theta_{I3}]$$

Additionally, the configuration of the lines is specified as follows:

- Line 1: From the Punta Evans Substation to the Bight Substation.
- Line 2: Parallel to Line 1, from the Punta Evans Substation to the Bight Substation.
- Line 4: From the Bight Substation to the School House Substation.

These data provide an overview of the electrical system's structure and are fundamental for state estimation. Voltage magnitudes and angles have been calculated with high accuracy, taking into account operational conditions and appropriate mathematical models. Measurements, subject to random errors modeled with Gaussian distribution and zero mean variance, show that increased variance corresponds to reduced accuracy. To assess the effectiveness of the state estimator, a linear algorithm was implemented using Python, providing detailed insights into the performance and accuracy of the estimator across various scenarios.

3 Results

This section presents the results of the state estimation analysis for the San Andrés Island electrical network. Table 3 shows the estimated values, the corresponding errors, and the comparison with expected values for voltage magnitudes and angles. The evaluation focuses on a single scenario, assessing the performance and accuracy of the state estimator under specific conditions. This scenario includes changes in network topology, and the results are analyzed to demonstrate how effectively the estimator performs in maintaining accuracy and reliability.

Values in Table 3 provide insight into the state estimator's performance, showing predominantly low percentage errors, indicating its effectiveness. However, higher percentage errors are observed in angle estimations (θ_{v1}, θ_{v2}, θ_{v3}), suggesting either angle estimation challenges or less accurate angle measurements compared to voltage magnitudes.

The estimator's performance was further evaluated across four scenarios, each representing unique operating conditions challenging the estimator's reliability and accuracy. These scenarios include changes in network topology, three-phase short

Table 3 State estimation obtained values

State variable	Expected value	Measured value by PMU	Estimated value	Error [%]
V_1 $[p.u]$	0.9893	0.9892	0.9892	0.01
V_2 $[p.u]$	0.9705	0.9692	0.9701	0.04
V_3 $[p.u]$	0.9658	0.9664	0.9655	0.03
θ_{v1} $[p.u]$	−0.5479	−0.5472	−0.5484	0.08
θ_{v2} $[p.u]$	−0.5672	−0.5680	−0.5680	0.14
θ_{v3} $[p.u]$	−0.5723	−0.5741	−0.5730	0.12

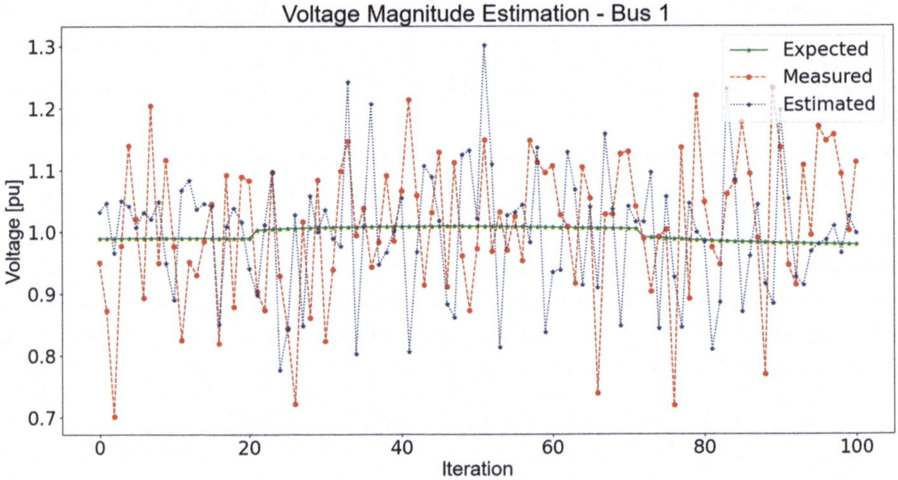

Fig. 4 Voltage estimation with variance of 0.1

circuits, load variations, and connection/disconnection of large-scale distributed generation, designed to comprehensively assess the estimator's performance under dynamic conditions.

To select the optimal variance value, a study evaluated three specific values: 0.1, 0.01, and 0.001, as shown in Figs. 4, 5, and 6. This approach aims to illustrate the dispersion of measurement errors associated with the studied event.

The analysis reveals that higher variance during the 1-s topological change period increases dispersion, challenging clear event observation. To ensure accurate interpretation across the 10-s simulation, a variance of 0.001 was selected over the usual 0.1 for phasor measurement units, prioritizing measurement accuracy and aiding visualization. This decision boosts confidence in data consistency and underscores the critical role of precise measurements in interpreting results. Simulations in PowerFactory employed a 10-s RMS approach to assess the state estimator's performance under dynamic conditions.

An event involving switches opening and closing on Line 4 assessed the linear state estimator's adaptability with PMUs to a new network configuration. Figure 7 provides a detailed representation of this scenario.

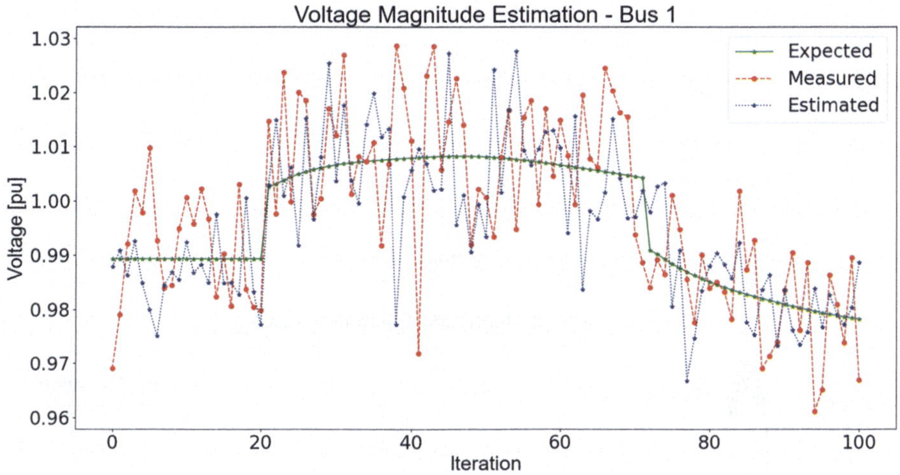

Fig. 5 Voltage estimation with variance of 0.01

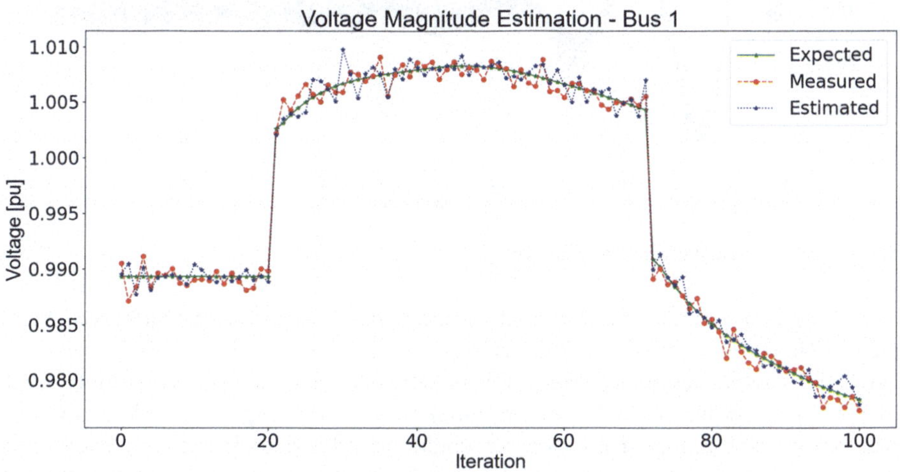

Fig. 6 Voltage estimation with variance of 0.001

During the simulation, a scenario was run in which Line 4 was disconnected at 200 ms and reconnected at 700 ms. It is noteworthy that this line connects Substation Bight (Node 2) with Substation School House (Node 3). The simulation was run for 10 s, for a total of 969 iterations. To visualize the results obtained, the first 100 iterations were selected and plotted, covering the critical period of this operation.

Figures 8, 9, and 10 show the estimation of voltages at the three nodes of the electrical network during the topological change scenario.

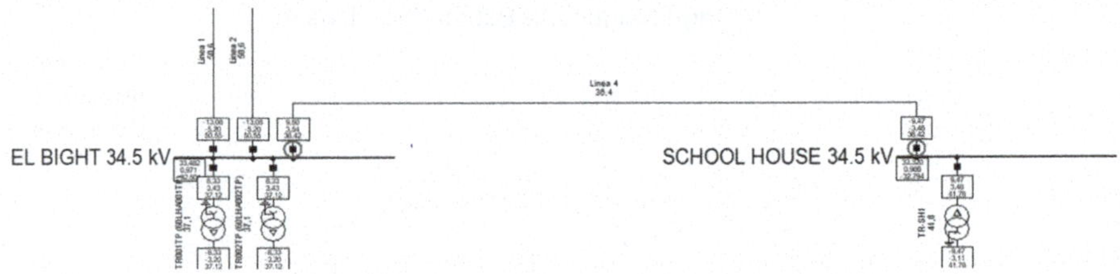

Fig. 7 Detailed representation of the disconnection of Line 4

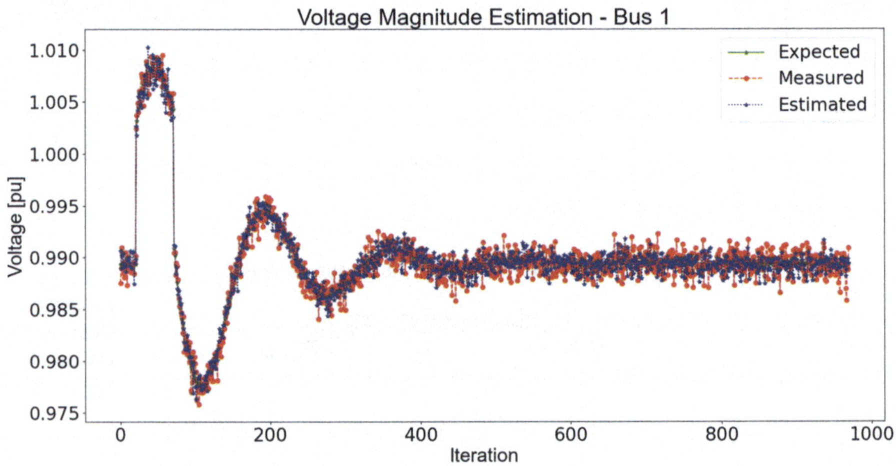

Fig. 8 Voltage magnitude estimation—Bus 1

For the first scenario, the variations of the voltage graphs can be analyzed. After the disconnection of Line 4 at 200 ms, Node 1, which is responsible for generation, had an initial value close to 0.990 p.u. When the line went out of service, it increased to 1.010 p.u. As for Node 2, with an initial value of 0.970 p.u., it peaked at 0.995 p.u. before stabilizing. This increase could be related to generation exceeding demand, as the disconnection of Line 4 left Node 3, with 5 associated distribution circuits, out of service.

Both Node 1 and Node 2 tend to stabilize with damped behavior until they return to their nominal value. Node 3, which remains without power after this disconnection, shows how its value drops to zero. After the line connection is restored, it returns to its nominal value.

After examining the patterns in the node voltages, it is equally essential to consider how the angles of voltage phasor vary. Figures 11, 12, and 13 show the phase angles at the three substations. These graphs provide a detailed insight into the behavior of system in this particular scenario.

In angle analysis, Nodes 1 and 2 show a peak around −0.57 to −0.52 p.u., stabilizing with a lower peak near −0.58 p.u. Node 3 initially at −0.49 p.u. peaks at 2.5 p.u. during Line 4 disconnection, quickly returning to −0.49 p.u. This

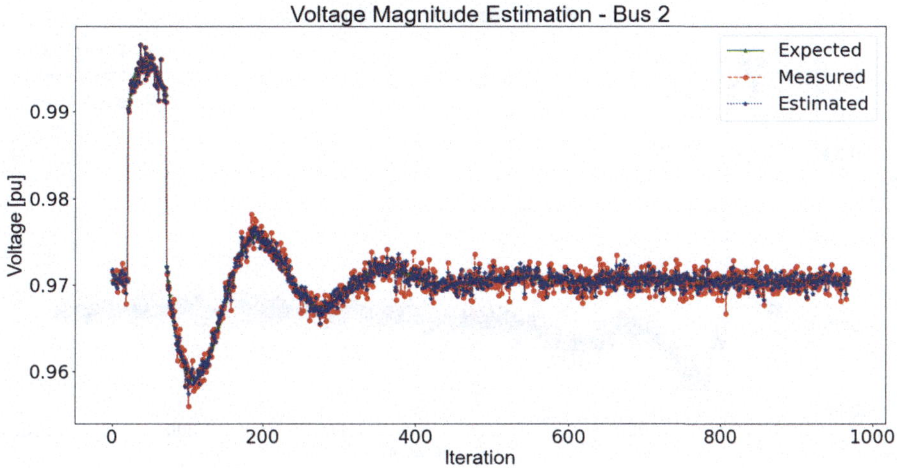

Fig. 9 Voltage magnitude estimation—Bus 2

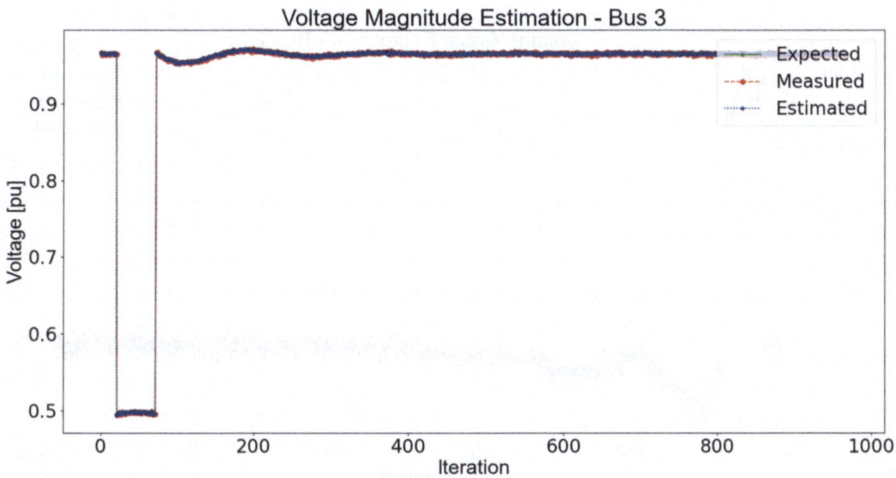

Fig. 10 Voltage magnitude estimation—Bus 3

underscores the estimator's adeptness in adapting to dynamic conditions and managing network topology changes effectively.

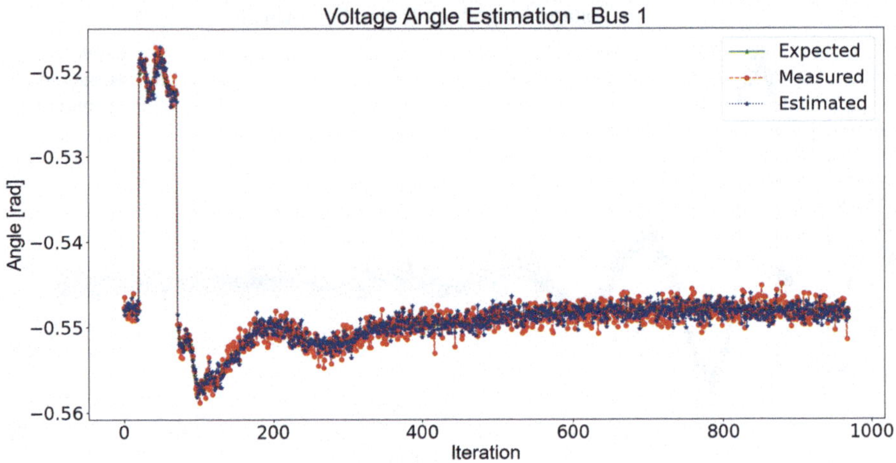

Fig. 11 Voltage angle estimation—Bus 1

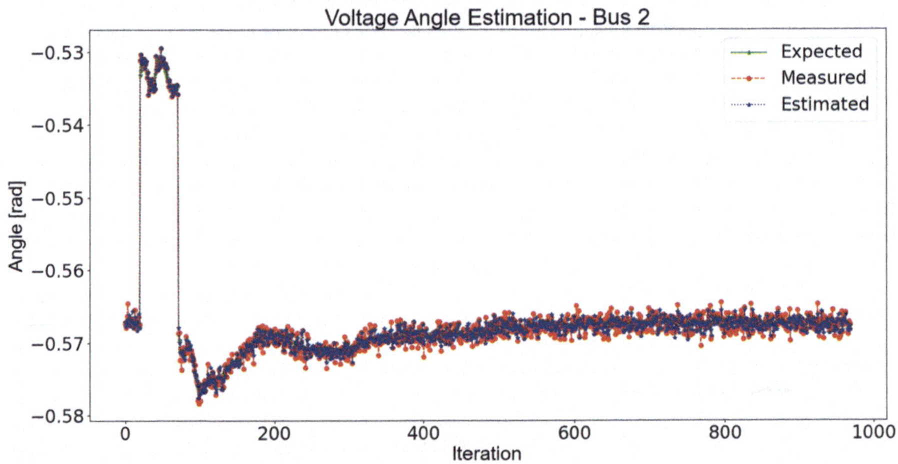

Fig. 12 Voltage angle estimation—Bus 2

4 Conclusion

Applications for the Wide-Area Monitoring System (WAMS) were selected through a process of prioritization and classification. This procedure included the evaluation of the cost-benefit ratio, which highlighted the convenience of starting the transition through monitoring applications. This choice is supported by the limited maturity of the current infrastructure on the San Andrés Island, where implementing control and protection applications based on phasors would involve a significant investment.

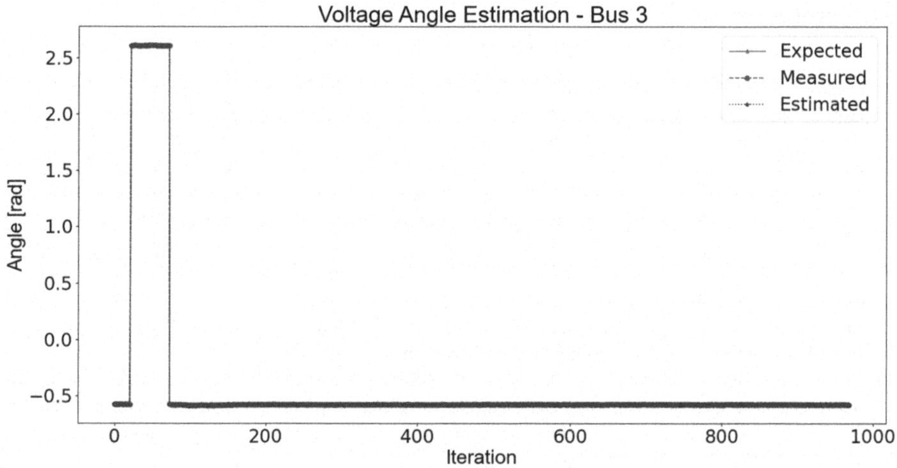

Fig. 13 Voltage angle estimation—Bus 3

The implemented state estimation strategy demonstrates its effectiveness in achieving accurate estimations with values very close to the expected voltage phasors at the bars of the San Andrés Island power grid. This approach was developed by using the network diagram and phasor measurements and adjusting the estimation calculations for a contingency event such as a topological change.

The state estimator maintains an error rate below 5% using a Python-based linear estimation method integrating voltage and current phasor measurements. Even under the most stringent conditions, which showcase optimal performance, it provides a clear view of system behavior. This approach, though nearly ideal, is crucial for understanding the system's maximum potential and emphasizes the importance of precise measurements for informed decisions and real-time adjustments. Accurate measurements enable the estimator to track system behavior effectively, including topological changes, facilitating informed and strategic decision-making.

The analysis of the topological change scenario reveals coherent behavior in the measurements obtained. This scenario is evaluated through the state estimator; this provides the opportunity to subject the system to different conditions and to evaluate the robustness and accuracy of the estimator. This analysis is important as it offers a deeper understanding of the behavior of the system, especially in the context of future changes. With the transition to a more distributed and dynamic environment in which conventional generation paves the way for renewable sources, greater robustness is necessary.

Acknowledgments This research was supported by Electrical Machines and Drives (EM&D) from Universidad Nacional de Colombia, Red de cooperación de soluciones energéticas para comunidades, code: 59384 and Project Think Green on the island of San Andres [BPIN: 2016000100002 EEDAS ESP].

References

1. Ángel, M., Fornié S., Matanza, J., Cossent, R.: El futuro de las redes eléctricas inteligentes
2. Aquino-Robles, J., Corona, L., Robles, A.: Recursos Energéticos Distribuidos frente a los sistemas eléctricos de potencia convencionales (2010)
3. Rivera, G., De La Torre, A.: Vista de Estrategias para la Implementación de un Sistema de Monitoreo de Área Extendida WAMS en el Sistema Nacional Interconectado del Ecuador (2013). https://revistaenergia.cenace.gob.ec/index.php/cenace/article/view/134/132
4. Phadke, A.G., Bi, T.: Phasor measurement units, WAMS, and their applications in protection and control of power systems. J. Mod. Power Syst. Clean Energy. **6**(4), 619–629 (2018). https://doi.org/10.1007/s40565-018-0423-3
5. Lozano, C.A., Castro, F., Ramirez, S.L.: Unidades de Medición Fasorial (PMU): Sistema de Posicionamiento Global - Adquisición de datos (2012)
6. Matavalam, A.R.R., Singhal, A., Ajjarapu, V.: Monitoring long term voltage instability due to distribution & transmission interaction using unbalanced μPMU & PMU measurements. In: 2020 IEEE Power & Energy Society General Meeting (PESGM). IEEE (2020). https://doi.org/10.1109/PESGM41954.2020.9282086
7. Zhu, Z., Fan, Z.: An efficient consumption optimisation for dense neighbourhood area demand management. In: IEEE International Energy Conference (ENERGYCON), 2016, pp. 1–5. IEEE (2016). https://doi.org/10.1109/ENERGYCON.2016.7514063
8. Lopez, J.A., Lu, C.N.: Adaptable system integrity protection scheme considering renewable energy sources output variations. IEEE Trans. Power Syst. **35**(5), 3459–3469 (2020). https://doi.org/10.1109/TPWRS.2020.2975280
9. Hunt, R., Technology Lingwei Zhan, Q., Julio Romero Aguero, O., Technology Yilu Liu, Q.: San Diego Gas & Electric Eric Udren, Quanta Technology Tariq Rahman (2021)
10. Arias, D.: Índice de detección de problemas de estabilidad de tensión en base a WAMS (2012)
11. UPME: Integración de las energías renovables no convencionales en Colombia (2015)
12. Jones, K.D., Thorp, J.S., Gardner, R.M.: Three-phase linear state estimation using phasor measurements. In: IEEE Power and Energy Society General Meeting. IEEE (2013). https://doi.org/10.1109/PESMG.2013.6672516

Fault Prediction in Wind Turbines: A Recurrent Neural Network Approach for IGBT Failure Detection

Génesis Vásquez, David Rosales, and Jorge Maldonado-Correa

1 Introduction

Wind turbines (WTs) play a significant role in harnessing the power of the wind in the field of wind power generation. However, the presence of anomalies or failures in these devices can compromise their performance, affect energy production, and increase maintenance costs [1]. Therefore, effective early anomaly detection methods are necessary to improve the operation of WT. On a global scale, wind power generation capacity has experienced exponential growth, reaching more than 3372 gigawatts by 2022 [2]. This growth has not only contributed to reducing emissions, avoiding the release of 270.9 tons of CO_2 in 2020, but has also generated more than 12.7 million jobs [3].

In the specific context of Ecuador, there has been steady progress integrating wind energy into its energy matrix. This growth has been supported by emblematic projects such as the Villonaco wind farm (VWF) in the province of Loja. The VWF is located in a complex terrain at an altitude of approximately 2720 m above sea level. It consists of 11 GoldWind GW70/1500 WT with direct drive technology and a permanent magnet synchronous generator with a rated power of 1500 kW. This facility has an estimated wind speed of more than 12 m/s/year and is considered one of the world's highest altitudes and most extreme wind power sites [4].

A WT is a complex system that converts wind energy into electricity. It comprises several elements, such as blades, rotor, gearbox, generator, tower, and controllers. The wind rotates the blades and rotor, driving the main shaft, accelerating the

G. Vásquez (✉) · J. Maldonado-Correa
Universidad Nacional de Loja, Loja, Ecuador
e-mail: genesis.vasquez@unl.edu.ec

D. Rosales
Instituto Superior Tecnológico Sudamericano, Loja, Ecuador

M. Espinoza-Andaluz et al. (eds.), *Congress on Research, Development and Innovation in Renewable Energies*, Green Energy and Technology,
https://doi.org/10.1007/978-3-031-88995-0_19

generator through the gearbox, converting the wind energy into mechanical and then electrical energy. The blade angles are adjusted to suit different wind speeds, and the yaw system aligns the turbine with the wind direction [1]. WT can face some failure causes that affect their performance and operability. Extreme weather conditions can produce structural damage despite their rugged design. In addition, natural wear and aging of mechanical and electrical components over time can result in decreased performance and, eventually, failures [1, 5, 6]. Detecting failures in WT at a practical level is an overly complex problem. In general, the tracking and monitoring of a wind farm is done using a supervisory control and data acquisition (SCADA) system, which stores the data from the installed sensors in the form of time series and stores relevant information regarding variables such as active power, reactive power, wind speed, voltage, current, temperature, among others [7]. The SCADA system monitors wind installations in real-time, but it is not able to identify failures, much less make decisions based on the measured variables [5, 7, 8]. Current methodologies tend to rely heavily on supervised approaches, which require labeled data sets, and in the real world, the data from a wind farm's SCADA system does not contain tag information when a fault has occurred [6]. The insulated gate bipolar transistor (IGBT) in the WT helps control the speed of the generator, which in turn controls the frequency of the alternating current produced [9]. Fluctuations in generator speed can cause changes in the frequency of the generated AC current, affecting the IGBTs switching. If the switching frequency of the IGBTs does not match the frequency of the AC current, overvoltages may occur, which could damage the IGBTs [1, 10, 11]. In this study, the focus is on the failures produced in the IGBT since, according to the operation and maintenance (O&M) data of the VWF, most of the recorded failures are related to damages in these semiconductor devices. In addition, in the existing scientific literature, failures related to the blades, rotor, gearbox, generator, converter, etc. are studied. However, only some studies have analyzed the problems related to this component in depth.

Recurrent neural networks (RNNs) have emerged as an alternative tool to address the challenges associated with early fault detection in WT. Unlike traditional neural networks, RNNs can process data sequences and capture temporal dependencies through feedforward connections in the network [1, 12]. This feature makes them suitable for modeling and predicting the dynamic behavior of complex systems such as WTs.

Numerous studies have applied machine learning (ML) models to real SCADA system data for anomaly detection, while others have been limited to comparing different ML models with simulated data. In this context, it is important to highlight the advantages and disadvantages of the main techniques used. RNN-gated recurrent unit (GRU) offers excellent ability to handle long-term dependencies and is computationally efficient, though it may be less powerful than LSTM in some very complex tasks. Long short-term memory (LSTM) is highly effective in capturing long-term dependencies and robust against the vanishing gradient problem but has higher computational complexity than GRU [13].

In [5], they conducted a study using a type of LSTM (long short-term memory) network and attention mechanism to detect and classify WT blade imbalance faults;

the proposed model achieved a fault detection accuracy of over 99%, indicating a highly effective ability to identify and classify these faults with high accuracy. In addition, in [14], a method for WT condition monitoring using SCADA data is proposed. This approach combines a convolutional neural network (CNN) and a bidirectional recurrent neural network (BiGRU) to capture multidirectional spatial-temporal features of the data. First, anomalous data are filtered out, and the validity of the information is improved. Then, input variables are selected and transformed into high-dimensional features using CNN. These features are fed into the BiGRU network with an attention mechanism to improve learning accuracy by highlighting important information. The approach is validated by detecting anomalous operations and faulty components in WTs using real cases in a wind farm.

Convolutional neural network (CNN) combined with bidirectional GRU (BiGRU) effectively captures multidirectional spatial-temporal features and improves learning accuracy but presents greater complexity in architecture and possible overfitting if not properly designed. Multiple stacked LSTM has a high capacity to capture non-linear and non-stationary relationships but comes with higher computational cost and risk of overfitting on small datasets. The choice of the RNN-GRU model for this specific application of IGBT fault detection is based on its robustness and efficiency. The GRU offers an optimal balance between the ability to capture complex temporal dependencies and computational efficiency, crucial for real-time processing of SCADA data. Moreover, its simpler architecture compared to LSTM makes it less prone to overfitting, especially relevant when working with limited or noisy datasets, as is common in industrial environments [15, 16].

The proposed method for fault detection involves an initial SCADA data preprocessing process to identify patterns or relationships between variables that can explain IGBT faults. This analysis allows the selection of the most relevant SCADA variables, which will be used to build an RNN-GRU model to predict new values of these variables. Subsequently, the Isolation Forest (IF) algorithm establishes a threshold to classify the possible failures based on the predictions obtained. The novelty of this approach lies in the specific combination of ML techniques to address IGBT-related faults in wind turbines, an area that has received less attention in existing literature. By focusing on IGBTs, critical components in wind turbine energy conversion systems, this method promises to significantly improve early fault detection, thus contributing to the overall efficiency and reliability of wind farms. The core contributions of our study are outlined as follows:

First, the proposed method combines supervised (RNN) and unsupervised (IF) learning approaches to detect potential IGBT-related faults. Second, a data set comprising real information from a SCADA system at a wind farm located in complex terrain is used. Third, the proposed RNN-GRU model is highly applicable and scalable to different types of time series data. Finally, detection threshold is determined in conjunction with IF, based on prior knowledge of the faults recorded by the O&M data.

The document is organized as follows. Section 2 presents the applied methodology. Section 3 shows the proposed fault prediction method. Section 4 shows the results obtained by testing the method using data sets from a real wind farm. Conclusions are drawn in Sect. 5.

2 Methodology

This section describes the strategy for IGBT fault prediction. The scheme of the proposed method is depicted in Fig. 1, where an RNN-GRU model is used as the prediction method. The data set belongs to a real wind farm located in the city of Loja, Ecuador. Then, data preprocessing and exploratory analysis of the dataset are addressed by obtaining a correlation matrix and kurtosis analysis to select the variables for the model. Based on the new data predicted by the RNN-GRU model, an unsupervised learning model is implemented to define the threshold at which new values are identified as potential failures. Finally, the evaluation and validation metrics of the proposed model are included.

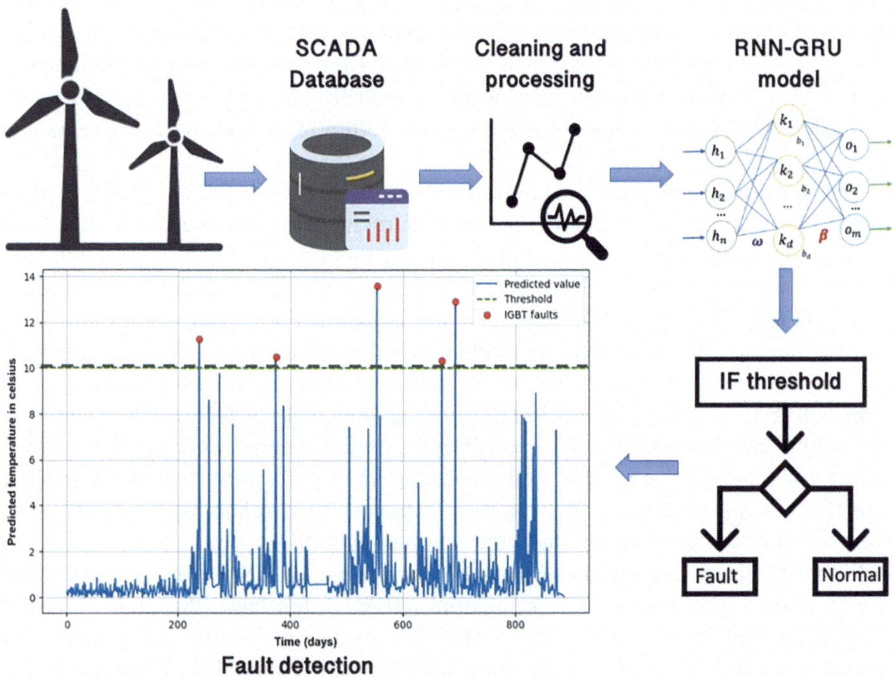

Fig. 1 Scheme of the proposed method

2.1 Data Processing

The data used in this study correspond to operation from the SCADA system of the VWF, which is in southern Ecuador, between UTM coordinates 693,030 E 9558392 N and 693,526 E 9556476 N. VWF is on complex terrain at an altitude of approximately 2720 a.m.s.l. [4]. These variables are subjected to preprocessing and exploration to identify pertinent patterns and characteristics that may be linked to IGBT failures. In the exploratory data analysis, missing/null values and outliers were identified, collectively comprising $<5\%$ of the total instances. Consistent with several studies in the current state of the art, these instances were eliminated to prevent their influence on subsequent models, considering they represent a negligible minority percentage. Subsequently, a slight imbalance was observed in the distribution of fault and non-fault data (45% and 55%, respectively). Consequently, random under sampling was implemented on the majority class to equilibrate the specified categories.

In addition to the SCADA data, information on O&M is available for the WTs at VWF. These data provide the following information: type of failure, date of failure, subsystems affected, and actions taken. This information is used to model and test whether the proposed methodology can predict the occurrence of faults in the data. RNN-GRU is a model to which several time series of different variables are entered as inputs to predict the variables for multiple time instants. To determine which variables to include in the model, a correlation and kurtosis analysis was carried out to determine the variables that have the best relationship with the behavior of the IGBT and that collaborate to obtain the best performance of the model.

2.2 RNN-GRU Model

The GRU workflow is like a basic RNN. It needs an input vector and a previous hidden state through gates at each time step to produce a new hidden state. This new state contains short- and long-term information and is transmitted to the next timestamp. Unlike LSTM, GRU has fewer parameters and is computationally more efficient, as it combines short- and long-term state vectors into a single vector, which simplifies its structure [16, 17]. In this work, an RNN with GRU cells was employed to predict SCADA variables related to IGBT faults. The architecture of the network was determined through a systematic process of experimentation and performance evaluation. The researchers opted for a single hidden layer after extensive testing revealed that additional layers did not significantly improve the ability to detect early defects. This decision aligns with the principle of parsimony in model selection, where simpler models are preferred when they perform comparably to more complex ones. The number of neurons in the hidden layer was set to 128 after a series of experiments with various configurations. This choice represents a balance between model complexity and performance. Smaller neuron counts (64 and 32) were tested but showed reduced predictive capability, while larger numbers (256 and 512) increased computational overhead without proportional performance gains. The 128 neuron configuration provided optimal learning capacity for the specific dataset and prediction task at hand.

For the output layer, the hyperbolic tangent activation function (tanh) was employed. This choice was motivated by tanh's stability during training, which facilitates more balanced gradients in backpropagation compared to alternatives like ReLU. The sigmoid function was selected for the activation gates due to its ability to regulate information flow accurately. Its output range (0 to 1) is particularly suited for the gating mechanism in GRU cells, effectively facilitating the learning process through error backpropagation. To prevent overfitting and enhance the model's generalization capability, the early stopping technique was implemented. This approach continuously monitors the model's performance on the validation set during training. A threshold was set where training stops if performance drops below 90% of the best observed validation performance. This threshold was determined empirically to strike a balance between allowing sufficient training and preventing overfitting. The batch size was set to 32 samples after experimenting with many sizes (16, 32, 64, and 128). This choice provided a good balance between computational efficiency and training stability. Smaller batch sizes led to noisier gradient estimates and longer training times, while larger batches reduced the model's ability to generalize effectively. A learning rate of 0.001 was selected after testing a range of values (0.1, 0.01, 0.001, 0.0001). This rate allowed for stable learning without significant oscillations during training. Higher rates led to unstable training and potential divergence, while lower rates resulted in excessively slow convergence.

These hyperparameters were not arbitrarily chosen but were the result of extensive experimentation and model tuning. It has been employed a combination of grid search and manual tuning, evaluating each configuration's performance in predicting new SCADA variable values. The final set of parameters represents the optimal configuration for the specific problem domain, balancing predictive accuracy, computational efficiency, and model generalization.

2.3 *Validation of the Prediction Model*

The training and validation data sets are for 1 year to ensure that the model can adequately capture seasonality. In this case, the data corresponds to the year 2019. January to November are used for model training, representing 80% of the data set, while December is used exclusively for validation, constituting the remaining 20% of the data for this year. This 80–20 split is a common practice in machine learning, balancing the need for sufficient training data with the requirement for a robust validation set. The choice to use December as the validation period allows us to assess the model's performance on future, unseen data, simulating real world application scenarios.

This approach also has certain limitations. The single month used for validation may not fully represent the variability present throughout the year, especially if that month has unique characteristics or patterns that are not present in other months. Although an alternative approach could be to randomly sample all months to create the validation set, this would also risk information leakage between the training and validation periods, which would overestimate the model performance. The model's performance is evaluated by analyzing the loss curves of both the training and validation sets. This allows us to determine how accurate the model is in predicting the actual values. In addition, metrics such as mean absolute error (MAE) and root mean square error (RMSE) compare actual and predicted values. These metrics provide valuable information about the quality of the model's predictions, making it easier to evaluate its performance.

3 Determination of IF Threshold

The proposed method defines a decision threshold to identify whether potential IGBT-related faults exist in the series of predicted values. The IF algorithm is used to determine the threshold. This unsupervised and non-parametric algorithm works well when the labeled data is not available. Outliers are identified by considering an isolation score threshold. IF identifies outliers in the data set by using random decision trees to isolate outliers by partitioning the feature space. By repeatedly dividing the dataset into smaller subsets, the algorithm can quickly determine which points are more accessible to isolate and, thus, more likely to be outliers [17].

Figure 2 depicts several thresholds to determine the threshold that identifies the highest percentage of anomalies. The value of 0.76 has been experimentally established based on the information of the percentage of failures recorded by the O&M data and the value of "anomaly score" given by IF.

IF sets an "anomaly score" value, calculated using the average depth reached by a sample within the constructed decision trees. The shallower the average depth a sample reaches, the more anomalous it is considered. This score is normalized by dividing it by the expected average depth value for random data and then converted to an anomaly score. Points are placed based on the anomaly score, where those farther away from the threshold are considered more anomalous.

The threshold was determined by an iterative experimental process, balancing the trade-off between false positives and false negatives. First, a range of potential threshold was determined based on the distribution of anomaly scores in the training data. For each potential threshold, we calculated the precision, recall, and F1 score on a held-out validation set. We then plotted these metrics against the threshold values to identify the point of optimal balance between precision and recall. The threshold of 0.76 was selected as it maximized the F1-score, providing the best compromise between identifying true IGBT-related anomalies and minimizing false alarms. It is important to note that this threshold may need periodic reassessment as the system's behavior evolves over time or if operational conditions change significantly.

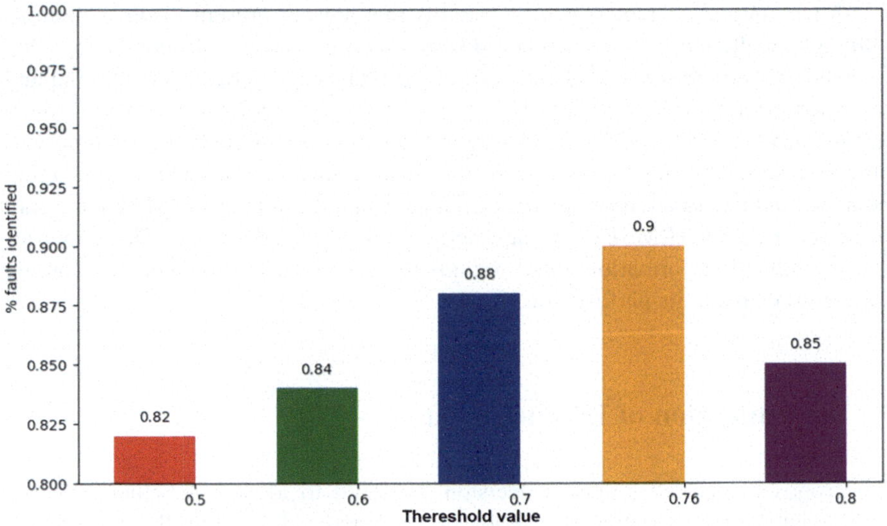

a) Determination of the threshold based on the percentage of detected failures.

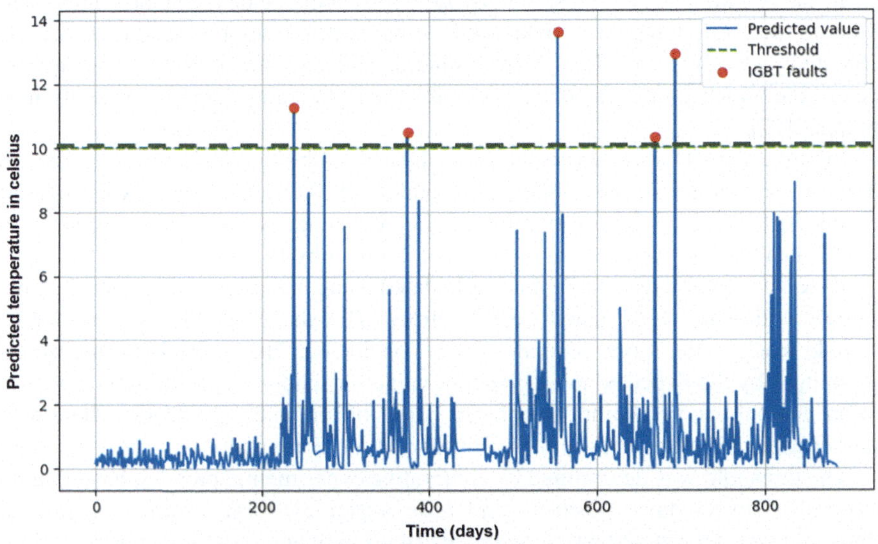

b) Detected IGBT faults using the threshold for the predicted variable igbt temperature.

Fig. 2 Threshold defined by IF and applied to the predicted variable IGBT temperature. (**a**) Determination of the threshold based on the percentage of defected failures. (**b**) Detected IGBT faults using the threshold for the predicted variable IGBT temperature

Values detected as anomalies should be verified if they are indeed IGBT-related anomalies. Once the threshold of 0.76 is defined, if the value obtained is less than this threshold, it is labeled as an IGBT failure in the future; otherwise, it is labeled as normal data. Figure 2 shows the predicted variable IGBT temperature applying the set threshold and detected possible IGBT-related faults.

4 Results

This section provides and evaluates the results of data preprocessing and variable selection to apply the proposed fault prediction method for real data from the CEV SCADA system. The correlation analysis in Fig. 3 provides an overview of the interactions within the WT SCADA system. A clear relationship is observed between IGBT temperature and environmental factors such as wind speed, indicating a direct response to atmospheric conditions. In addition, negative correlations with variables related to blade angle suggest the existence of automatic adjustments in generator speed to compensate for changes in reactive power and blade orientation.

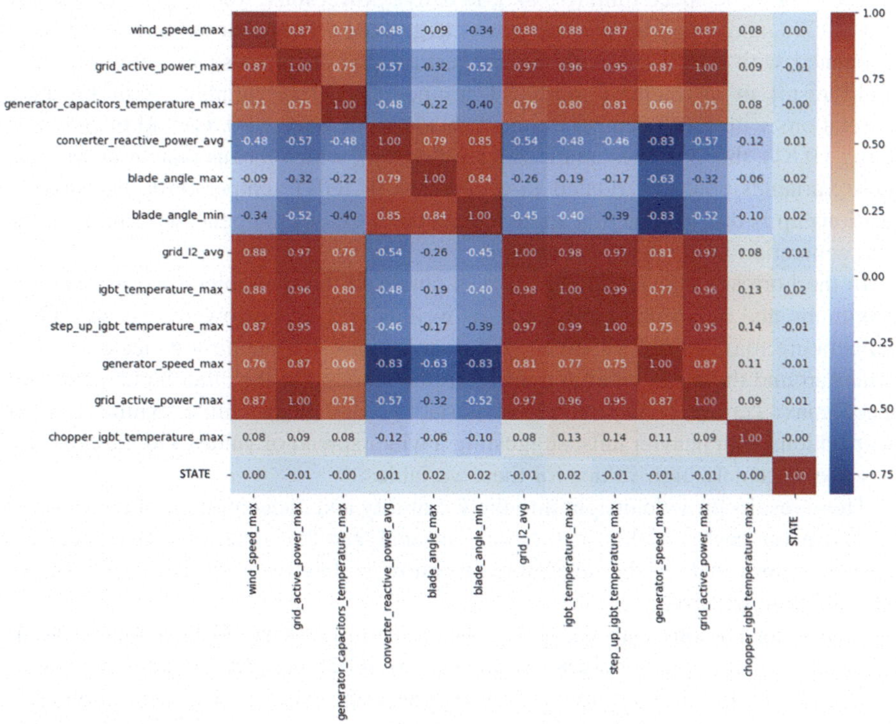

Fig. 3 Correlation matrix of SCADA variables

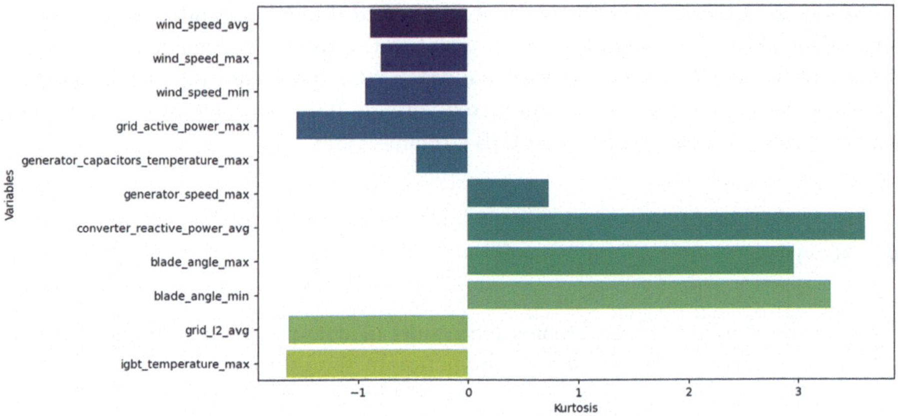

Fig. 4 Analysis of the kurtosis of the SCADA variables

The IGBT temperature variable was highlighted as the main source of information about the temperature fluctuations experienced by the IGBT, and which variables are related to these changes. Among the variables studied, the variables with positive correlations with IGBT temperature are wind speed avg. (0.78), wind speed max. (0.77), wind speed min. (0.77), grid active power max. (0.85), generator speed max. (0.76), and converter reactive power avg. (0.81). These variables are positively associated with IGBT temperature, suggesting that an increase in these variables could result in an increase in IGBT temperature. The variables with negative correlations with IGBT temperature are: blade angle max. (−0.64) and blade angle min. (−0.81), this indicates an inverse relationship, which means that an increase in these variables could be related to a decrease in IGBT temperature. The inverse relationship is due to how this angle affects the speed of the generator and thus the heat generation in the electrical system.

In the kurtosis analysis in Fig. 4, it is observed that variables such as average, maximum, and minimum wind speed, as well as grid active power max. and IGBT temperature max, exhibit relatively flat distributions, indicating a concentration of values around the mean without many extreme values. On the other hand, generator speed max, certain converter-related variables, and blade angle exhibit sharper distributions with heavier tails, suggesting a wider spread of values and the presence of some extremely high values for these variables.

These results provide insight into the variability and concentration of data around the mean for each variable in the data set analyzed and determine that there are extreme values that can be determined to know if they occurred before, during or after a failure occurred.

In the correlation analysis, it has been identified that IGBT temperature is influenced by multiple variables. An increase in IGBT temperature does not necessarily indicate an IGBT failure. Therefore, the proposed approach focuses on early fault detection by considering both current and future values of IGBT temperature-

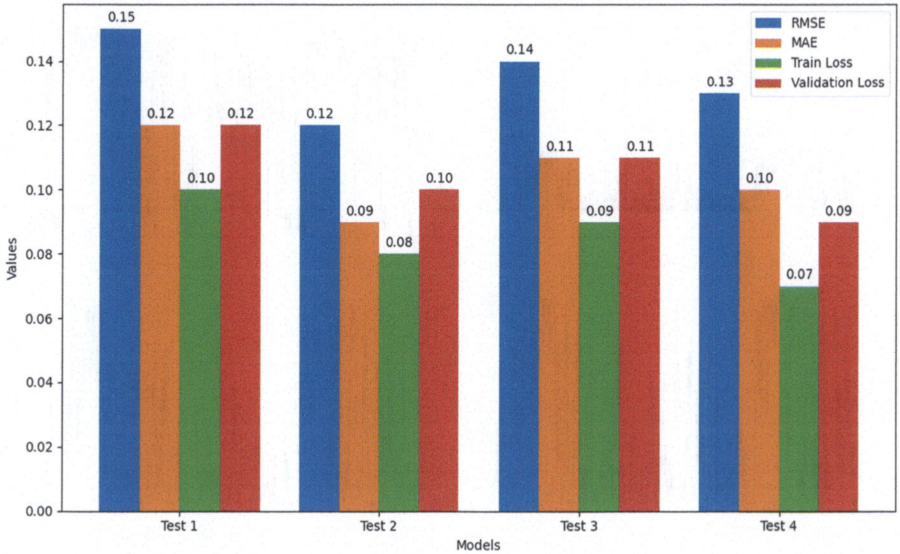

Fig. 5 Values obtained from different models for the RNN-GRU

related variables, applying the RNN-GRU model to predict future values, and detecting potential failures by determining a threshold using the IF model.

Figure 5 presents the RMSE, MAE, training loss, and validation loss values of the different models tested for data prediction, varying the hyperparameters in each case. The RMSE and MAE values are low in all models, indicating that the models have made predictions with relatively low errors. The "Test 2" model has the lowest RMSE (0.12) and MAE (0.09) values, suggesting that it is the most accurate model. The training and validation loss values are consistent, indicating that the model does not suffer from significant overfitting. Although the "Test 4" model does not have the lowest RMSE, it shows the lowest training and validation losses, indicating good generalization ability without overfitting. The analysis suggests that the "Test 2" and "Test 4" models are the most promising. "Test 2" offers the best prediction accuracy, while "Test 4" shows an excellent balance between training and validation losses. The "Test 4" model also shows the lowest losses in both metrics, suggesting that it has a good balance between fit and generalization.

After the experiments, it was determined to apply the "Test 4" model to make the predictions. Figure 6 shows the actual measurements and predictions for the four important SCADA variables from November to December. There is a remarkable closeness between the predicted values and the actual data, showing the model's predictive capability. The IF threshold is applied to the new data. Figure 6 also shows the outliers detected and labeled as possible IGBT faults.

With the faults detected by the model, it is possible to check whether the anomalies detected correspond to events recorded in the O&M data. Figure 7 shows two possible faults detected for the last days of November. The extreme

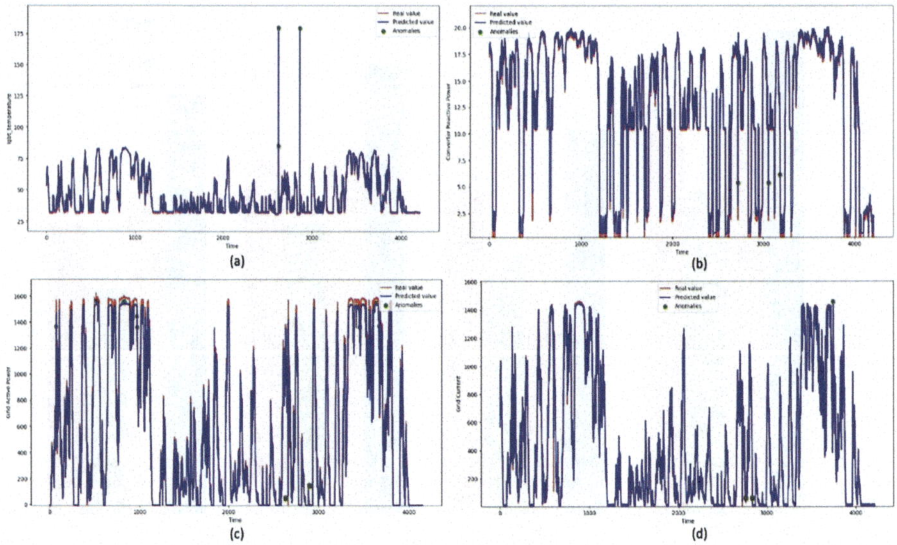

Fig. 6 Faults detected by the method at the predicted values

values recorded in the IGBT temperature suggest a possible anomaly. However, when analyzing the other variables, such as active power and grid current, it is observed that both reach a value of 0 at the time of fault detection. In addition, the converter's reactive power is below the normal range.

Figure 7 shows that three faults associated with the IGBT were reported, occurring on November 15, 19, and 21. However, the proposed method only manages to detect the failures on November 19 and 21. Despite attempts to adjust the threshold, the model fails to detect the remaining fault, suggesting that it may be related to another variable not considered in this initial analysis. However, the results of the method were promising when the tests were extended to multiple samples from different WTs.

To provide further insight into the performance of the RNN-GRU model, it is worth noting that the different test cases (Test 1, Test 2, Test 3, and Test 4) were designed to evaluate the impact of varying hyperparameters on the model's performance. Specifically, the hyperparameters that were varied across the test cases include the number of hidden layers, the number of units in each layer, the activation function, and the dropout rate. These changes in hyperparameters influenced the model's ability to learn complex patterns in the data, its capacity to generalize to new data, and its susceptibility to overfitting. For instance, Test 2, which achieved the lowest RMSE and MAE values, employed a configuration with two hidden layers, 128 units in each layer, and a dropout rate of 0.2. In contrast, Test 4, which showed the best balance between training and validation losses, used a configuration with three hidden layers, 256 units in each layer, and a dropout rate of 0.3. These differences in hyperparameter settings allowed us to evaluate the robustness of the model and identify the most effective configuration for predicting IGBT temperature

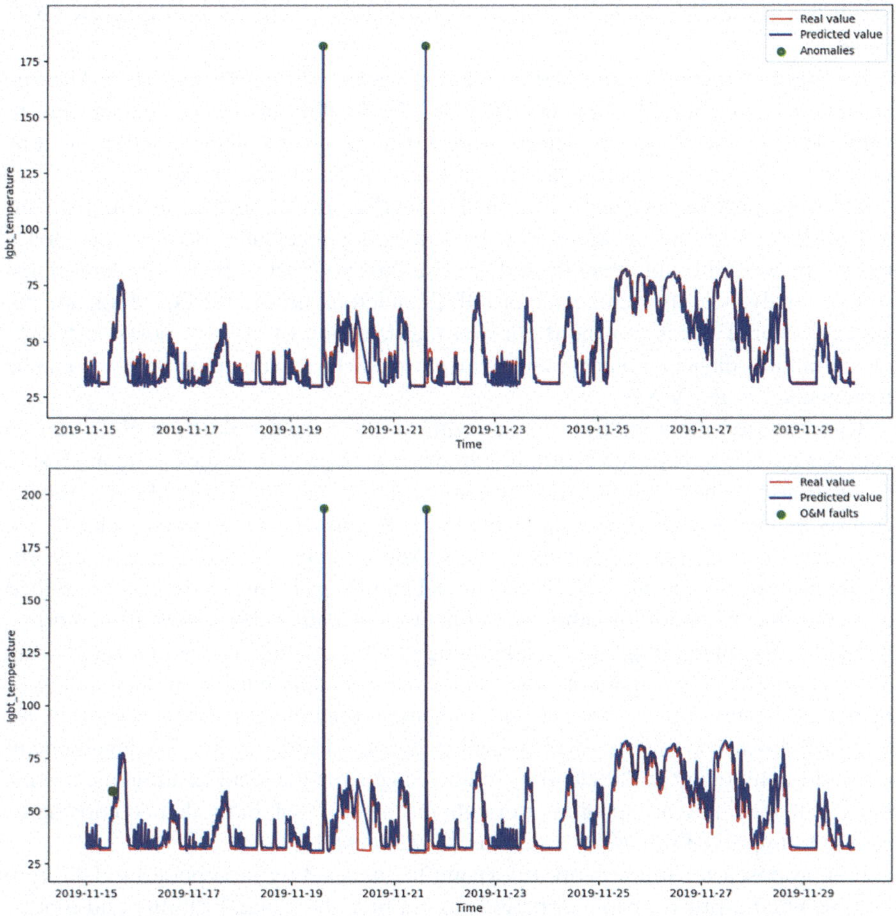

Fig. 7 Comparison between failures from the O&M record and the failures recorded by the proposed method

values. By comparing the performance of each test case, we were able to select the most promising model (Test 4) for making predictions and detecting potential IGBT faults and to gain a deeper understanding of the trade-offs between model complexity, accuracy, and generalizability.

5 Conclusion

This study presents a method to detect failures associated with IGBTs in WTs using predicted values from the SCADA system. It is based on a statistical analysis to identify the SCADA system variables that explain the IGBT behavior. In addition, it uses information from the O&M log. It applies the Isolation Forest (IF) algorithm to establish a threshold to identify faults in the values predicted by the RNN-GRU model.

During data preprocessing, the correlation matrix was examined, and a kurtosis analysis was performed. This revealed that IGBT failures occur mainly due to component overheating. This temperature rise is linked to other variables such as wind speed, converter reactive power, grid current, and blade angle.

The hyperparameters selected for the RNN-GRU model resulted in a model with an RMSE of 0.13 and an MAE of 0.10, indicating acceptable performance and a balance between fit and generalizability. The threshold set at 0.76 was determined from the analysis of the percentage of IGBT-related failures in the O&M log and the application of IF. This threshold allowed the detection of 90% of possible IGBT-related failures days in advance, with which a predictive maintenance system can be implemented in the VWF.

The proposed methodology for detecting IGBT-related failures in WTs can be extended to detect other types of faults in wind turbines. For instance, by analyzing the relationships between SCADA system variables and other critical components, such as generators, gearboxes, and blades, it is possible to identify patterns and anomalies that can indicate potential faults. Additionally, the use of machine learning algorithms, such as the RNN-GRU model employed in this study, can be applied to predict the behavior of other components and detect deviations from normal operation. By integrating this methodology with existing condition monitoring systems, wind turbine operators can gain a more comprehensive understanding of the overall health of their turbines and implement predictive maintenance strategies to reduce downtime and increase overall efficiency. Furthermore, the development of a multi-component fault detection system can enable the identification of complex faults that may not be apparent through the analysis of individual components, leading to improved reliability and reduced maintenance costs.

It is proposed for future work to expand the data set by incorporating data from multiple wind farms over longer periods to improve the generalizability and robustness of the model. In addition, alternative algorithms such as LSTM models could be explored to compare their performance relative to the RNN-GRU model developed in this research. Added to this, a system for real-time fault detection could be developed and integrated with existing SCADA systems for practical application, extending the methodology to simultaneously monitor multiple critical components, providing a more comprehensive assessment of the wind turbine status. In attention to the experimental obtaining of the threshold for FI, the development of adaptive thresholds that adjust for seasonal variations or specific turbine characteristics to improve the accuracy of fault detection is also proposed. These future directions are intended to refine the methodology, improve its practical applicability, and broaden its scope in the field of wind turbine predictive maintenance, ultimately resulting in more efficient and reliable wind energy systems.

References

1. Gao, Z., Liu, X.: An overview on fault diagnosis, prognosis and resilient control for wind turbine systems. Processes. **9**(2), 300 (2021). https://doi.org/10.3390/pr9020300
2. IRENA: Renewable capacity statistics (2021)
3. WWEA: World wind energy association. Statistics News (2023)
4. Maldonado-Correa, J., Vásquez, G., Paccha-Herrera, E., Cuenca, J., Rojas, M., Solano, J., Valdiviezo, M., Samaniego, C., Valarezo, J.: Analysis of the annual energy production of the Villonaco Wind Farm, pp. 73–87. Springer, Cham (2023). https://doi.org/10.1007/978-3-031-26813-7_5
5. Chen, H., Liu, H., Chu, X., Liu, Q., Xue, D.: Anomaly detection and critical SCADA parameters identification for wind turbines based on LSTM-AE neural network. Renew. Energy. **172**, 829–840 (2021). https://doi.org/10.1016/j.renene.2021.03.078
6. Maldonado-Correa, J., Valdiviezo-Condolo, M., Viñan-Ludeñam, M., Samaniego-Ojeda, C., Rojas-Moncayo, M.: Wind power forecasting for the Villonaco Wind Farm. Wind Eng. **45**(5), 1145–1159 (2021). https://doi.org/10.1177/0309524X20968817
7. Wen, W., Liu, Y., Sun, R., Liu, Y.: Research on anomaly detection of wind farm SCADA wind speed data. Energies. **15**(16), 5869 (2022). https://doi.org/10.3390/en15165869
8. Chen, H., Ma, H., Chu, X., Xue, D.: Anomaly detection and critical attributes identification for products with multiple operating conditions based on isolation forest. Adv. Eng. Inform. **46**, 101139 (2020). https://doi.org/10.1016/j.aei.2020.101139
9. WWEA: Central Eólica Villonaco: Ícono del aprovechamiento de los recursos renovables no convencionales. https://www.celec.gob.ec/noticias/central-eolica-villonaco-icono-del-aprovechamiento-de-los-recursos-renovables-no-convencionales/
10. Huang, X., Chang, W., Zheng, T.: Study of the protection and driving characteristics for high voltage high power IGBT modules used in traction convertor. In: IEEE 10th Conference on Industrial Electronics and Applications (ICIEA), pp. 1330–1334. IEEE (2015). https://doi.org/10.1109/ICIEA.2015.7334314
11. Mahmoud, M., Azzedine, B., Tayeb, A., Djamal-Edine, C.: Wind energy conversion system control robustness based on current analysis of IGBT open-circuit fault. Przeglad Elektrotechniczny. **96**(10), 14–24 (2020). https://doi.org/10.15199/48.2020.10.03
12. Liu, Z., Zhang, L.: A review of failure modes, condition monitoring and fault diagnosis methods for large-scale wind turbine bearings. Measurement. **149**, 107002 (2020). https://doi.org/10.1016/j.measurement.2019.107002
13. Encalada-Dávila, Á., Moyón, L., Tutivén, C., Puruncajas, B., Vidal, Y.: Early fault detection in the main bearing of wind turbines based on gated recurrent unit (gru) neural networks and SCADA data. IEEE/ASME Trans. Mechatron. **27**(6), 5583–5593 (2022). https://doi.org/10.1109/TMECH.2022.3185675
14. Xiang, L., Wang, P., Yang, X., Hu, A., Su, H.: Fault detection of wind turbine based on SCADA data analysis using CNN and LSTM with attention mechanism. Measurement. **175**, 109094 (2021). https://doi.org/10.1016/j.measurement.2021.109094
15. He, Y., Liu, J., Wu, S., Wang, X.: Condition monitoring and fault detection of wind turbine driveline with the implementation of deep residual long short-term memory network. IEEE Sensors J. **23**, 13360–13376 (2023). https://doi.org/10.1109/JSEN.2023.3273279
16. Yang, T., Teng, J., Li, C., Feng, Y.: Wind turbine fault detection and diagnosis using LSTM neural network. Chin. Control Conf. (CCC). **39**, 4042–4047 (2020). https://doi.org/10.23919/CCC50068.2020.9188709
17. Al Farizi, W., Hidayah, I., Rizal, M.: Isolation forest-based anomaly detection: a systematic literature review. In: 8th International Conference on Information Technology, Computer and Electrical Engineering (ICITACEE), pp. 118–122 (2021). https://doi.org/10.1109/ICITACEE53184.2021.9617498

Index

M. Espinoza-Andaluz et al. (eds.), *Congress on Research, Development and Innovation in Renewable Energies*, Green Energy and Technology,
https://doi.org/10.1007/978-3-031-88995-0

The manufacturer's authorised representative in the EU is Springer
Nature Customer Service Centre GmbH, Europaplatz 3, 69115 Heidelberg,
Germany. If you have any concerns regarding our products, please
contact ProductSafety@springernature.com

Printed and bound by CPI Group (UK) Ltd, Croydon, CR0 4YY
30/09/2026
02225596-0006